LIBRARY
WARWICKSHIRE
COLLEGE

MANAGING DISEASES *in* GREENHOUSE CROPS

William R. Jarvis
Agriculture Canada
Harrow, Ontario

APS PRESS
The American Phytopathological Society
St. Paul, Minnesota, U.S.A.

To Jo and Sarah

Reference in this publication to a trademark, proprietary product, or company name is intended for explicit description only and does not imply approval or recommendation to the exclusion of others that may be suitable.

Library of Congress Catalog Card Number: 92-71055
International Standard Book Number: 0-89054-122-1

Second printing, 1993

Printed in the United States of America on acid-free paper

The American Phytopathological Society
3340 Pilot Knob Road
St. Paul, Minnesota 55121-2097, USA

Contents

PART TWO

Strategy for Disease Control 87

Preface

> To affect, yea to effect their owne deaths, all living are importun'd. Not by Nature only which perfects them, but by Art and education which perfects her. Plants quickned and inhabited by the most unworthy Soule, which therefore neither will, nor worke, affect an end, a perfection, a Death. This they spend their Spirits to attaine; this attained, they languish and wither. And by how much more they are by Mans industry warm'd, and cherisht, and pamper'd, so much the more early they climbe to this perfection, this Deathe.
>
> JOHN DONNE, Paradox 1

Between 1980 and 1990, enormous strides were made in the environmental technology available to growers of greenhouse crops, with matching increases in production potential. Little, if any, of this technology was aimed primarily at the reduction of losses from insects and diseases, but it is possible to manipulate a crop and its environment specifically for that purpose. Together with advances in biological control of diseases, this greenhouse technology offers the grower opportunities to reduce the very costly input of pesticide applications and soil sterilization. At the same time, it can considerably reduce the escape of pesticides, fertilizers, and water into the environment outside the greenhouse and can ensure less pesticide residue on food and florist crops.

The degree of production control that can be achieved reaches its peak with cress seedlings (*Lepidium sativum* L.) grown as a salad crop in the United Kingdom, where the only hand that touches the seed and the plant is the consumer's. Every step of seeding, growing with correct light, temperature, and fertigation, and then packing is fully automated under computer control. It is assembly-line production. Most other crops can be produced with the aid of similar technology, but an appreciable amount of skilled labor is also necessary.

Sick plants can be detected electronically, but nothing has yet matched the computer in the experienced grower's head in monitoring abnormal growth of plants. There has been introduced the concept of the "speaking plant": the plant "speaks" to a computer through monitors that detect strains set up in the plant by environmental and disease stresses. The computer is programmed to respond to signals from the monitors by altering the environment, for example, by turning on the irrigation, warming the soil, opening the ventilators, or moving a thermal blanket over the crop canopy. Computer performance, however, can be considerably improved by continual intuitive feedback from the experienced grower. Perhaps, after all, horticulture is still more an art than a science.

Be that as it may, horticultural technology has a lot to offer in developing strategies whereby crops may escape disease, even in the presence of pathogens, and diseased crops may be manipulated so that losses in yield and quality are minimized. Most greenhouse production is very cost- and labor-intensive, and the cost of traditional pesticidal control at the first appearance of symptoms or simply in a preset program of insurance sprays or sterilization of media can have considerable impact on profitability.

This book explores the physical and chemical backgrounds of greenhouse crop production to see how modern technology may be modified to effect disease control. Some of the modifications may create environments that are less than optimum for productivity, but most of them are less expensive than traditional and more direct action against pathogens. Most of them also result in a more rational and reduced use of pesticides and energy-profligate processes, such as soil steaming. Some compromise is inevitable in the economics of crop production and the protection of crops from both diseases and invertebrate pests.

The reduction in the use of pesticides that can be realized in high-technology greenhouses scarcely qualifies crops as organically grown, but vegetables, fruit, flowers, and foliage plants grown in such environments certainly have lower pesticide residues. In addition, lower pesticide use makes for a safer and pleasanter workplace. Already, many tomato growers produce crops entirely without pesticides, relying solely on biological control of insect and mite pests and environmental control of diseases.

The opportunities for disease control are many and varied. Many rooting media, such as rock wool, perlite, and nutrient solutions in the nutrient film technique, are free of pathogens from the outset, and soil and soil mixes can be sterilized or, better, pasteurized to retain some biological control. The greenhouse environment can be monitored and controlled by microprocessors. Seed and planting material can be introduced into the greenhouse in a very healthy state, even pathogen-free, in the case of micropropagated plantlets. Biological controls can start with a relatively clean slate in an environment that can be manipulated to their advantage. Pesticides and other means of direct action against pathogens can be used more sparingly, integrated fully into the overall protection strategy. Full integration also includes tactics for insect and mite control, many of which rely increasingly on biological control.

During the review process, it became increasingly obvious that it is not true to say that a greenhouse is a greenhouse, no matter where in the world. The topography and macroclimate outside the greenhouse clearly have great effects on the environment inside. Some of these effects, such as shifts in light quality with latitude, are recognized and quantified, but I suspect that many others are not. This book, therefore, may be regarded as a generalized account in which due allowance must be made for local variations. Here, the experience of the grower and the extension agent will compensate for my shortcomings.

The literature I have chosen to illustrate various facets of the biology of greenhouse crop diseases is very much a personal selection, which other people may think is inappropriate or insufficient. I have tried, however, to give a lead into the literature on particular topics by what I think arc key papers. My selection is by no means exhaustive.

I hope this book will stimulate growers and horticulturalists to think in terms of crop protection, and exponents of crop protection to think how their science an be integrated smoothly into economic production practices.

Acknowledgments

It is a pleasure to acknowledge the help I have received from many scientists, extension officers, and, not least, experienced growers in Europe and North America. Many people gave me illustrations and unpublished data or discussed special problems with me or helped in other ways. These people include Andrea Buonassisi, Eric Champagne, Ken Cockshull, Adrian Cornelisse, B. N. Dhanvantari, Gillian Ferguson, Clem Fisher, André Gosselin, Michael Graham, Shalin Khosla, Jane Kochan, Tom Papadopoulos, Cathy Peace, C. C. Powell, David Rogers, Patricia Tilcock, James Traquair, Graham van der Hage, Chr. von Zabeltitz, Priva (U.K.) Ltd., and Vary Industries Ltd. Dr. R. S. S. Fraser and Rhona Floate kindly allowed me to use the specialist library of Horticulture Research International, Littlehampton, West Sussex, England.

In particular I thank my technicians, Susan Barrie, Ken Slingsby, and Harry Thorpe, and also Susan Duransky, who patiently and skillfully processed every single word several times. I am specially indebted to Ann Chase and John Paul Jones, who read the manuscript critically and constructively and gave me new insight into problems and solutions outside my own field. I am also indebted to the publications staff of the American Phytopathological Society, who guided me through the technicalities of book production.

Finally, none of this would have been done without the support of my wife, Jo, and daughter, Sarah.

Introduction

The purpose of the greenhouse is to extend the cropping season of tender vegetables, fruit, and ornamental crops and to protect them from adverse environmental conditions, such as low temperatures and precipitation, and from certain pests, such as rodents and birds (Hanan et al 1978; Mastalerz 1977).

The so-called greenhouse effect is achieved, in part, by the capture of solar energy, which is received by the earth mostly in short wavelengths, between 300 and 475 nm. Almost half of that energy is reradiated outward by soil and plants, in infrared wavelengths between 3,500 and 25,000 nm. Glass is transparent to wavelengths between 350 and 2,400 nm, transmitting 90% of light at 350 nm from direct sunlight and diffuse sky radiation, but it is opaque to thermal reradiation of wavelengths greater than 3,000 nm.

This is one reason for the greenhouse effect, but only 22% of the increase in temperature in a glasshouse can be attributed to this radiation entrapment (Bot 1983; Businger 1963). The rest is accounted for by conduction and convection in the enclosed air mass, which has less turbulence than the outside air. Convection transfers energy in two forms: the so-called *sensible heat*, which is transferred directly by the rising and mixing of warmed air, and *latent heat*, which is transferred indirectly when water is converted into vapor with no change in temperature. At 20°C this latent energy, about 2.45×10^3 $J \cdot g^{-1}$, is about 600 times the energy required to raise the temperature of water from 20 to 21°C. Conduction and convection largely account for the warming of a polyethylene-covered greenhouse, despite the high transparency of this material to infrared radiation.

A large amount of energy (60–70% of the total received by the plants) is dissipated by the transpiration of the crop; almost all of the remainder is reradiated or is lost by convection and conduction. Only 1% is utilized for photosynthesis, which occurs in the visible range of the spectrum in wavelengths between 400 and 700 nm, with blue (470 nm) and red (650 nm) being used the most effectively. The energy balance of a greenhouse is summarized in Figure 1.

Air exchange with the outside is restricted, and so water vapor transpired by the plants and evaporated from warm soil tends to accumulate, creating a low vapor pressure deficit (high humidity). Therefore, inside the greenhouse, the environment is generally warm, humid, and wind-free. This environment promotes the excellent growth of most crops (Cockshull 1985; Maher and O'Flaherty 1973), but it is also ideal for the development of diseases caused by bacteria and fungi (K. F. Baker and Linderman 1979; Bewley 1923; M. N. Cline et al 1988; Fletcher 1984) and for the activity of viruliferous insects (Hussey et al 1967). Groundbed crops are rarely rotated, and so soilborne pathogens tend to increase in population unless the soil is sterilized. But complete sterilization by fumigation or by heat, usually from steam, removes most of the soil microflora, and pathogens reentering a totally sterilized soil have little or no competition (Cook and Baker 1983). Pasteurization by heating soil to about 70° C, however, kills most plant pathogens and leaves a microflora that is usually competitively disease-suppressive. The limited microflora of some soil substitutes, such as rock wool and vermiculite, and of nutrient film systems does not always provide sufficient competition for pathogens, and large inoculum potentials present impediments to disease-free production (Zinnen 1988). Other media, such as peat and composted hardwood bark, on the other hand, suppress

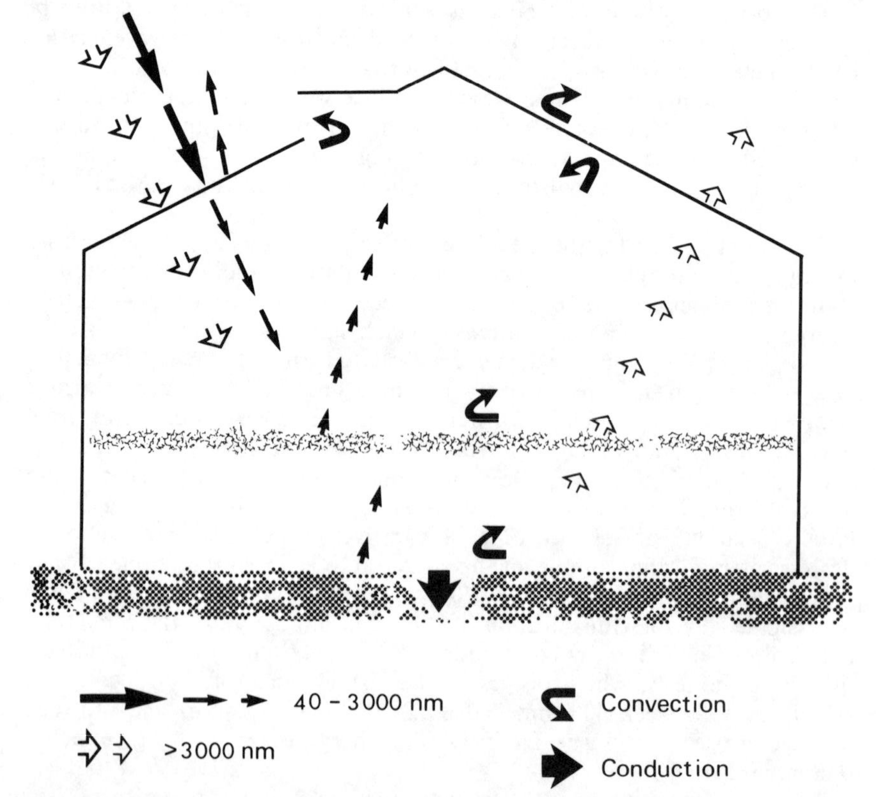

Figure 1. The energy balance of a greenhouse. Redrawn from Damagnez (1981).

some diseases biologically or chemically (Hoitink and Fahy 1986). Once infested, however, these media are difficult to disinfest.

Greenhouse production requires very high levels of economic investment, and space is at a premium. Therefore host density becomes a very important factor in pathology (Burdon and Chilvers 1982), not least for its implications in microclimate regulation and the ease with which pathogens move from plant to plant (Bravenboer 1974).

Because of the precision of control that greenhouse growers can exercise over the environment, irrigation, and nutrition, crops are being pushed into chronic stress. Whereas in 1975 a yield of 16 $kg \cdot m^{-2}$ of tomatoes was considered acceptable, yields of 36 $kg \cdot m^{-2}$ were not unusual in 1988, and over 40 $kg \cdot m^{-2}$ was possible (Anonymous 1990; Jarvis 1989). The partition of assimilates to the developing fruit from other tissues of the plant means that these tissues become relatively more susceptible to such pathogens as *Alternaria* and *Fusarium* spp., classified by Horsfall and Dimond (1957) as low-sugar pathogens. Other tissues enriched by translocated carbohydrates become more susceptible to high-sugar pathogens, such as *Botrytis cinerea* Pers.:Fr. Similarly, the susceptibility of tissues varies with physiological age and with position on the plant, as a reflection of their gross carbohydrate status (Grainger 1968).

For the majority of pathogenic fungi and bacteria, infection is usually accomplished in a film or drop of water on the plant surface. Unless temperature, humidity, and ventilation are well regulated, this surface water can persist in the greenhouse until infection becomes assured.

Most greenhouse crops are labor-intensive and for long periods require daily attention to tying, pruning, harvesting, and other routine operations. The risks of spreading pathogens on fingers, tools, and machinery are added to the risks from accidental wounds and from the exposure of relatively large areas of subepidermal tissues (usually susceptible) by pruning. In hydroponic systems and quite often in groundbed soils, osmotic stress—the so-called fertilizer burn—befalls roots. In many crops, the margin for error in calculating fertilizer rates is very small, and damage from excesses is rapid (Graves 1983; Jensen and Collins 1985), although in soil there is often a buffering from the adsorption of nutrients by soil components, which reduces or delays root damage and infection by opportunist pathogens.

Greenhouses are designed to protect crops from many adverse conditions, but most pathogens are impossible to exclude. Windblown spores and aerosols containing bacteria readily enter doorways and ventilators; soilborne pathogens enter in windblown dust and adhere to footwear and machinery; aquatic phycomycetous fungi are present in irrigation water from ponds, wells, and ditches (Hendrix and Campbell 1973); and insects transmit viruses and mycoplasmas from crops and weeds outside to crops inside, and they may well carry fungi and bacteria, too. The cucumber bacterial wilt pathogen, *Erwinia tracheiphila* (Smith) Bergey et al, is frequently brought into greenhouses by two cucumber beetles (*Diabrotica undecimpunctata* Mann. and *D. balteata* LeConte) (J. G. Leach 1964). Fungus gnats (Sciaridae) are often associated with high infestations of *Pythium* spp. in cucumbers in rock wool production systems. Indeed, larvae of *Bradysia impatiens* (Johanssen) can transmit oospores of *Pythium* spp. (Gardiner et al 1990). Often seeds and planting material such as transplants, cuttings, bulbs, and rhizomes are sources of pests and diseases and must be carefully inspected and rejected or treated.

Once inside a greenhouse, pathogens may be very difficult to eradicate. For example, the root-knot nematode *Meloidogyne incognita* (Kofoid & White) Chitwood survives under concrete pathways and at depths of up to 150 cm in sandy groundbed soils (Bird 1969; P. W. Johnson 1975; P. W. Johnson and McKeen 1973), well out of reach of steam and fumigants (P. W. Johnson 1975; P. W. Johnson and McKeen 1973). Tomato mosaic virus survives in many places (Broadbent 1960), including dry clothing, on which it survives for at least 2 years (J. Johnson and Ogden 1929). Some pathogens with soilborne sclerotia, such as *Phomopsis sclerotioides* van Kesteren (causing cucumber black root rot) and *Pyrenochaeta lycopersici* R. Schneider & Gerlach (causing tomato corky root rot) build up in population even with periodic groundbed sterilization (Ebben and Last 1975; Last et al 1969).

It is evident that greenhouses do not afford the protection from insect pests and pathogens that might be supposed. Nevertheless, the degree of control that can be achieved over limiting environmental factors, such as temperature, humidity, and ventilation, and over nutrition and water supplies enables disease-escape procedures to be devised, given adequate knowledge of the epidemiology and autecology of pathogens. When such measures are integrated in soilless systems, with biological control of some diseases and insect pests, with resistant germ plasm, and with modern techniques of applying pesticides at very low but effective rates, very good control of diseases, insects, and nematodes becomes highly practicable. Indeed, in many cases, tomato growers have not needed to apply pesticides, except sometimes soil sterilants, relying on biological control of insects, disease-resistant germ plasm, manipulation of the environment to prevent infection by foliar pathogens, and high standards of hygiene to reduce inoculum levels.

PART ONE

Greenhouse Production Systems: The Pathologist's View

Greenhouse structures are essentially light scaffolding covered by sheet glass, fiber glass, or one or more of several plastic materials, with a range of energy-capturing characteristics, all designed to maximize light transmission and heat retention. Crops are grown in groundbed soil, usually heavily amended with peat or farmyard manure; in pots containing soil or soil mixtures or soil substitutes; and in hydroponic systems, such as sand or rock wool culture and flowing nutrient systems without a matrix for the roots.

Most crops are labor-intensive and energy-demanding for considerable periods of winter cropping. They normally require, therefore, a high level of technology to obtain adequate economic returns on investment. Most of this technology is designed primarily for optimizing crop production, and historically not enough attention has been paid to exploiting and amending production technology for the control of pests and diseases.

Well-grown and productive crops are generally less susceptible to diseases, but in many cases a compromise has to be made between optimum conditions for economic productivity and conditions for disease escape. For example, many of the fuel-saving procedures discussed by Hurd and Sheard (1981) run contrary to phytopathological advice.

Greenhouse climates are essentially warm, humid, and wind-free (Hussey et al 1967), ideal for the development of many pests and diseases. How these climates may be regulated to avoid epidemics is a major concern of plant pathologists. Soil is the source of many pathogens, but the cost of soil sterilization is so high—it is probably the most expensive single item of production costs—that growers have sought alternatives to growing crops in groundbed soil. This has led to the development of soilless substrates, such as rock wool, and the nutrient film technique (Cooper 1979; Graves 1983; Jensen and Collins 1985; Molyneux 1988).

Consideration of the physical factors limiting greenhouse crop production has to take into account those also affecting the incidence and severity of crop diseases. The control of these diseases lies largely in understanding pathogen autecology and disease epidemiology in the greenhouse environment and designing rational disease-escape measures to be used together with resistant germ plasm and direct pesticidal control (Bravenboer 1974).

Greenhouse Structures and Equipment

CHAPTER 1

Siting and Form

For optimum crop production, greenhouses are designed to capture the maximum incoming radiant energy (Aldrich et al 1967; Critten 1984; Hanan et al 1978; Manbeck and Aldrich 1967; Mastalerz 1977). They should be sheltered from damaging high winds (Spek 1972; Woodruff 1954), unshaded, and oriented in a north–south direction (Aldrich et al 1967; Hanan 1970b; Mattson and Maxwell 1971). Within a greenhouse, row crops are usually oriented north–south, and benches are usually sited to make the most economical use of the available area. In all cases, it is desirable to exploit natural air movement for ventilation and air circulation within crops, and for this reason greenhouses are best oriented parallel to the winds prevailing at the time of full cropping (J. N. Walker and Duncan 1974b) and away from still-air topography (Plate 1). These effects can be modeled, and the model can be used to predict investment in construction, covering, and heating costs on new sites (Sase et al 1984). Heat loss varies with wind velocity and inversely with relative humidity.

Drainage of cold runoff water into the greenhouse must be prevented in order not to chill the groundbed and not to bring in waterborne pathogens, such as *Erwinia carotovora* (Jones) Bergey et al (Pérombelon and Kelman 1980); *Pythium* spp. (Hendrix and Campbell 1973); and *Olpidium radicale* Schwartz & Cook, the vector of melon necrotic spot virus (Tomlinson and Thomas 1986), and *O. brassicae* (Woronin) P. A. Dang., the vector of lettuce big vein virus (Tomlinson and Garrett 1964).

The greenhouse is best surrounded by at least 10 m of grass, kept weed-free, or bare gravel, concrete, or asphalt. Weeds are reservoirs of viruses and viruliferous insects (Harris and Maramorosch 1980) and of many bacterial and fungal pathogens. Similarly, field crops botanically related to those in the greenhouse are best avoided or at least kept as far away as possible.

Greenhouse Structures

Most commercial greenhouses are of ridge-and-gutter roof construction, to obtain maximum capture of solar energy, and are built on as light a scaffolding as possible, to reduce structural shading.

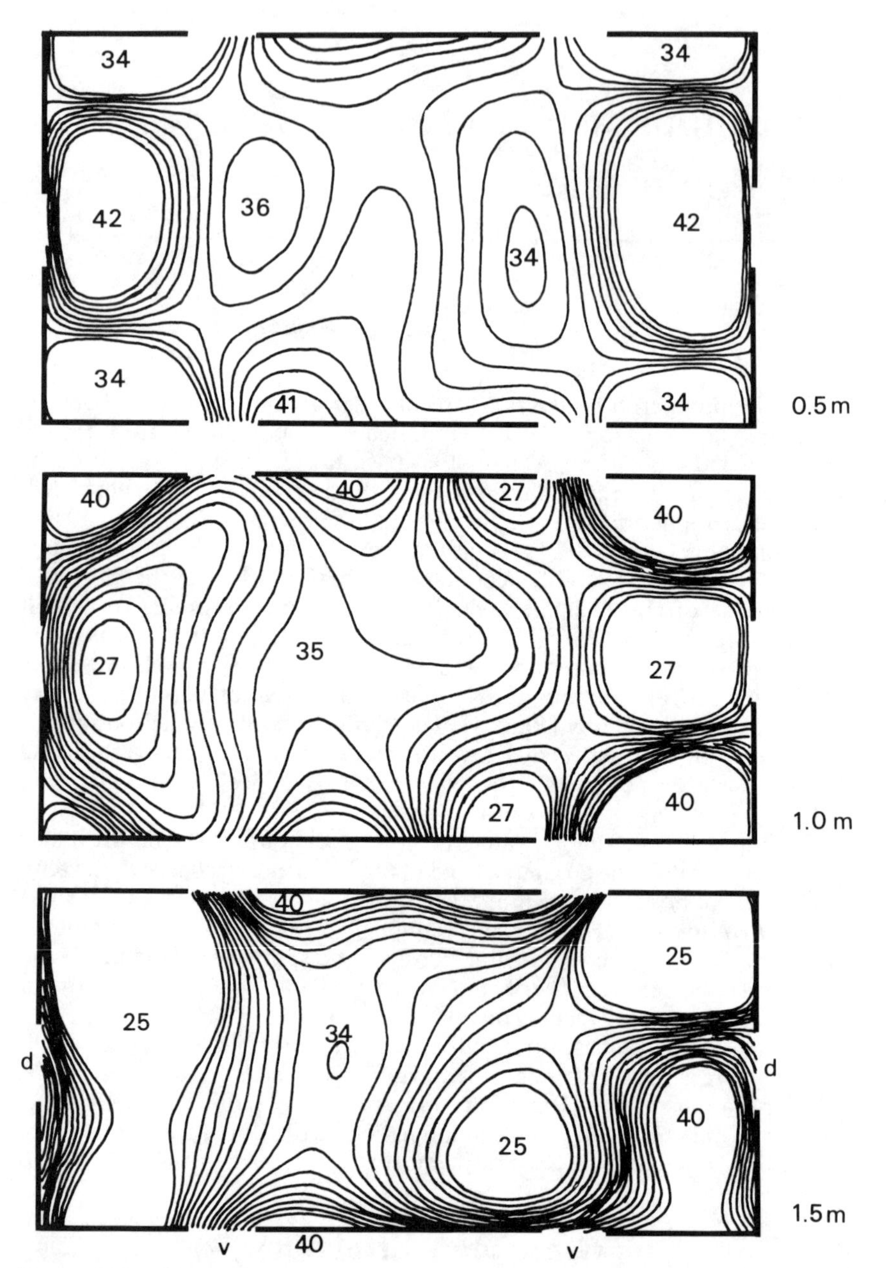

Figure 2. Isotherms in a polyethylene tunnel greenhouse at heights of 0.5, 1.0, and 1.5 m. The patterns are considerably influenced by the doorway (d) at each end and ventilation openings (v) in the sides. Redrawn from Baeten et al (1985).

Depending on the heating system, there is a steep vertical temperature gradient in a greenhouse (Carpenter and Bark 1967a,b; Hanan 1958), which is altered to various extents by air circulation and ventilation systems (Carpenter 1969; Carpenter and Bark 1967a,b; Duncan and Walker 1973; Hanan 1958; J. N. Walker and Duncan 1973a,b, 1974a). Since it is desirable to maintain crops under as uniform a temperature as possible, a relatively large headspace is required, to accommodate the hot air as well as hardware such as supplementary lights, thermal blankets, and carbon dioxide burners or emitters. This headspace is best provided by the tall Venlo type of greenhouse, which also enables tall, trained crops, such as tomatoes, cucumbers, and peppers, to bear a relatively greater photosynthetic area vertically per unit of floor area and permits more leaf pruning near the base of the stem for better air circulation (Plate 2).

Horizontal temperature gradients also exist (Figure 2) and must be corrected by suitable air circulation systems (Bark and Carpenter 1969; Carpenter 1969; Carpenter and Bark 1967a,b; Duncan and Walker 1973; Koths and Bartok 1985; J. N. Walker and Duncan 1973b, 1974a) and by the insulation of basal walls in cold climates (Plate 3) (Jarvis and Nuttall 1981).

In order to obtain better air circulation, heat distribution, and light penetration, a given floor area is better covered by multispan structures than by several single isolated greenhouses.

Covering Materials

In order to create the greenhouse effect, the covering material must be transparent to incoming solar radiation with wavelengths in the range of 400–3,000 nm and opaque to the long waves (longer than 3,000 nm) reradiated back into space by plants and soil (Damagnez 1981; Geiger 1966; Monteith 1972). Glass is an ideal material for this purpose, whereas polyethylene was listed by Trickett and Goulden (1958) as far from ideal (Table 1).

Commercial greenhouses are covered either with soda–lime–silica glass of various weights, sometimes toughened for snow and hail resistance, or with various plastic materials, commonly polyethylene, sometimes reinforced

Table 1. Transmission of light at 600 nm by some greenhouse covering materials and ultraviolet cutoff wavelengths[a]

	Transmission at 600 nm (%)	Ultraviolet cutoff[b] (nm)
Glass	90	320
Polymethyl methacrylate	87	290
Polyvinyl chloride	87	300
Polyethylene	82	230

[a] Adapted from Trickett and Goulden (1958).
[b] Wavelength at which transmission is 10% or less.

with glass fiber. The various materials have different energy transmission properties, as well as physical properties that affect durability and wettability (Aldrich et al 1967; Bowman 1972; Chandra 1982; Critten 1984; de Halleux et al 1985; Germing 1985; Manbeck and Aldrich 1967; Nijskens et al 1984; Trickett and Goulden 1958; Waaijenberg 1985; J. N. Walker and Slack 1970). The properties of several materials were summarized by Hanan et al (1978), Mastalerz (1977), and Trickett and Goulden (1958).

In general, a glasshouse with gaps purposely left unsealed between panes is ventilated naturally, particularly when it is oriented parallel with the prevailing wind. It has been calculated that there are 2.5 air changes per hour in a glasshouse with 61- $\times$ 61-cm panes overlapping by 1 cm. This, however, was also calculated to leak almost 26.4×10^7 $J \cdot h^{-1}$ in a 6.2-$km \cdot h^{-1}$ wind (Whittle and Lawrence 1960). Morris (1959) considered that leaks accounted for up to 10% of energy losses. Nevertheless many growers consider an unsealed greenhouse to be superior in many ways in crop production. Of concern to the pathologist is the leakage of humid air, generally a desirable factor. Plastic-covered houses almost always have a more humid climate and are more difficult to ventilate. Typically they have more problems with water-dependent diseases, such as gray mold and the downy mildews, than naturally leaky glasshouses (Calvert and Hand 1975; Watkinson 1975).

In a warm, humid greenhouse atmosphere, water condenses on glass and plastic when the temperature outside is low. A material as infrared-transparent as polyethylene may approach the opacity of glass in that range when water forms a film on it (de Halleux et al 1985; Trickett and Goulden 1958). Some plastic coverings are treated so as to induce water to film and run off rather than form beads. Condensed water in films on the inside of the roof adds considerably to high humidity in the greenhouse (Hurd and Sheard 1981).

Some plastics are phytotoxic. Hardwick et al (1984) noted that dibutyl phthalate (DBP), used as a plasticizer in sealing strips, was emitted at up to 2,010 $pg \cdot L^{-1}$ in the air of a greenhouse, enough to cause discoloration and necrosis in cabbage seedlings (*Brassica oleracea* L.) within 3 wk. Hardwick et al (1984) also suspected that other phytotoxins from plastic hoses, pots, and wire insulation, for example, could contaminate the air and water.

Heating and Cooling Systems

Greenhouses are heated naturally by solar radiant energy, some of which may be reflected by clouds and reflective surfaces on the ground. It enters the greenhouse as short-wave radiation and is converted to heat on absorption by plants, soil, benches, and walkways (Morris 1959). Less than 5% of solar radiation is utilized in photosynthesis and growth; the surplus energy is dissipated by thermal reradiation, reflection, transmission, convection, and evaporative cooling (which alone dissipates 70–90% of incident energy). In order to avoid thermal damage, this last factor is the single most important one that must be controlled by the greenhouse manager.

For much of the year in most latitudes, solar energy is sufficient for both light and heat, but for out-of-season cropping, additional heat must be supplied to the plant from burning fossil fuels, from electricity, or from stored solar energy (Short et al 1976; von Zabeltitz 1976a).

During the main growing period of most crops, the major problem is not one of heating but of cooling the greenhouse on sunny days, compounded with heating it at night or at least conserving heat, in order to obtain maximum productivity without spoiling crop quality.

Heating Systems

Numerous artificial heating systems and circulatory systems are available to the disease-conscious grower. They may use hot water or low- or high-pressure steam in pipes variously finned to increase radiation to the air; forced air, often ducted in perforated polyethylene tubing; above-crop infrared heaters, variously fueled; or hot-water pipes and electric heating cables in soil and on benches. The uniform distribution of heat depends largely on the placement of the heating elements, fans, and ducting (Duncan and Walker 1973; Hanan 1958; J. N. Walker and Duncan 1973b, 1974a). Excess heat can be expelled from the greenhouse simply by ventilating hot air masses through roof and side ventilators, with the assistance of fans.

There are many designs for solar-heated houses that store energy in the soil or in various other media, principally brine or concrete (Airhart 1984; Bartok and Aldrich 1984; Dale et al 1984; Garzoli and Shell 1984; Short et al 1976; von Zabeltitz 1976c). These are ideally linked to hot-water heat exchangers and air-conditioning systems.

Energy conservation at night or on cold sunless days is achieved by various temporary insulation techniques (Short et al 1976; Stickler 1975), but not without problems. Watkinson (1975), for example, noted an increased incidence of ghost spot, caused by *Botrytis cinerea* Pers.:Fr., on tomatoes in a polyethylene-lined greenhouse where ventilation was restricted.

Cooling Systems

Radiant energy flux entering the greenhouse can be reduced by painting the roof and insolated sides with opaque paints, variously tinted (Goldsberry and Wolnick 1966); by cloth, plastic, or lath shades; or by water films (Morris et al 1958). Of particular interest to the pathologist are evaporative systems of cooling; they include the fan-and-pad system, whereby water is evaporated by blowing or sucking air through wet fibrous pads with low-speed high-volume fans, and low- and high-pressure misting systems (Carpenter and Willis 1957; Davidson 1953; King 1963; Krug 1963; Mastalerz 1977). In the simplest system, plants may be sprayed with water (Seemann 1962), but many water-dependent pathogens may easily infect them, depending on how long the plants remain wet, either because they have been sprayed directly or because dew has formed on them at the lower temperature in a humid atmosphere. High-pressure fogging systems generating water drops on the order of 0.2 μm in diameter, which evaporate before they alight on the plant, are to be preferred (Mabbet 1986; Mee 1977; Musselman et al 1985).

Cooling a crop is accomplished in three ways: by permitting radiative cooling, usually at night; by ventilating to promote evaporative cooling of transpiring leaves and soil (Bailey 1985; Goedhart et al 1984; King 1963; Nederhoff 1984; Sase et al 1984; van de Vooren and Strijbosch 1980; J. S. Wolfe and Cotten 1975); and by evaporative cooling of a fog or a wet surface (pad cooling) to lower the temperature of the ambient air. Evaporative cooling, however, is more difficult to achieve in tropical and subtropical areas, because the ambient air already has a very low water vapor pressure deficit. Sometimes crops are cooled by water sprays, but from the pathologist's point of view this invites several water-dependent bacterial and fungal pathogens, such as *Botrytis cinerea* and downy mildews. Less efficient is indirect cooling by running cold water over the roof of the greenhouse (Morris et al 1958).

Root-zone cooling in potted plants can be achieved by circulating cold water through pipes at or just below the bench surface. The pipes are usually embedded in channels in insulating material, such as expanded polystyrene, doubling as hot-water pipes when needed.

Fogging. For meteorologists a fog is an aerosol of water drops condensed on dust particles as the result of the cooling of humid air. For horticulturalists a fog is made by forcing water at high pressure (600–4,000 kPa) through very fine nozzles (0.15 mm in diameter or less) (Davidson 1953; Mabbet 1986; Mee 1977; Musselman et al 1985). This forms fog droplets about 0.2 μm in diameter, with about 100 drops per cubic centimeter (Plate 4). In order not to block the nozzles with particles and precipitated salts, the water supply must be filtered, and calcium and iron must be kept in solution by pretreatment with chemicals appropriate to the salts dissolved in the water. In the long run, there is a risk that salts will be deposited on plants by evaporation, but the damage is insignificant, compared with that from spraying or misting.

The prime effect of fogging is to cool the air evaporatively, extracting about 2.45×10^3 $J \cdot g^{-1}$ of heat energy. The cooler air is denser and, in falling, contributes to convective air movement and the movement of carbon dioxide through the crop. If the fogging nozzles are correctly placed above the crop, most of the droplets evaporate in the air and never reach the plants. Thus, although the air is humidified, the plants do not become wet. Plant stress is thereby relieved without inviting water-dependent fungal and bacterial pathogens. There is some diminution in light levels during fogging, but it is small in comparison with the loss from shading by whitewash or blinds. Most of the light transmitted through fog is in the visible range.

Fogging provides an even distribution of temperature and humidity throughout the house and crop and reduces the need for supplementary fans for air circulation. Seasonal fluctuations in humidity are leveled out, and there is less need for irrigation. The corrosion of structures is reduced, unless high humidities are maintained unrelieved for long periods.

Misting. *Mist* is no longer a valid meteorological term, having come to mean different things in different countries, from light fog to drizzle. In horticultural terms a mist is an aerosol obtained by forcing water at moderate pressure (100 kPa) through fine nozzles or from fine holes in the rose of an irrigation hose. The name is commonly applied to the aerosol in propagation units for rooting cuttings. Considerable caution must be used with misting in order not to invite water-dependent pathogens and to avoid crop damage, but it is a better system than overhead irrigation for relieving plant stress and for irrigating and applying foliar nutrients and pesticides. Mist drops do not evaporate entirely in the air, and many land on the plants and the ground. More water is used in misting than in fogging, and so more energy, in the form of latent heat of vaporization, is needed to evaporate it.

Pad cooling. In areas of low to moderate humidity, air can be cooled and incidentally humidified by passing it over or through a pad of fibrous material soaked in water or over an open water surface (Giacomelli et al 1985; Hanan et al 1978). The excessive humidity that this introduces is a problem that requires a compromise solution, depending on the crop and the likelihood of disease.

This system is cheap to install and run but is relatively inefficient in providing a uniformly cooler environment, especially in a humid climate. It sets up large temperature and humidity gradients in the greenhouse, and the fast-moving air that flows through the evaporative pad may damage plants near the inlet. Because of the steep gradients set up, it is difficult to control the fans with a thermostat or humidistat set in the crop. In order to accommodate the flow of cooled air through the greenhouse, an equivalent outlet must be provided, which is wasteful of a carbon dioxide–enriched atmosphere.

Overhead irrigation. From the phytopathological point of view, overhead irrigation should not be done solely for the purpose of cooling a crop; the risks of spreading pathogens and facilitating infection are too great. There are occasions when overhead pesticide spraying may be necessary, but again there are risks of inviting and spreading pesticide-resistant pathogens.

Roof cooling by water films. In the summer months, a very simple method of reducing greenhouse temperature is to run a film of water down the roof (Morris et al 1958). There is only a very slight reduction in transmitted light.

Air Circulation and Ventilation

Air circulation is essential in a greenhouse, to provide a uniform distribution of temperature, humidity, and carbon dioxide, and ventilation is required to replenish carbon dioxide (unless it is artificially supplied) and to remove accumulating toxic gases, such as ammonia from the soil and ethylene from the soil as well as ripening fruits. It also removes excess

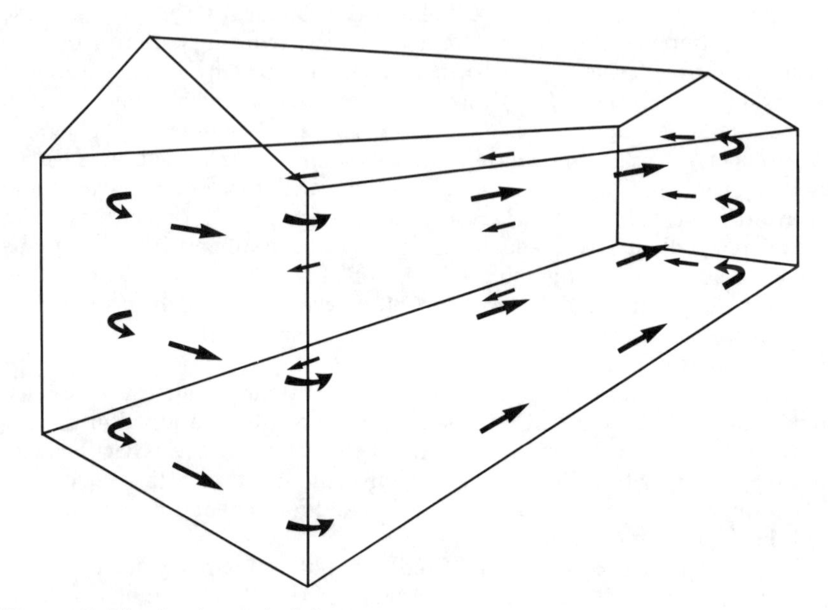

Figure 3. The horizontal airflow system moves large masses of air around the greenhouse.

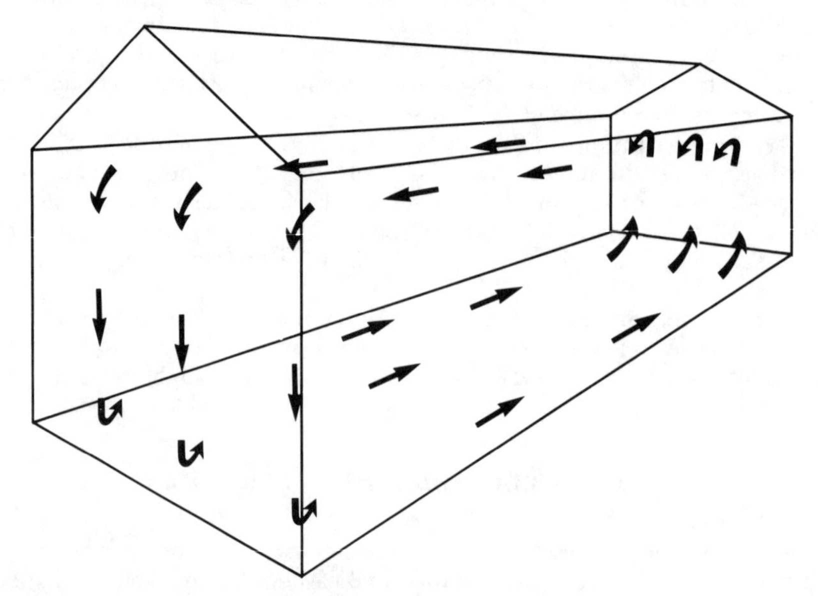

Figure 4. The vertical circulation system moves air through the crop more efficiently than the horizontal airflow system and reduces condensation on the underside of the roof or thermal blanket.

moist air and overheated air when the solar heat gain is very high (Hanan et al 1978; J. N. Walker and Duncan 1973a,b, 1974a). The desired rates of change can be calculated for computer use (Bot 1983; Goedhart et al 1984; Nederhoff 1984) and modeled (Albright 1984; Bot 1980; Tachibana and Minagawa 1984).

In order to exploit natural ventilation, greenhouses should not be located in unduly sheltered places, such as hollows and areas in the lee of trees and buildings, and they should be oriented parallel to the prevailing wind (J. N. Walker and Duncan 1974b). Wind effects can be modeled (Sase et al 1984).

Several air circulation systems are available (Koths and Bartok 1985; J. N. Walker and Duncan 1973a). The fan-jet system uses perforated plastic ducting in the roof space (J. N. Walker and Duncan 1973b, 1974a) inflated by a fan, so that warm air in the roof space is pushed downward to displace cooler air below (Plate 5). This system is not efficient for moving air around the plants (Koths and Bartok 1985).

Some vegetable growers, wanting good air circulation at the bases of deleafed stems, install a version of the fan-jet system with the ducts near the ground and between the rows (Plate 6) (von Zabeltitz 1976b).

The turbulator system uses vertical fans to stir the air above the crop and move it toward the eaves. Several fans are required, and they still do not move air through the canopy very well. Moreover, cold spots are caused below the fans.

Much more efficient for disease management is the horizontal airflow system (Koths 1983; Koths and Bartok 1985), which moves large, coherent air masses horizontally around the greenhouse (Figure 3). The fans, sufficient in number to maintain an air movement of at least 12 $m^3 \cdot min^{-1}$, are situated just above the crop. This system reduces temperature stratification, scrubs

SUMMARY

Greenhouse Structures and Equipment

In order to construct an environment conducive to crop productivity but inimical to pathogen activity, a number of physical factors must be considered: the siting, shape, and covering material of a greenhouse determine its solar energy entrapment, and this energy can be conserved by such devices as solar storage ponds and thermal blankets. Cooling the crop depends mostly on evaporative cooling from the crop itself or from water surfaces. Efficient and uniform distribution of heat in the crop and the removal of excess heat and water vapor depend on efficient ventilation systems, such as horizontal and vertical airflow systems. All climate control systems for the most part concentrate on avoiding wetting the crop, so as not to invite infection by water-dependent pathogens.

the canopy of humid air, utilizes carbon dioxide more efficiently, reduces oxygen stratification at the leaf surface (Burrage 1971; Waggoner and Shaw 1953), and keeps irradiated plants cooler (Carpenter and Nautiyal 1969; Hanan 1965, 1970a,b). Similarly, heat is transferred to plants losing heat by radiation on clear, cold nights, and so the risk of attaining the dew point at the plant surface is reduced. Natural leaks through unsealed overlaps in glasshouses are relatively small with horizontal airflow.

The vertical circulation system, a variation of the fan-jet system, is even better in mixing the air. It moves air masses along the roof space, downward at the end of the greenhouse, and then through the crop (Figure 4). Resistance from the crop results in a relatively low air speed across the whole width of the house. All the air is moved from the headspace; consequently, this warm air transfers heat to plants near the ground and under benches and keeps the air moving gently through the canopy. Condensation on the underside of the roof or a shade screen is avoided (G. van der Hage, personal communication).

Growing Systems

CHAPTER 2

Except for very rare examples of aeroponics, in which plants are grown with their roots bathed in a misted nutrient solution (Soffer and Berger 1988), commercial ornamental and vegetable crops are grown in a solid or liquid medium providing water, nutrients, oxygen, and support to the roots. The most common medium is soil, often heavily amended with organic manure, composts, peat, and other materials to provide nutrients and improve its physical structure. There is a wide variety of soilless media, ranging from straw bales, peat, sand, gravel, perlite, and vermiculite to synthetic substances, such as rock wool and polyurethane foam, to the flowing nutrient solution used in the purely hydroponic nutrient film technique, in which the roots are supported only on the base of a gully containing a shallow layer of the nutrient solution.

Solid media can be contained in bags laid on the ground or in pots of a wide variety of sizes, shapes, and materials. The pots are set on benches, also of widely different designs.

The physical and chemical properties of all these materials have profound implications for the susceptibility of crops to diseases and for disease control.

Soil Groundbeds

A soil groundbed (known as a border in the United Kingdom) is almost always an area of natural field soil over which a greenhouse is built. The groundbed water table therefore coincides with that outside, and this often brings problems. Unless drainage is arranged to run outward from the greenhouse, the influx of cold water from outside cools roots near the outside walls, allows the entry of waterborne pathogens, such as *Erwinia* spp. and *Pythium* spp., and brings into the greenhouse excess fertilizer and herbicide runoff from adjacent fields and sometimes industrial wastes or toxic fumes.

The relative amounts of mineral sands, gravels, and clays and organic materials determine the physical properties of the soil and the availability of nutrients (Bunt 1961, 1976; de Boodt and Verdonck 1972). Drainage and oxygen availability are also important. Groundbeds are usually amended annually with a wide range of materials to improve those physical properties and to modify nutrient availability by adjusting the soil pH. Common amendments are farmyard manure, peat, straw, peanut hulls, corncobs, spent mushroom compost, and other crop residues; also added are composted sawdust, hardwood bark, and municipal sludge. Not infrequently pathogens are imported on these materials, along with herbicide residues and industrial toxins, such as heavy metals. Some of these problems were reviewed by Hoitink and Fahy (1986) and are discussed elsewhere. Organic mulches reduce soil compaction during work routines, reduce evaporation from the soil, contribute carbon dioxide during decomposition, and add organic matter to the soil.

Depending on color, structure, and water-holding capacity, groundbeds retain or reradiate various amounts of solar energy, maintain different temperatures by evaporation, and warm and cool quickly or slowly. For example, if a crop of tomatoes follows lettuce in the early spring, a soil groundbed with a high impedance to heat exchange may take several weeks of insolation to achieve an adequate temperature for tomato roots (Damagnez 1981). Conversely, after a summer tomato crop, the soil may take several weeks to fall to the optimum for lettuce again.

The thermal properties and water potentials of soils are considerably modified by mulching with organic materials or plastic film. Originally used to decrease soil splash on fruit, straw mulch on tomato groundbeds was found to insulate the soil in late winter and early spring (Jarvis et al 1983), exacerbating Fusarium crown and root rot, caused by *Fusarium oxysporum* Schlechtend.:Fr. f. sp. *radicis-lycopersici* W. R. Jarvis & Shoemaker, a disease more severe in cool soil. Organic mulches are increasingly being replaced by plastic film, commonly white polyethylene, wholly or partially covering the soil. Polyethylene reflects heat (Plate 7), which is occasionally excessive. It reduces evaporation from the soil and hence keeps down the humidity of the greenhouse.

Peat Bags

Peat is a common soil substitute, either alone or mixed with vermiculite, polystyrene beads, perlite, or terra-cotta pellets. It has a high water-holding capacity and a high anion exchange capacity, and it maintains a fair structure for oxygen exchange during cropping. When peat is mixed with vermiculite, which has a high cation exchange capacity, the mix has an excellent buffering capacity, which reduces the risk of accidental overfertilization. Polystyrene beads added to the mix improve aeration (Bunt 1961, 1976; de Boodt and Verdonck 1972).

Peat and peat mixes are commonly used in pots and in white polyethylene bags laid on polyethylene sheets on the ground on the sides of shallow gullies to facilitate drainage. Slits are made in the bag about 2.5 cm above its base to establish its water table. Peat and peat mixes in bags and pots warm up quite quickly in the sun.

Peat contains a wide variety of microorganisms (Dickinson and Dooley 1967; Kavanagh and Herlihy 1975), including pathogenic *Pythium* spp. in commercial bales (Favrin et al 1988; Kim et al 1975; Robertson 1973). Kim et al (1975) also found *Fusarium* spp. in all of 52 commercial peats they examined but not *Verticillium* or *Rhizoctonia* spp. Couteaudier et al (1985) isolated *F. oxysporum* f. sp. *radicis-lycopersici* from peat compost imported into France from the Netherlands. Peat therefore cannot be assumed to be free of pathogens. On the other hand, Tahvonen (1982a,b) and Wolffhechel (1989) reported that some batches of peat were disease-suppressive. Light peat, from the upper layers of the peat bog, contained *Streptomyces* spp. and *Trichoderma viride* Pers.:Fr., well-known antagonists of fungi. Dark peat, from lower layers of the bog, was not disease-suppressive. Jarvis (1977b) reported that sterilized peat was highly conducive to Fusarium crown and root rot of tomato transplants, but an unsterilized Canadian peat suppressed the disease very well.

Rock Wool and Other Inert Substrates

Rock wool, manufactured of spun molten rock, is sterile and chemically inert. Vermiculite, perlite, polystyrene, and a foamed phenol–formaldehyde complex are also heat-sterilized during manufacture, and all but vermiculite are also inert.

Planting blocks are made of rock wool with the fibers running vertically, for capillary water movement, and are wrapped in plastic around the sides. In rock wool slabs the fibers run horizontally. Slabs are encased in polyethylene bags with planting and drainage holes.

In areas where sawdust is plentiful, vegetables are often grown hydroponically in vertical or horizontal bags of sawdust (Plate 8). Sawdust is scarcely inert, but excellent crops are raised in it. In British Columbia, the sawdust is usually from Douglas fir (*Pseudotsuga menziesii* (Mirb.) Franco) and western hemlock (*Tsuga heterophylla* (Raf.) Sarg.).

All of these substrates, in relatively small masses above the ground, respond quickly to changes in ambient temperature. None needs sterilization except if reused.

In the United Kingdom wheat straw bales have traditionally been used for cucumber and tomato production (Plate 9). The bales isolate the root system from soil infested with root disease fungi and nematodes, and they warm up relatively quickly in the early part of the season. Similarly, cucumbers have often been planted on ridges.

Polymer gel modules are another inert substance used as a substrate. In a paper sandwich in a plastic sleeve, a thin layer of dry gel swells in a nutrient solution to form a layer about 10 cm thick, which serves as a root matrix. Yields of tomatoes and cucumbers from plants grown in this substrate are comparable to those from plants grown in rock wool (Anonymous 1989; Flaherty 1989). Nutrient checking is more difficult in polymer gel modules, but disposal after use is easier.

Nutrient Film Technique

There are many commercial and homemade versions of equipment for the nutrient film technique (NFT), in which nutrients in solution flow continuously or intermittently by ebb and flow over roots in shallow gullies

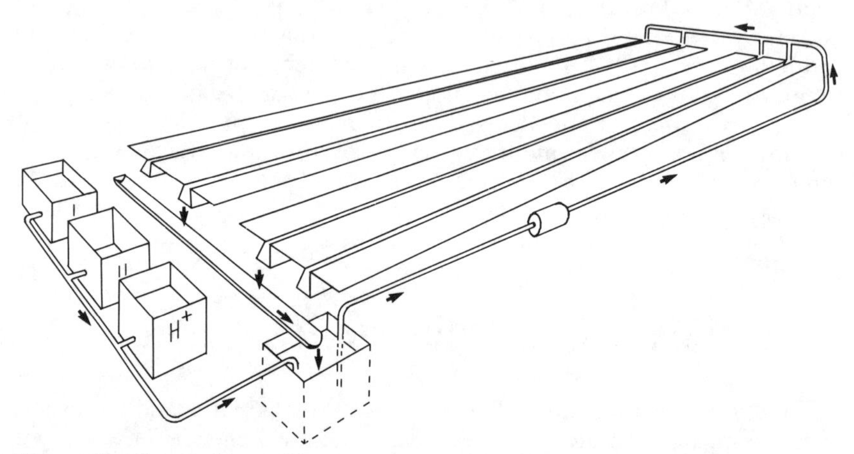

Figure 5. The nutrient film technique, with the typical storage of stock solutions of calcium nitrate (I) and other nutrients (II) and nitric or phosphoric acid (H^+). The stock solutions are mixed and then circulated, flowing down the sloped gullies containing the root systems of plants.

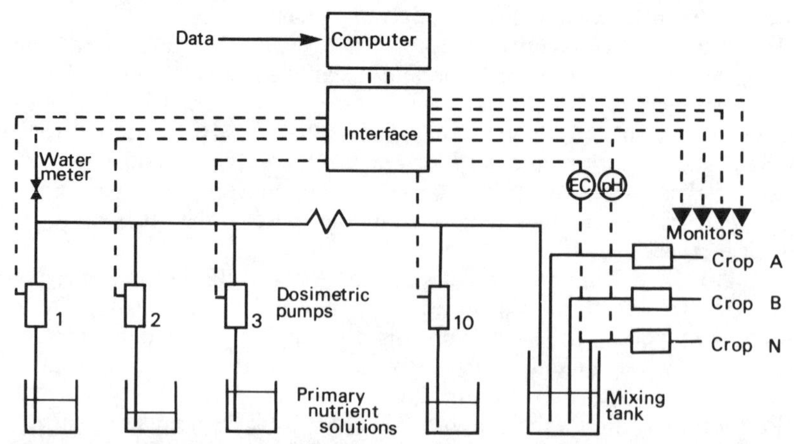

Figure 6. Schematic diagram of a multinutrient injector system. A computer responds to sensors that measure electrical conductivity (EC) and pH and switches dosimetric pumps to dispense nutrient solutions in a mixing tank and thence to crops or greenhouse zones. The computer is also programmed to dispense nutrients according to crop age and several environmental parameters. Redrawn from Papadopoulos and Liburdi (1989).

Table 2. Typical nutrient stock solutions for vegetables produced by the nutrient film technique[a]

Stock solution[b]	$g \cdot L^{-1}$
Solution A	
Calcium nitrate	75
Solution B	
Potassium nitrate	90
Monopotassium phosphate	30
Magnesium sulfate	60
Iron chelate (15% Fe)	3
Manganese sulfate	0.4
Boric acid (omitted for cucumbers)	0.24
Copper sulfate	0.08
Zinc sulfate	0.04
Ammonium molybdate	0.01

[a] Adapted from I. Smith (1988).
[b] Prepared for dilution to 1:100.

or on benches (Graves 1983; Jensen and Collins 1985). Typically, these gullies of folded polyethylene sheet are laid down in parallel rows at a slope of 1–2% (Figure 5), either on the floor of the greenhouse or on benches. The nutrients are supplied from two stock solutions, one of calcium nitrate, the other of all other salts (Table 2). The two solutions are mixed after dilution to avoid precipitation of calcium phosphate. A third stock solution of nitric or phosphoric acid is used to adjust pH and to supply a small amount of nitrate or phosphate. The nutrient supply is governed by an electrical conductivity sensor, which measures total dissolved salts. A pH electrode measures acidity. Stock solutions are pumped according to total salt and pH determinations, and the whole process is usually controlled by microprocessors. Papadopoulos and Liburdi (1989) developed a rather more advanced version of NFT nutrient regulation in which each component is separately monitored and metered into the system (Figure 6).

Aeration is very important in NFT, and nutrient flow rates should not be less than $2\ L \cdot min^{-1}$, with a free fall of solution back into the reservoir.

Notwithstanding the absence of soil from the main NFT system, soilborne pathogens are sometimes a problem (Zinnen 1988), usually when plants are raised in starter blocks made of soil or a soil mix.

Benches and Containers

Benches isolate potted plants from the soil and from soilborne inoculum carried by water splash. They enable plants to be heated or cooled more uniformly and provide easy access for cultural operations. These operations are further facilitated by moveable benches or by containers on rollers, which can be moved on the benches. A worm drive system can be arranged to open the spacing between containers progressively as the crop matures (Zinnen 1988).

For the convenience of workers, benches are 80–90 cm high and not more than 90 cm wide, to limit the distance to reach across. Benches with open tops of wooden slats or welded-wire fabric are recommended, to enable air to move upward through the plant canopy, with enough space between containers to permit free flow (Plate 10). Adequate air circulation is difficult on solid benches, whose sole advantage is the accommodation of capillary or ebb-and-flow subirrigation (Plate 11).

The structure of a bench is important in the distribution of heat from heaters below or alongside it (B. M. Jenkins et al 1988). Solid or crowded benches tend to keep the air warmer near the floor than in the growing zone, with the coolest air 25–90 cm above the canopy. There are wide variations in temperature 15 cm above the bench surface. Temperature distribution is improved with open-mesh benches and spaced-out containers permitting air movement upward through the canopy. Carbon dioxide and humidity gradients are also lessened.

Bench heating is usually recommended for the production of potted plants, although too warm a root zone may induce guttation at night in some cultivars of begonia and peperomia, causing leaf damage (van Weel 1984). This problem could not be solved by air movement or by lessening the difference between night and day temperatures. Only raising the electrical conductivity in the root zone was effective. Bench heating can also create optimum conditions for warm-season pathogens during the winter.

Microclimate at the Bench

Within the general microclimate of the greenhouse are submicroclimates imposed by the thermal properties of the bench design, pot material, and potting mix, which influence plant growth and diseases. Bowman (1972) investigated the effect of some of these factors on root-zone temperatures during nocturnal radiation (Table 3). Because of radiation and evaporation from potting composts, the root-zone temperature differs from that of the ambient air, although convection tends to minimize the difference. The lowest temperatures occurred in clay pots on a solid bench under a polyethylene greenhouse cover; evaporative cooling, stagnant cooled air, and

Table 3. Average dawn temperatures (°C) in John Innes potting compost in different pots and on different benches under two greenhouse coverings[a,b,c]

	Glass covering		Polyethylene covering	
Bench	Clay pots	Black polystyrene pots	Clay pots	Black polystyrene pots
Slatted capillary	12.9	14.5	13.2	13.6
Wire mesh	12.9	14.5	12.4	13.8
Grit on corrugated asbestos	12.7	13.4	11.6	12.8

[a] Adapted from Bowman (1972).
[b] Ambient air was controlled at 16.7°C.
[c] Least significant difference = 0.3 ($P < 0.05$).

intensified radiation were additive cooling effects under these conditions. The highest temperatures occurred in polystyrene pots on a wire bench in a glasshouse; here the effects were additive and tended toward the ambient air temperature.

Bunt and Kulweic (1970) found the contents of clay pots to be about 1°C cooler at night than those in plastic pots and 4°C cooler under high-intensity radiation. They preferred plastic pots for tomato propagation in winter and clay pots in summer. At night, pot temperature was negatively correlated with the difference between the greenhouse temperature and that of the outside air. Within clay pots there was a marked horizontal temperature gradient, especially in dry soil, caused by evaporative cooling from the porous clay surface. The soil in black pots was generally about 4°C warmer than that in white pots.

Air Movement

The movement of air through the canopy induces thigmomorphogenesis (Biro and Jaffe 1984; Jaffe and Biro 1979), resulting in a thickening of the stem and a shortening of the internodes and thus a sturdier plant. Plants shaken to induce thigmomorphogenesis are more resistant to Fusarium wilts (Shawish and Baker 1982). The boundary layer is also reduced in disturbed plants, and evaporation is increased. Thigmomorphogenesis is induced by ethylene produced in the internodes, and it is possible that ethylene originating from bacterial activity in the soil (A. M. Smith and Cook 1974) and passing through the canopy adds to this effect. Ethylene, however, may predispose plants to other pathogens, among them *Botrytis cinerea* Pers.:Fr. (W. H. Smith et al 1964). Continuous, gentle movement of air upward through the leaf canopy is obtained by using wire mesh benches and transplant trays with ventilation holes alternating with plant cells (Plate 12). Adequate spacing of pots on the benches also aids ventilation.

Ventilation upward through the canopy prevents temperature gradients, facilitates active transpiration, and evaporates inoculum drops (Blackman and Welsford 1916). Containers vary widely in size, shape, color, and material, and these characteristics impart different thermal and moisture-holding properties. In addition, transplants for groundbed growing are often produced in blocks of peat, various soil mixes, and rock wool, all with different thermal properties. In general, peat and soil blocks are not good starting materials for hydroponic production in nutrient film systems or on rock wool or similar slabs. Chance contamination of these blocks by *Pythium* spp. usually leads to significant losses from damping-off when the blocks are set out in the production house.

Crop Spacing and Training Systems

Managing the spatial relationships of plants to permit effective air movement and ventilation, as well as to reduce stressful interplant competition for light, water, and nutrients, is a necessary part of disease escape.

Spacing

Interplant competition in vegetable crops was discussed by Frappell (1979) and in a theoretical way by Charles-Edwards (1982), but most commercial spacings have evolved empirically to obtain maximum productivity per unit of area of the ground or bench. For example, optimum plant spacings for productivity have been worked out by trial for cucumbers by Bakker and van de Vooren (1984b), for tomatoes by Fery and Janick (1970), for chrysanthemums by Janick and Durkin (1968), and for carnations by Durkin and Janick (1966). In the last case, flower production decreased with increasing plant density in a linear log-log relationship. There was also an edge effect, with plants on the edge of the bench more productive than those in the middle. Flower quality declined linearly with increasing density. In compromise, overall productivity could be increased by increasing the density of plants at the edge of the bench.

Usually potted plants are spaced by the grower's experience according to the season, the sizes of the plants and the pots, and the available bench

SUMMARY

Growing Systems

Greenhouse crops are grown in various soils and soilless media whose physical and chemical properties are adjusted to obtain maximum productivity. These properties, such as heat conservation, water-holding capacity, fertilizer levels, and pH, can also be manipulated to reduce the amount of inoculum of pathogens and the probability of infection.

Systems for growing crops in the greenhouse vary widely in complexity. The most common rooting media are soil and various soil mixtures incorporating peat, vermiculite, perlite, and several other materials usually added to modify soil structure. In order to avoid the costs of soil disinfestation, many growers use soilless substrates, such as rock wool, with nutrients added intermittently or continuously in solution. The nutrient film technique dispenses with a porous rooting medium and bathes roots in a flowing nutrient solution. In general, this technique, which was designed to improve the precision of crop nutrition as well as to avoid risks of soilborne diseases and the cost of soil sterilization, also confers relative freedom from diseases. Disease epidemics in nutrient film systems can usually be attributed to poor hygiene.

The thermal and gas exchange properties of rooting media affect the growth of roots as well as the activities of pathogens. Peat, a common rooting medium either alone or in mixtures, often suppresses pathogen activity, depending on its origin. On the other hand, several pathogens, including *Fusarium* and *Pythium* spp., have been found in commercial peat.

space. As the season progresses, interplant distances have to be increased, and they can be adjusted automatically on moveable benches (Zinnen 1988). Occasionally, the required spacing is marked out on the bench (Plate 10).

Training

In order to get the maximum photosynthetic area per unit of ground area, indeterminate vegetable crops, such as tomatoes, cucumbers, and peppers, are trained on strings attached to support wires 2.5–4 m high. Cucumbers are grown with various spacings between and within rows, depending on the time of year and the cultivar, giving densities of 1.4–1.8 plants per square meter. They are grown in a number of training systems, including the Guernsey arch (Plate 9), vertical and inclined cordons, and the umbrella (Figure 7). The Guernsey arch ideally requires ventilation under the arch as well as regulation of the microclimate above the canopy, in order to prevent the persistence of water on the plants.

The design of benches is important in affecting the ventilation of seedling trays and potted plants. In order to escape infection from pathogens that require surface water for germination and entry into the host, potted plants on benches should be adequately spaced and well ventilated, preferably by through-the-bench ventilation. In general, this method of ventilation encourages good plant growth and reduces infection from water-dependent pathogens. Evaporation of inoculum droplets is enhanced, and thigmomorphogenesis is induced by the movement of air through the canopy, making transplants sturdier and more resistant to certain diseases. Ethylene-induced thigmomorphogenesis, however, affects susceptibility to pathogens; it decreases the susceptibility of tomatoes to Fusarium wilt, for example, but it may make them more susceptible to *Botrytis cinerea* unless inoculum droplets are evaporated quickly.

Very little is known of the effects of plant spacing and training systems on the incidence of diseases in tall crops grown in groundbeds. Generous spacing of potted plants on the bench decreases infection from water-dependent pathogens, but there is very little evidence on the effects of within- and between-row spacing, training and pruning systems, and plant habit on disease incidence in row crops in greenhouses. A priori reasoning suggests that dense plant communities would have a microclimate conducive to foliar diseases and the rapid spread of pathogens from plant to plant.

Traditional indeterminate tomato culture requires the stems to be lowered periodically from their support on an overhead wire so that the lower stems, devoid of fruit and leaves, lie along the ground. In straw mulch, they lie in a microclimate conducive to infection by pathogens such as *Botrytis cinerea* (Plate 13). The situation is improved somewhat by plastic sheet mulch, but in a significantly better technique the stems are run along horizontal wires 30–40 cm above the floor and parallel to the row or are supported on hooked pegs stuck in the ground.

Habit

Generally, scant regard is given to the role of the habit of greenhouse plants in disease escape. Rather, the emphasis is on form in foliage plants, on the number and quality of flowers in florist crops, and on productivity in fruit and vegetable crops. Consequently, there have been few attempts, if any, to breed cultivars with an open, less leafy habit to provide a drier microclimate.

In lettuce, Abdel-Salem (1934) found gray mold (caused by *Botrytis cinerea*) to be more prevalent in cabbage types with the outer leaves spreading to the ground than in the upright cos types. The microclimate beneath the spreading leaves remains humid and gives *B. cinerea* and also *Sclerotinia*

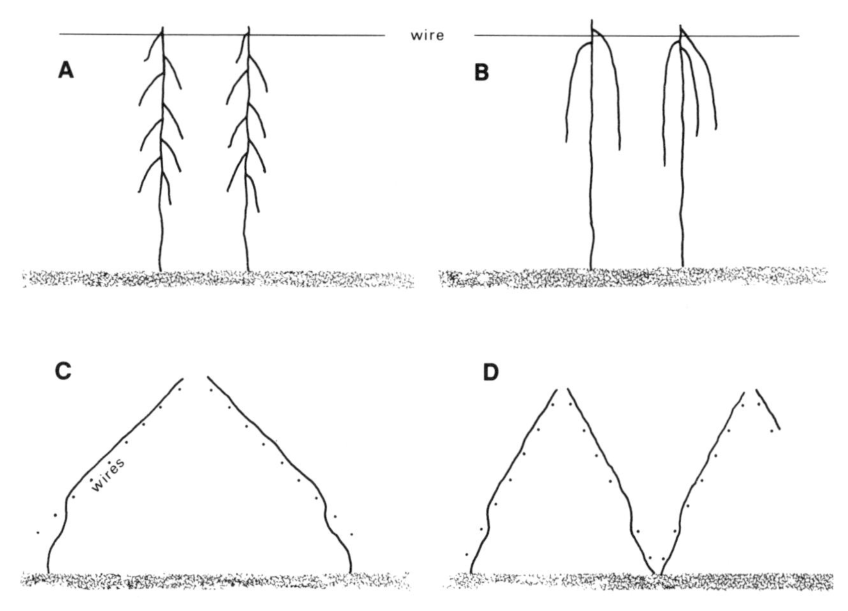

Figure 7. Training systems for cucumbers. **A,** Vertical cordon. **B,** Umbrella. **C,** Guernsey arch, supported on horizontal wires. **D,** Inclined cordon. In a Venlo-type greenhouse the top support wire is about 4 m above the ground, creating a relatively foliage-free stem area. In a traditional house, the top wire is about 2–2.5 m high. Redrawn from Anonymous (1983).

sclerotiorum (Lib.) de Bary and *S. minor* Jagger more time to achieve infection.

Plant habit is also a function of nutrition. Overfertilized plants, especially those overfertilized with nitrogen, usually have vigorous vegetative growth with leaf canopies difficult to ventilate and keep dry. Water-dependent pathogens are thus favored.

Pruning

The number of fruit permitted to form on cucumber, tomato, pepper, and similar crops determines the vegetative vigor of the plants, and so there has to be some compromise between productivity and physiologically stressing the plants to a point at which new flower development is inhibited or delayed. This stress might explain the attack of cucumber stems by *Penicillium* (Jarvis and Barrie 1988) and *Aspergillus* spp. (S. D. Barrie, unpublished observation), hitherto not regarded as virulent pathogens. A similar stress can be induced in some tomato cultivars if too many fruit per cluster are permitted to develop. W. R. Jarvis and H. J. Thorpe (unpublished) advised some tomato growers to remove one or two clusters of fruit to prevent the collapse of plants infected by *Fusarium oxysporum* f. sp. *radicis-lycopersici*, which typically occurred at the stress of first fruiting (Jarvis 1988). Some yield was sacrificed in order to save the plants, which survived and yielded from the remaining clusters.

The same sort of compromise has to be made in removing yellowing, senescent leaves from the stems of indeterminate vegetable crops. They contribute little to photosynthesis, and their removal facilitates ventilation of the lower stem area. In order to improve ventilation, as well as to save on labor costs, some growers are overzealous in leaf removal and so reduce yield. Their crops, however, generally escape gray mold.

The Greenhouse Environment and Crop Protection

CHAPTER 3

Radiation, water relations, carbon dioxide exchange, and nutrition govern all of the processes in crop production and pathogenesis, and the skill with which environmental factors are manipulated in the greenhouse determines productivity and profitability.

During the last half of the 20th century, horticultural science in the greenhouse has advanced from manual and largely empirical adjustments of temperature, ventilation, nutrition, irrigation, and carbon dioxide enrichment to advanced, elaborate modeling of physical and physiological processes and monitoring and control of environmental factors by microprocessors (Bot 1983; Challa et al 1988; M. N. Cline et al 1988; Jarvis 1989; Ting et al 1989; Zinnen 1988).

Deficiencies of solar radiation are made up for by heating systems and supplementary lighting, and excesses are countered mainly by ventilation to expel large masses of hot air, as well as by various direct cooling systems.

Humidity, or more accurately water vapor pressure deficit, in the crop's environment represents in a sense the summation of water relations and radiation, and as such its measurement and control figure largely in crop productivity.

When a crop is ventilated to cool it, large amounts of water vapor are expelled, and so the rate of evapotranspiration from the crop and the rate of evaporation from the soil are directly affected. Carbon dioxide enrichment is achieved at the expense of some restriction in ventilation and hence an increase in humidity in the greenhouse. High humidity restricts transpiration and water and nutrient uptake, which are also governed by air and soil temperatures.

There is, therefore, a highly complex interaction of many factors in crop production, and these factors also affect the activities of plant pathogens and insect pests. The delicate balancing of the environment to achieve maximum and economical production is increasingly being aided by microprocessor evaluation of data and control of operations.

Economics has a high priority in climate control (Bogemann 1980; Challa et al 1988; J. S. Wolfe 1970), because it almost always involves the expenditure of energy. Some of this energy may be recoverable.

Algorithms for Productivity

Udink ten Cate (1980) and Udink ten Cate and Challa (1984) described a hierarchical computerized system for describing crop growth, utilizing the concept of the "speaking plant" (Udink ten Cate et al 1978), and Hashimoto et al (1984) showed how the speaking plant responds to its environment in its carbon dioxide utilization and leaf temperature as affected by radiation, heating system temperature, and cloudiness outside.

Udink ten Cate and Challa (1984) recognized three levels of hierarchy (Figure 8), with the outputs from the lower levels providing the inputs to the upper levels. The ultimate output variables comprise such factors of economic output as yield, earliness, and quality, and the initial input variables are such things as planting material, fertilizers, pest and disease control, labor, climate (inside and outside the greenhouse), and cultural practices.

The hierarchical system implies that inputs and outputs of subsystems (levels) can be formulated in measurable terms and that feedback from higher to lower levels can either be neglected or be measured exactly. Feedback that can be measured forms the criterion for the presence of measurable interaction.

By this criterion, the greenhouse climate subsystem can be isolated from the plant growth system, and it constitutes the first level of the hierarchy (Figure 8). Its output is a set of variables characterizing the greenhouse climate—temperature, humidity, carbon dioxide concentration, and so on. Its input is the set of controls that alter climate—heating, ventilation, and carbon dioxide enrichment. There are uncontrollable disturbances at this first level, such as wind speed, radiation, and outside temperatures. The greenhouse climate subsystem is controlled by establishing a feedback loop.

The processes that determine the daily growth of the plant—photosynthesis, storage of assimilates, respiration, transpiration, and so on—constitute the second level of the hierarchy (Figure 8). The disturbances at this level include irrigation, pruning, and fertilizer supply. Since some of the factors at this level are not easy to measure, they have to be estimated or modeled. The controls applied at this level also affect those at the first level.

The third level (Figure 8) concerns season-long processes, such as flowering and fruiting, morphogenesis, senescence, and the effects of pests, diseases, and nutritional imbalances. Most of these effects are measurable; the others must be estimated. Some are also determined by first-level factors, especially pests and diseases, and therefore there is interaction between first-level controls and third-level processes.

Finally, there has to be an integration of these processes and other disturbances, such as the application of pesticides.

There now comes the problem of manipulating the inputs and controls of the hierarchical system into producing maximum yields (in which the control of pests and diseases has a major positive effect) at a minimum economic input. In this manipulation, the experience of the grower is a

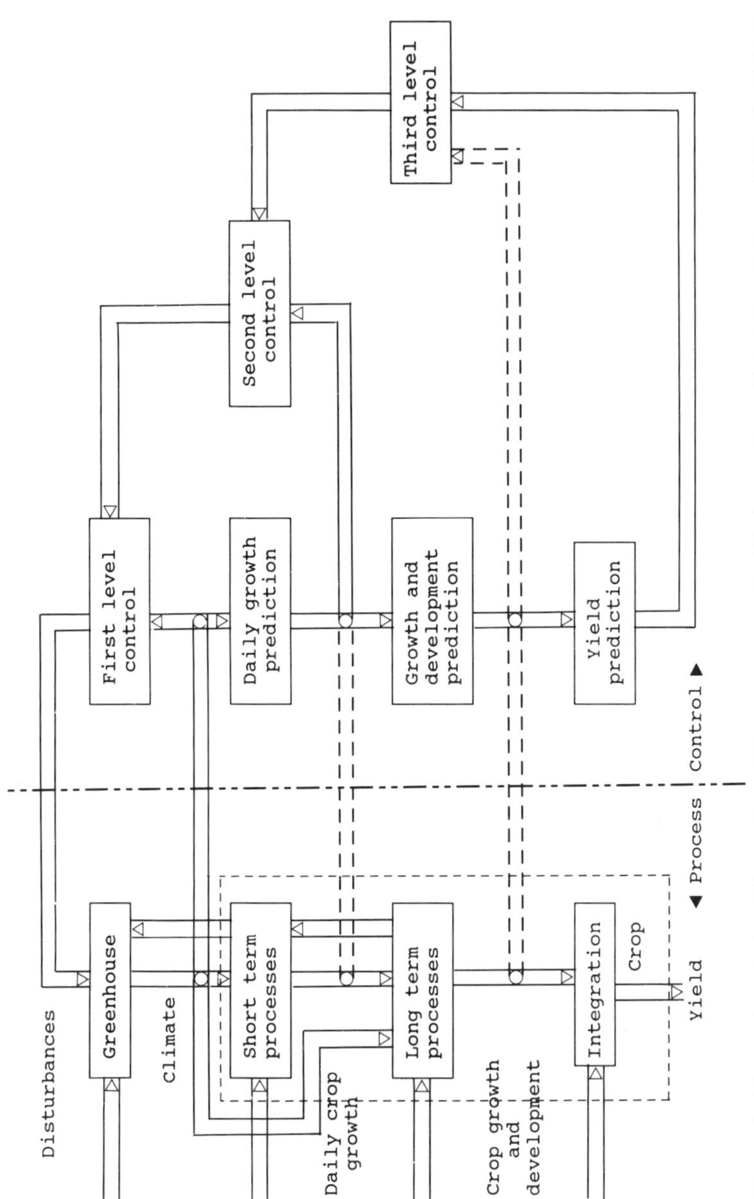

Figure 8. Hierarchy of algorithms describing crop growth. Redrawn from Udink ten Cate and Challa (1984).

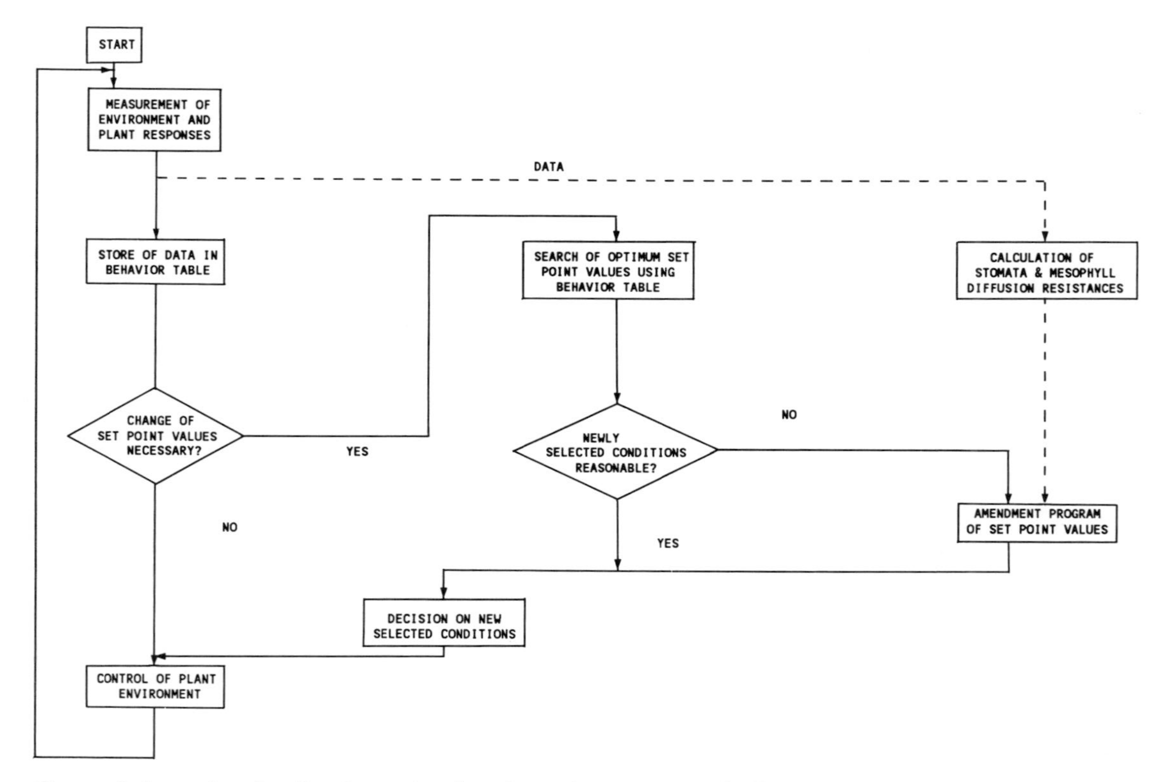

Figure 9. Learning feedback mechanism for enhancement of climate control in the greenhouse. Redrawn from Harazono et al (1984).

major influence, together with the largely empirical blueprint production systems; in the learning control procedure of Harazono et al (1984), based on a method of Kozai (1976), the system is developed by a process of modeling alternating with empirical adjustment.

Harazono et al (1984) investigated two second-level factors (sensu Udink ten Cate and Challa 1984)—transpiration rate and net assimilation rate—as determined by four first-level processes—air temperature, vapor pressure deficit, carbon dioxide concentration, and light intensity—in cucumber leaves in a wind tunnel. Measurements of these parameters were taken every 20 s and averaged every 3–10 min over 3- to 8-day experiments. The data were stored in a "behavior table," which was modified as the experiments progressed by a learning feedback mechanism (Figure 9), becoming increasingly more sophisticated as data and experience accumulated. Notwithstanding an apparent adverse effect of frequent changes in the environment, the net assimilation rate did improve over 8-day learning and adjustment periods. This, in fact, is what many growers do with a blueprint and a microprocessor, adjusting controls by trial and error, guided by their experience (green fingers) as growers (Biemond 1989; Harazono et al 1988; Hashimoto et al 1981, 1984; P. Jones et al 1988; Kano and Shimaji 1988; Kurata 1988).

Several determinant processes have been investigated, although not all in a learning control procedure, and these studies provide the basis for integration into a composite program such as the knowledge-based and hierarchically distributed on-line control system described by Hoshi and Kozai (1984). In this case, the hierarchy was based on time—minute-by-minute, hourly, and daily or weekly assessment of data. The practicalities and pitfalls of a system to control energy input were described by Takakura et al (1984) and van Zinderen-Bakker et al (1984).

Algorithms for Disease Escape

In order to be able to avoid diseases or to control them in the greenhouse it is necessary to be able to detect and measure the environmental parameters that determine the many factors affecting pathogenesis in its broadest sense. These factors include, among others, temperature, humidity, and wind speed as affecting spore dispersal; the likelihood of dew; the duration of surface wetness at various temperatures; the conditions of ventilation necessary to evaporate water films; the rate of soil warming to temperatures not optimum for root pathogens; and the maintenance of adequate soil moisture and osmotic status to promote compensatory root growth in the presence of root nibblers.

In the mind of the grower, the profitability of crop productivity from minimum energy, materials, and labor inputs is the overriding concern. In some sectors of the industry, unless there is a disastrous epidemic, a proportion of the crop is written off to diseases and insects—in practice about 5% seems wholly acceptable, and advisers are usually called in only when losses near 10–15% are threatened. Pathologists would not agree with this complacency when it would appear that control measures not requiring pesticides or sterilants often involve relatively simple regulation of the environment. Perhaps some of the deviations in crop growth algorithms to accommodate adjustments made in the interest of disease escape would be economically unacceptable to growers, but there seem to have been no feasibility studies.

Algorithms and models describing the growth of a few crops are in an advanced state and written for hierarchical decision making by computer (Anonymous 1988a; Challa and van de Vooren 1980; Ditner et al 1985; Garzoli 1985; Hesketh et al 1986; P. Jones et al 1988; Kano and Shimaji 1988; Kurata 1988; Maher and O'Flaherty 1973; Parlitz 1984; Saffell 1985; Takakura et al 1984; Tantau 1980, 1984, 1985; Udink ten Cate and Challa 1984). The computer monitors many environment sensors and on the basis of these analyses issues signals that turn on heating, switch on fans, and open ventilators in sequences that could prevent diseases as well as give maximum productivity (Ditner et al 1985). In practice, however, there may well have to be economic compromises between maximum productivity and yield losses resulting from disease-escape measures (Augsburger and Powell 1986). For example, closing the ventilators to maintain a high concentration of carbon dioxide usually reduces the water vapor pressure deficit and increases the risk of dew formation and consequent infection by water-dependent pathogens. The grower has to choose between that and opening the ventilators, risking a gray mold epidemic, for example. Even in this case, there are degrees of risk; the petal flecking caused by *Botrytis cinerea* Pers.:Fr. on roses may be economically more important than ghost spots on tomato fruit caused by the same pathogen, but a single gray mold lesion on a tomato stem could kill the whole plant. A grower of Easter lilies will certainly decide that it is more important to get most of the crop ready for the Easter market than to delay it by disease-escape measures.

Although much research needs to be done in greenhouse crops (Udink ten Cate 1980; van der Borg 1980), the effects of pests and diseases on growth can be measured, modeled, and calculated by their effects on the carbon balance (Bloomberg 1979; Boote et al 1983; Rouse 1988; Rouse et al 1985). A leaf spot disease affects the photosynthetic area (leaf area index); diseased leaves shade other leaves and may produce a toxin that affects the photosynthetic ability of the remaining leaves. Similarly, the carbon allocated to roots damaged by nematodes may be insufficient to replace damaged roots. The same may be true in the case of damage by the root-nibbling *Pythium* spp. (Stanghellini and Kronland 1986), in which plant collapse is rather sudden in the absence of macroscopic root symptoms. De Long and Powell (1988) initiated work on a deterministic model for powdery mildew of rose using modified leaf wetness recorders; high levels of leaf wetness were recorded over a 9- to 12-h period during powdery mildew epidemics.

Temperature

The literature on the effects of radiation and temperature on plant growth and crop productivity is vast and has been comprehensively reviewed by

Businger (1963), Campbell (1977), Charles-Edwards (1982), Eastin et al (1969), and Garzoli and Blackwell (1973), among many others. Algorithms and models describing radiation and temperature relations in greenhouses, without regard for pests and diseases, have been described by Critten (1983), de Koning (1988a,b), Giniger et al (1985), Manbeck and Aldrich (1967), and van der Post et al (1974).

Leaf, flower, and fruit temperatures are of particular phytopathological importance. About 99% of radiation absorbed by foliage is reradiated in infrared wavelengths or lost by conduction, convection, and evaporative cooling. At night, leaves can be 1–3°C cooler than the ambient air. If transpiration is restricted, however, they can be as much as 21–28°C warmer than the ambient air (O. F. Curtis 1936; Gates 1980; Shull 1936). With good transpiration, heat is lost by evaporative cooling, which may account for the dissipation of 65% of absorbed radiant energy. Leaf temperatures are considerably reduced, therefore, by increased rates of airflow (Carpenter and Nautiyal 1969). Flower temperatures are related to flower colors (Hanan 1965, 1970a); red carnations, for example, absorb more energy than white ones and can become warmer than the ambient air and white flowers. It would seem likely that fruits that change color when ripening, such as tomatoes and peppers, would also have different energy-absorbing characteristics and hence different temperatures. These differences become important when the dew point is considered.

Fruits, too, may have skin temperatures on the order of 10–12°C higher than the ambient air temperature. Smart and Sinclair (1976) detected a gradient of 3°C across an insolated grape berry. Fruit temperatures were related to the flux density of absorbed radiation, the size and thermal conductivity of the fruit, and convective heat transfer, which in turn depended on wind speed. Similarly, Schroeder (1965) found a temperature gradient within watermelon and cantaloupe fruits exposed to the sun, as well as diurnal variations, which diminished in amplitude toward the center of the fruit. The temperature of the pericarp 1 cm below the skin of red tomato fruits rose from about 20 to over 50°C by noon in air that rose from about 26 to 37°C in the same period. Green tomato fruits were 4–8°C cooler than red fruits throughout most of the same day.

It is important to recognize that a crop is not always grown at a temperature that is optimum for vegetative growth but rather may be grown at a temperature that is optimum for the production of flowers or fruit. These temperatures are not necessarily the same. Moreover, production temperatures change over the season and may differ in different geographical regions. In Ontario, for example, tomato plants grow best vegetatively at 25°C maintained over a 24-h period, but fruit production, depending on the light and the cultivar, is best at air temperatures of about 19–21°C during the day and 17–18°C at night. A minimum soil temperature of 14°C is recommended. In order to raise hardy transplants, there also has to be a chilling period (Table 4) (I. Smith 1988).

Temperature regulation, therefore, has to be very exact; for the early part of a spring tomato crop additional heat has to be provided, together with supplementary lighting to control seedling growth (I. Smith 1988).

As the season progresses, the problem is not one of heating but of cooling the crop. At the same time, ventilation for cooling is very wasteful of carbon dioxide in an enrichment program.

Incoming radiation largely determines the temperature relations of the greenhouse atmosphere, the root zone, and the plant (Challa and van de Vooren 1980; Ditner et al 1985; Garzoli 1985; Maher and O'Flaherty 1973; Parlitz 1984; Saffell 1985; Takakura et al 1984; Tantau 1980, 1984, 1985; Udink ten Cate and Challa 1984), and so the radiometer (Monteith 1972) has an overriding hierarchical position in many of the decisions made by a microprocessor. Most greenhouse crops are grown to a blueprint plan requiring separate day and night air temperatures and minimum soil temperatures (M. N. Cline et al 1988; I. Smith 1988; Zinnen 1988). However, several investigators, for example, Bailey (1985), Challa and van de Vooren (1980), Critten (1983), de Koning (1988a,b), Hurd and Graves (1984), and Parlitz (1984), have suggested that temperatures, as well as other factors, should be integrated over daily periods to obtain optimum averages.

Thermistors and thermostats have fast responses to changing temperatures, but there is considerable inertia in heating and cooling systems and in temperature changes in large masses of air and soil. It is fortunate that this is so, since gradual changes are better for plant growth. In computer programs, therefore, desired temperature changes in a blueprint program are made in small increments over an hour or two. This helps avoid the rapid rises in temperature early in the morning that might result in dew on slow-to-warm plants.

A temperature program written to cover the development of a crop over a season with a very wide range of incoming and outgoing radiation, and taking phytopathological concerns into account, can be very complicated indeed. Some of the parameters of pathogenesis may not be known in some cases.

Heating Systems

There is a very wide range of greenhouse heaters, variously sited and adapted for different uses in different crops (Hanan et al 1978; Mastalerz 1977). Growers have many choices, governed by the degree of temperature control required, whether priority is given to temperature control in the root zone or control in the ambient air, the uniformity of heating throughout the greenhouse and the location of the heaters in the crop (often heating pipes double as railway lines for work platforms in cucumber and tomato crops, as shown in Plate 14), the quickness of response to changes demanded by the temperature sensors, and the change of temperature between night and day (Table 4).

Space heaters (Plate 15), burning natural gas, propane, fuel oil, or occasionally fluidized sawdust or coal dust, have a quick response time, are relatively efficient in that they can be fitted with flue-gas heat exchangers, and can be used to augment carbon dioxide levels. This type of heater is usually combined with a fan for air circulation. A louvered exhaust fan

Table 4. Recommended air temperatures for raising tomato transplants and for fruit production in Ontario[a]

Growth stage	Light	Air temperature (°C)[b,c] Day	Night
Seed germination	Not critical	24	24
2 wk after cotyledon expansion	Maximum available	10–13	10–13
After pricking out	Good	21	18
	Poor	19	17
Flowering and fruiting[d]	Good	21	18
	Poor	19	17

[a] Adapted from I. Smith (1988).
[b] A minimum soil temperature of 14°C is recommended.
[c] Air temperatures can be 1–2°C lower for cold-tolerant cultivars (e.g., Vendor). A reduction of 1–2°C causes catface in fruit of very vigorous cultivars (e.g., Ohio CR 6).
[d] Minimum temperatures for flowering and fruiting.

and an oxygen inlet are also required, to prevent an excessive amount of carbon dioxide from collecting in the greenhouse.

Direct radiant heat can be provided by infrared heaters, which are usually mounted above the crop. They can be fueled in various ways and are frequently electrical (Hanan et al 1978; Knies et al 1984). It is difficult, however, to obtain uniform heating over large areas, but infrared heaters are relatively cheaper to operate than space heaters, by some 6–12% (Knies et al 1984). On the other hand, infrared heating is difficult to control by computer and air temperature sensors alone. A combination of infrared sensors and air temperature sensors solves this problem (G. Van der Hage, personal communication).

Systems of hot-water pipes, variously finned to improve heat transfer, maintain very uniform temperatures but are slower than steam systems in responding to changes called for by sensors, an advantage in avoiding dew points. They are excellent for root-zone heating—pipes are inserted in gullies in expanded polystyrene or other insulating material (Plate 16).

Solar energy can be collected in brine or some other heat sink and supplied to the greenhouse through a heat exchanger, which is ideally linked to a hot-water circulatory system (Airhart 1984; Bartok and Aldrich 1984; Dale et al 1984; Garzoli and Shell 1984; Jewett et al 1984). For hydroponic systems, electric immersion heaters are usually placed in the reservoir. However, care is needed in the selection of a suitable heater—nickel toxicity has been attributed to the corrosion of heater elements. The nutrient gulleys in nutrient film systems are often insulated to minimize heat loss to the underlying soil or bench.

In the root zone, temperature sensors are best sited near the roots, whereas for ambient air temperatures they should be in the crop canopy in an aspirated box, shielded from the sun and direct heating units.

Humidity and Vapor Pressure Deficit

In order that a plant may maintain active growth and production, it must be able to transpire freely during photosynthesis. Transpiration is only possible if water is able to leave the plant by evaporation (Gates 1980) and, for short periods, by potentially deleterious guttation (L. C. Curtis 1943; O. F. Curtis 1936). Only if the plant can absorb water through its roots can it make good the water lost by transpiration and guttation and used in metabolic processes.

In a greenhouse, water is in two forms: liquid water in the soil, hydroponic systems, and plant cells and gaseous water vapor in the air and the intercellular spaces of the leaves. It is as a gas that water is exhausted from the plant through the stomata.

The amount of water vapor that a given volume of air can hold is dependent on temperature (Figure 10). When that limit is reached, the air is saturated; water molecules in the air return as liquid water to the soil as fast as water molecules leave the liquid for the air. At that equilibrium, the plant can lose no more water vapor, and transpiration stops; guttation of liquid water then begins if root pressure is still osmotically pumping water up the plant (Bradfield and Guttridge 1984; L. C. Curtis 1943).

There are three or four ways of expressing the amount of water vapor in the air, and because this is what limits transpiration, it is important to understand them. Water vapor has a mass, and so air moisture can be expressed as grams of water per kilogram of air, which is *specific humidity*. It can also be expressed as a density, in grams of water per cubic meter of air, which is *absolute humidity*. The expression most familiar to growers is *relative humidity* r (or RH), the ratio between the mass of water vapor in the air and the mass it can hold at the saturation point.

Relative humidity is

$$r = \frac{x}{x_s} \times 100$$

expressed as a percentage, where x is the amount of water vapor in a mass of air that holds an amount x_s ($g \cdot kg^{-1}$) at saturation, at the same temperature. Like the saturation point, relative humidity is temperature-dependent; warming humid air lowers its relative humidity.

Relative humidity can also be expressed in terms of density: $r \approx (q/q_s) \times 100$, where q and q_s are the water vapor densities ($g \cdot m^{-3}$) in subsaturated and saturated air, respectively, at the same temperature.

It can also be given in terms of vapor pressure: $r \approx (p/p_s) \times 100$, where p and p_s are the partial pressures exerted by water vapor in subsaturated and saturated air, respectively, at the same temperature.

The bar is the common unit of measurement of gas pressure, but it is more correct to use the SI unit pascal (Pa). A pascal is equivalent to 10^{-5} bar and is 1 $N \cdot m^{-2}$. At 20°C and 50% RH, $p_s = 2.34$ kPa, and $p =$ 1.17 kPa. At 30°C and 50% relative humidity, $p_s = 4.24$ kPa, and $p =$ 2.12 kPa.

Since transpiration is limited by the decreasing water vapor pressure deficit as saturation is approached, it is more meaningful to express air moisture in terms of this deficit between the maximum water vapor pressure (at saturation) and the vapor pressure at some point below saturation (Anderson 1936). This is the water vapor pressure deficit, and it is expressed as a pressure in pascals (Grange and Hand 1987; Snyder and Shaw 1984).

The mathematical relationship between vapor pressure deficit and relative humidity is

$$\text{vpd} = 0.6108e^{17.269T/(T+237.3)}(1 - \text{RH}/100)$$

where vpd is the vapor pressure deficit (kPa), T is the temperature (°C), and RH is the relative humidity (%) (Snyder and Shaw 1984). Vapor pressure

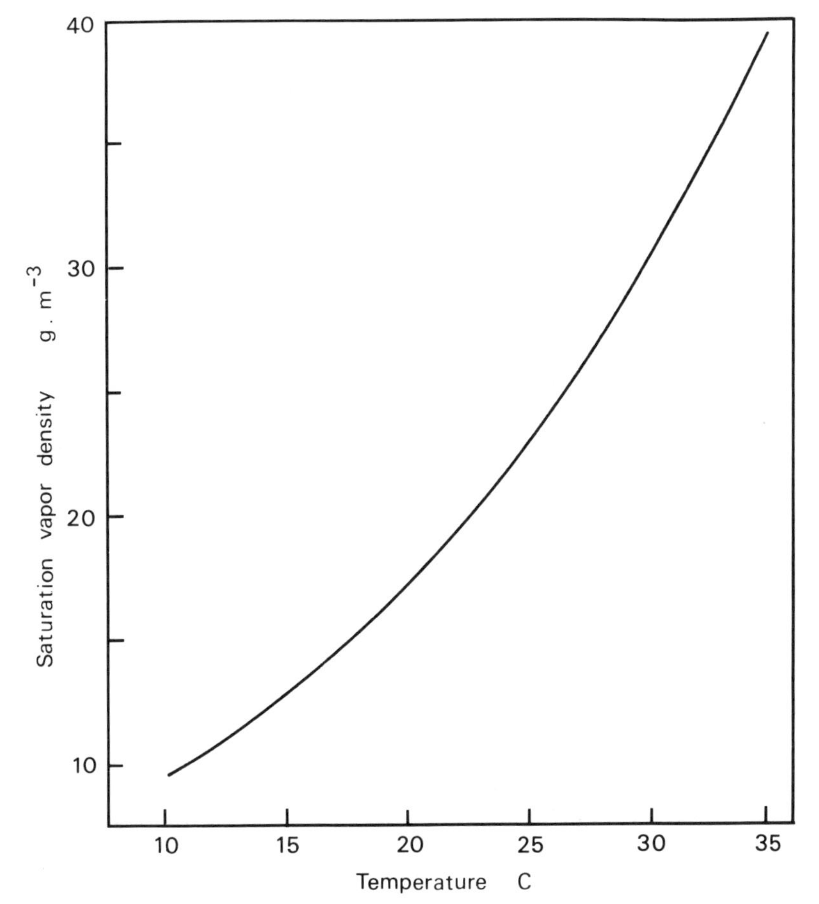

Figure 10. The amount of water vapor that a given volume of air can hold is temperature-dependent.

Table 5. Vapor pressure deficits (kPa) at selected temperatures and relative humidities

Temperature (°C)	Relative humidity (%)							
	99	95	90	85	80	70	60	50
15	0.017	0.085	0.171	0.256	0.341	0.512	0.683	0.853
20	0.024	0.117	0.234	0.351	0.468	0.701	0.935	1.170
25	0.032	0.158	0.317	0.477	0.633	0.950	1.269	1.584
30	0.042	0.212	0.424	0.636	0.849	1.273	1.697	2.122
35	0.056	0.281	0.562	0.843	1.124	1.687	2.252	2.811

deficit is a measure of the "drying power" of the air (Grange and Hand 1987). Table 5 gives some approximate values.

Measurement of Vapor Pressure Deficit

In practical terms, vapor pressure deficit is measured by the rate of evaporation of water into the atmosphere. This is most easily done with a psychrometer, or wet-and-dry-bulb thermometer, which determines the fall in temperature in a thermometer whose bulb (or thermostat) is wrapped with a water-soaked wick and is cooled by the evaporation of the water (Figure 11), while it measures the temperature of the ambient air with a dry thermometer bulb nearby. This difference in temperatures represents

Figure 11. An aspirated psychrometer contains wet-bulb (right) and dry-bulb (left) temperature sensors. Relative humidity can be calculated from their readings. The wick around the wet-bulb sensor is kept wet by distilled water from a reservoir.

the latent heat of vaporization, the amount of heat lost in the form of energy used to evaporate the water. Both bulbs are placed in a louvered box, the Stevenson screen, to shield them from the sun and other heaters, and air is aspirated through the box at a fixed rate to ensure evaporation from the wick and to sample a large volume of air. Since dissolved salts affect the rate of evaporation, distilled water must always be used. Relative humidities and vapor pressure deficits can be read from several published

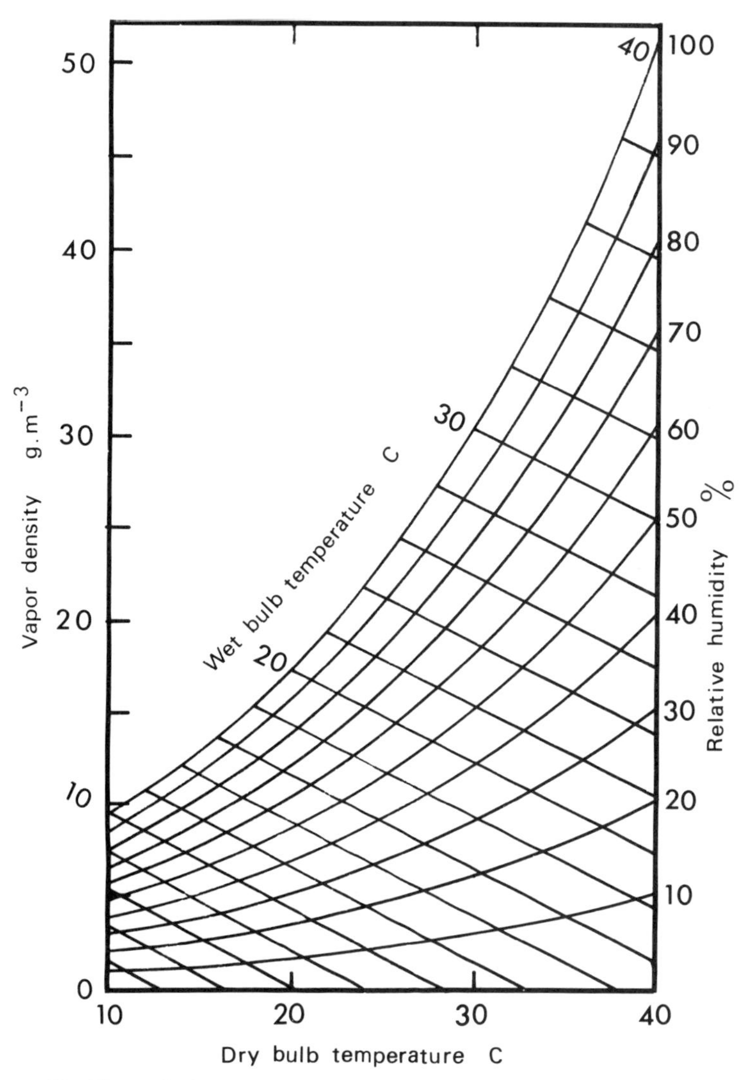

Figure 12. The relative humidity and dew point can be read from a chart when the wet- and dry-bulb temperatures are known. Redrawn from Campbell (1977).

tables of wet-and-dry-bulb readings (Forsythe 1964; Weast 1968) with a final accuracy of ±2% RH (Monteith 1972) or from graphs such as Figure 12 (Campbell 1977).

Another instrument commonly used in measuring humidity is the hygrograph, in which a hygroscopic material, usually bundles of human hair, expands in decreasing vapor pressure deficits. These movements are magnified by a series of levers in a barograph, usually calibrated in relative humidities. This type of hygrograph has an accuracy of only ±5–10% and requires frequent calibration (Monteith 1972).

A plastic wafer impregnated with lithium chloride responds to a change in humidity by changing its electrical resistance. This sensor has an accuracy of about 3% and requires frequent cleaning and recalibration.

In all cases, the accurate measurement of very low vapor pressure deficits is difficult.

Dew points can be measured (±1°C) by noting the temperature of a progressively cooling silver surface on which dew is formed. The instrument, however, is rare in greenhouses.

Various types of instruments have been catalogued by Monteith (1972).

Humidity and Crop Production

There have been relatively few studies on optimum humidities for crop production (Bakker 1984a,b, 1989; Bakker and van de Vooren 1984a; Bakker et al 1987; Cockshull 1985; Grange and Hand 1987). In general, humidity within a wide range—relative humidities of 55–95%, corresponding to vapor pressure deficits of 1.0 to 0.2 kPa—has little effect on the physiology and development of horticultural crops. Vapor pressure deficits higher than 1.0 kPa induce water stress; that is, plants lose water by transpiration faster than it can be replaced by their roots. Consequently yields decline. On the other hand, atmospheres with low vapor pressure deficits, with the concomitant likelihood of dew deposition, invite water-dependent pathogens. Low vapor pressure deficits may also lead to growth abnormalities in lettuce, cabbage, and chicory and to localized calcium deficiencies, which cause calcium scorch on tomato leaves and blossom-end rot in tomatoes and pepper fruits.

Root pressure, the osmotic pumping of water by root cells, is a function of the salinity of the rhizosphere solution and atmospheric humidity (Grange and Hand 1987). It is stimulated by high humidity at night, although not at vapor pressure deficits of less than 0.2 kPa. Prolonged high humidity, together with root pressure, may lead to glassiness, or veinal tipburn, in which the intercellular spaces in the mesophyll are saturated, similar to vitrification in tissue-cultured plants (Pierik 1988). The deposition of wax on the leaf surface may also be impeded (E. A. Baker 1974).

Bakker (1984a,b) and Bakker et al (1987) determined the effects of prolonged high humidity on the growth and productivity of cucumber. There was a significant negative correlation of vapor pressure deficit by day with total yield in all but one experiment. Fruit quality, as determined by dark green color, was reduced by high 24-h average humidity. Symptoms of calcium deficiency were correlated with the 24-h average humidity. Bakker et al (1987) concluded that high humidity should be maintained by day but that high 24-h average humidity should be reduced to avoid loss of fruit quality. No data were given on diseases in this work. Similar results were obtained with sweet pepper (*Capsicum annuum* L.) (Bakker 1989).

Photosynthesis is little affected over a wide range of vapor pressure deficits (Grange and Hand 1987). Pollination is similarly little affected (Grange and Hand 1987) at deficits within the range of 1.0 to 0.2 kPa. If the humidity is very high, tomato pollen is less likely to shed from the anthers, and at more than 1.0 kPa it tends not to adhere to the stigma (Kretchman 1968; Picken 1984); 0.7 kPa seems to be the optimum vapor pressure deficit for tomato pollination (Kretchman 1968).

High humidity can exacerbate the damage from pollutants in the atmosphere, such as sulfur dioxide, polypropylene, and ethylene (McLaughlin and Taylor 1981). High humidity, however, may be desirable to generate root pressure to move calcium into the plant (Collier and Tibbits 1984) and when entomopathogenic fungi are used to control insects (R. A. Hall et al 1982).

Evaporation and Ventilation

In order to maintain active transpiration, maintain moderate leaf and fruit temperatures, dry out wet plants and soil, and remove excessive atmospheric humidity, there has to be evaporation of water. This can occur only with adequate air movement through the crop and adequate ventilation of the greenhouse.

The energy required to evaporate liquid water is considerable; about 600 times more energy is needed to evaporate 1 g of water than to raise its temperature from 20 to 21°C. At 20°C, this energy amounts to $2.453 \times 10^3\ J \cdot g^{-1}$, and it must be supplied as heat from the immediate surroundings of the evaporating water, which therefore cool. This heat is the latent heat of vaporization. Conversely, when water vapor condenses to liquid, it gives up heat—the latent heat of condensation—and the temperature of the surroundings rises. It is the latent heat of vaporization that cools the air when fog droplets evaporate; this is the basis of fogging cooling systems. Water condensing on or evaporating from a leaf therefore brings about marked local changes in temperature at the leaf surface, enough to affect the relative activities of pathogens and their commensal microorganism.

If transpiration is impeded, leaf temperature can rise to as much as 45°C on a hot day (Gates 1980; Shull 1936); conversely, transpiration is reduced in a wind, because the leaf temperature falls by up to 10°C, with a marked drop in the vapor pressure of the leaf (O. F. Curtis 1936).

Evaporation is determined by wind speed as well as by humidity gradients. Thornthwaite and Holzman (1939) gave this relationship:

$$E = \frac{\rho k^2 (u_2 - u_1)(q_1 - q_2)}{\{\ln[(z_2 - d)/(z_1 - d)]\}^2}$$

where E is the rate of evaporation; ρ is the density of the air; k is von Karman's constant, derived from water vapor diffusivity; $u_2 - u_1$ is the difference in wind speed at two convenient heights z_2 and z_1; $q_1 - q_2$ is the difference in specific humidity at heights z_1 and z_2; and d is the zero-plane adjustment, the effective height of the evaporating surface.

If we assume that the wind speed immediately at the leaf surface is zero and that the saturation deficit is also near zero, then evaporation is a product of the specific humidity of the ambient air at saturation and the speed of its mass movement. To evaporate an inoculum drop, therefore, relatively dry air has to be passed over the leaf at a significant rate. By the same token, this also maintains a healthy transpiration rate.

Close to surface of the leaf, or the boundary layer (Sutton 1953), most water vapor exchange occurs by diffusion (Burrage 1971); outside the boundary layer, forced convection occurs. There is a vapor pressure gradient over a transpiring leaf (Frampton and Longrée 1941; Ramsey et al 1938; Thut 1939). The depth and gradient of the boundary layer depends on surface topography: a hairy leaf has a deeper boundary layer than a smooth one (Woolley 1964), and a steeper humidity gradient and a greater wind speed are required to enable forced convection to evaporate water as close to the leaf as possible.

In a greenhouse, the rate of evaporation depends on several factors (Bakker and van de Vooren 1984a); they include the difference between the saturation vapor pressure at an evaporating surface (for example, the surface of a mesophyll cell in a leaf) and the vapor pressure of the air in the greenhouse. This is a water vapor pressure gradient, and the steeper the gradient, the faster molecules of water leave the liquid for the air. There must also be a supply of energy for the latent heat of vaporization, which comes from radiant solar energy, from the greenhouse heating system, and from the plant or other wet surfaces. There must be air movement, too, the effect of which is to steepen the vapor pressure gradient if the moving mass of air is less than saturated.

Practical Control of the Environment

Controlling the greenhouse environment runs from manual opening and closing of ventilators to highly sophisticated sensing and switching by microprocessor. Most greenhouses have thermostats to switch the heating system on and off, and they can control temperature by opening and closing ventilators. However, the plant pathologist would say that it is equally important to use heating and ventilation to control humidity. Algorithms seldom recognize this or allocate humidity control a sufficiently high priority in the hierarchical system of control.

Virtually all of the many computer programs written to control the greenhouse environment do so to provide the optimum environment for productive growth (Parlitz 1984; Seginer 1980; Seginer and Raviv 1984; Takakura et al 1984; Tantau 1980; Udink ten Cate and Challa 1984; van de Vooren 1980; van de Vooren and Strijbosch 1980; van der Borg 1980), and several incorporate economic considerations (Bogemann 1980; Challa and van de Vooren 1980; Krug and Liebig 1980). Many are derived from an amalgamation of models (Udink ten Cate and van de Vooren 1984) with empirical modifications (Amdurskey 1980; Bot 1980; Harazono et al 1984), but few, if any, address the special problems of monitoring or manipulating the climate for disease control. Few, if any, algorithms are written primarily to control diseases and insect pests.

The construction of an algorithm for disease control depends primarily on a thorough knowledge of the biology of infection and secondarily on a knowledge of the inoculum potential in the rhizosphere or, in the case of airborne foliage pathogens, in the phylloplane and perhaps at some distance from the plant as well.

There may have to be compromises between maintaining the environmental optima for production and avoiding the optima for infection by a given pathogen known to be present. For example, humidity and temperatures approaching the dew point are primary environmental conditions affecting diseases of the shoot. Both temperature and humidity can be controlled in a number of ways, and the control algorithm may consider them in an order of priority determined by such factors as economics, geography, the season, and the stage of crop development. Then the environmental parameters may include the time of day, incoming radiation, wind speed and direction, shading, heating, fogging (either for cooling or for increasing humidity), irrigation, ventilation, and carbon dioxide levels. For root diseases, important parameters include radiation, the frequency of irrigation, the temperature of irrigation water, matric and osmotic potentials, and the pH of the nutrient solution.

Once algorithms have been written for the many environmental processes that determine crop productivity as well as freedom from disease, the microprocessor can assume most of the monitoring and operational control of the environment. As Kozai (1976) and Harazono et al (1984) have pointed out, the development of a profitable, productive cropping system depends on an alternating process in which modeling of optimal subsystems is successively modified by empirical knowledge drawn from the grower's experience. Not only humidity and temperature, which have been discussed at length, but also nutrition and carbon dioxide enrichment can be treated in this way. Since the environment has diurnal and seasonal variations, and plants, insects, and microorganisms have biological clocks (Follett and Follett 1981), an astronomical clock should be included in the equipment linked to the microprocessor. The monitoring and control of the environment is illustrated in Figures 13 and 14.

Since events of primary phytopathological significance take place at the plant surface, it is logical that environmental conditions be monitored there, and not, as often happens, in the ambient air some distance from the plant.

Most of the systems controlling the greenhouse environment can monitor and switch independently. Before the microprocessor became available, however, the integration of information on, for example, incoming radiation, temperature, and carbon dioxide levels was largely a matter of intuition and trial and error on the part of the grower, and it had to be modified

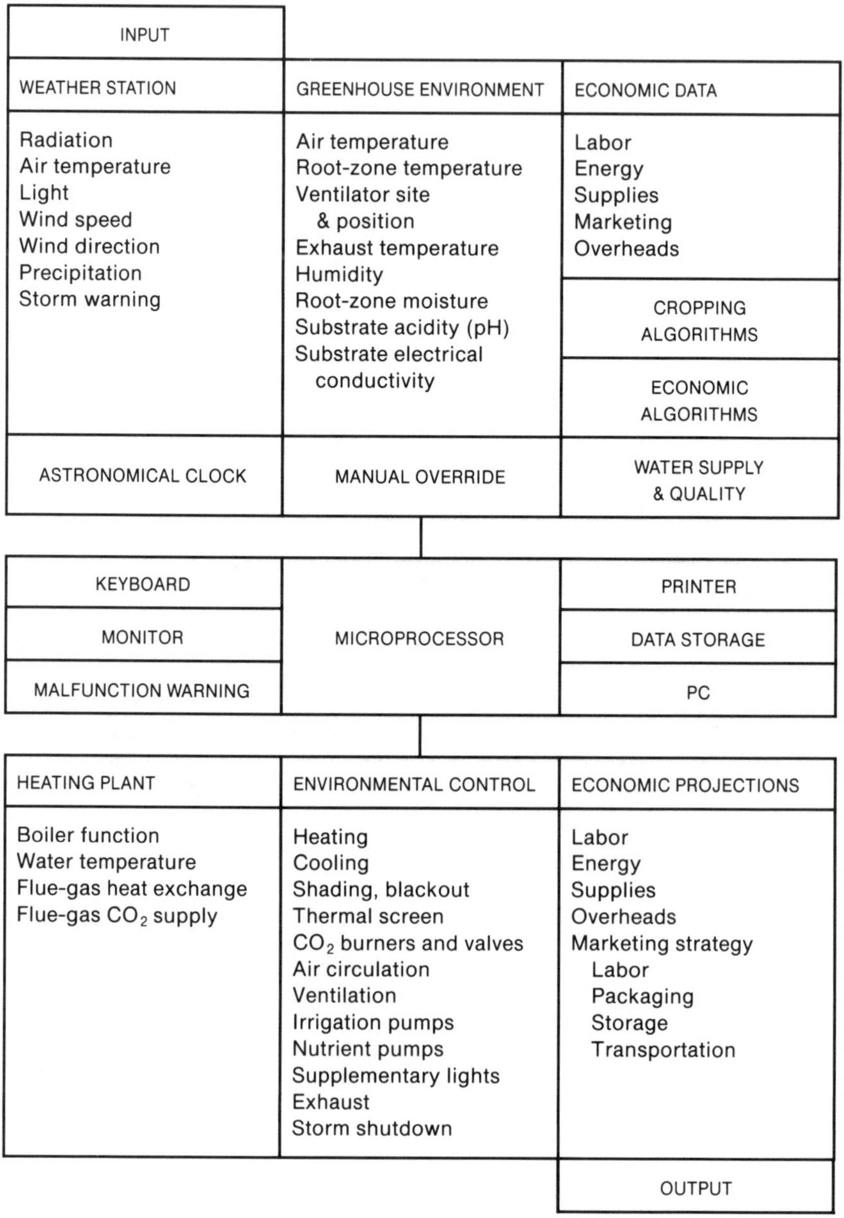

INPUT

WEATHER STATION	GREENHOUSE ENVIRONMENT	ECONOMIC DATA
Radiation Air temperature Light Wind speed Wind direction Precipitation Storm warning	Air temperature Root-zone temperature Ventilator site & position Exhaust temperature Humidity Root-zone moisture Substrate acidity (pH) Substrate electrical conductivity	Labor Energy Supplies Marketing Overheads
		CROPPING ALGORITHMS
		ECONOMIC ALGORITHMS
ASTRONOMICAL CLOCK	MANUAL OVERRIDE	WATER SUPPLY & QUALITY

KEYBOARD	MICROPROCESSOR	PRINTER
MONITOR		DATA STORAGE
MALFUNCTION WARNING		PC

HEATING PLANT	ENVIRONMENTAL CONTROL	ECONOMIC PROJECTIONS
Boiler function Water temperature Flue-gas heat exchange Flue-gas CO_2 supply	Heating Cooling Shading, blackout Thermal screen CO_2 burners and valves Air circulation Ventilation Irrigation pumps Nutrient pumps Supplementary lights Exhaust Storm shutdown	Labor Energy Supplies Overheads Marketing strategy Labor Packaging Storage Transportation

OUTPUT

Figure 13. Monitoring and control of the environment can be coordinated through a microprocessor. Economic information can be incorporated into the decision-making process, and additional economic information can be derived in part from environmental and crop growth data.

almost daily as the season progressed. The integrated algorithms developed for optimum crop growth require the rapid decisions that the microprocessor can make. The microprocessor can also sense that the dew point is about to be reached, for example, and can issue signals for corrective action to be taken, in this case switching on heating, opening the ventilators, expelling moist air, reclosing the ventilators, and pulling across the thermal screen. This action can be taken without disturbing the overall daily (or nightly) average temperature and with the minimum expenditure of energy consistent with profitability. Before the microprocessor, expelling large masses of warm, moist air could be very expensive and comparable in cost to the application of pesticides to achieve control of, say, *Botrytis cinerea* or *Bremia lactucae* Regel in lettuce (W. M. Morgan 1983a, 1984a,b, 1985).

Radiometer	
Astronomical clock	
Manual override	
Ventilator position indicator	Ventilators
Wind speed sensor	(windward, leeward)
Wind direction indicator	
Rain sensor	Storm warning
Snow sensor	
Light meter	Supplementary lighting
	Blackout screens
	Shading
Outside thermistor	
Ambient air thermistors	Heating
	Exhaust
	Air circulation fans
Wet-bulb thermistors	Pad cooling
	Roof cooling
Infrared thermistors	Fogging
Psychrometers	Thermal screens
Root-zone thermistors	Nutrient heating
	Root-zone heating
	Irrigation water heating
Water meter	
Water supply EC meter	
Nutrient EC meters	Stock solution pumps
Individual nutrient sensors	Individual nutrient pumps
pH meter	Acid pump
	Nutrient circulation pump
Tensiometers	Irrigation pumps
	Carbon dioxide burners
Carbon dioxide sensors	Flue-gas heat exchanger
	Oxygen inlet

Figure 14. Microprocessor control of the greenhouse environment. Input from sensors (left) initiates one or more control actions (right). EC = electrical conductivity.

The integration of sensors and activators is summarized in Figure 14. The master unit that controls the environmental control system can also be used with a personal computer with printer and graphics to monitor and control other operations, such as seeding, marketing, and determining labor requirements.

Temperature

Temperature can be measured quite accurately with various instruments, and the microprocessor can maintain a running average, which is required by several algorithms. Ambient air temperatures are usually sensed in a Stevenson screen, but they may have a rather distant relationship to the temperatures that determine the success or failure of infection of shoots and roots. Because of differences between the temperature of the ambient air and the temperatures of plant parts, the siting of sensors in a Stevenson

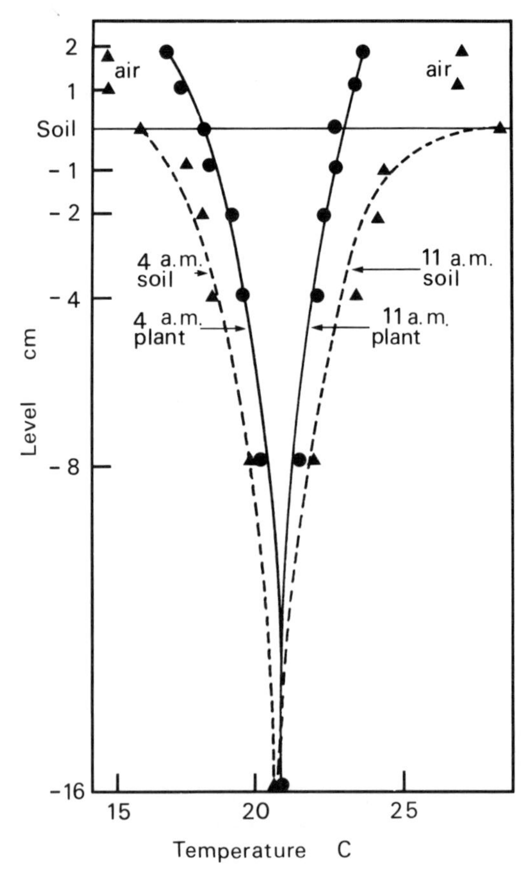

Figure 15. Temperatures in and 2 cm from the stem and roots of a corn plant. Redrawn from Waggoner and Shaw (1953).

screen 1.5 m above seedlings on the floor of a greenhouse (Plate 17) is an illogical way of managing the environment of the plants.

Leaf, flower, and fruit temperatures are particularly important, since these structures are the entry points for very many pathogens. For reasons discussed previously, these temperatures may differ significantly from the ambient air temperature. There are also wide differences between the temperatures of different parts of the same shoot of a plant, between plants of different size, shaded and unshaded, and between shoots and roots (Figures 15 and 16) (Waggoner and Shaw 1953).

The advantages of infrared thermometry in monitoring the environment of shoot pathogens, therefore, seem to be compelling. Amsen (1980) considered leaf temperature to be a valid parameter for control of the environment. From the phytopathological point of view, the root-zone temperature is equally important, and for a good estimate of this, several sensors must be sited in the root zone to determine the maximum, minimum, and average temperatures. Again, the microprocessor is able to produce a running average.

Humidity and Dew Point Control

Control of the water vapor pressure deficit of the air remains rather crude, and it is rather easier and cheaper to decrease the deficit than to

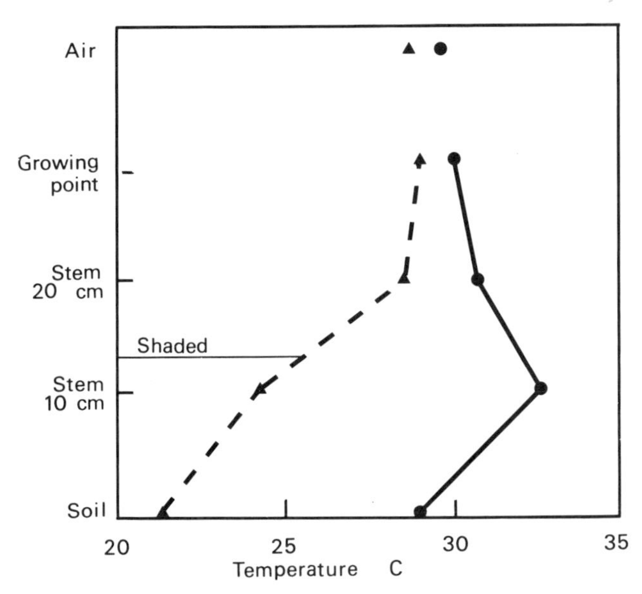

Figure 16. Temperatures in and near the stems of two tomato plants: a small plant, 30 cm tall (continuous line), and a large plant, 45 cm tall (dashed line), growing in peat soil in August, with the sun obscured. The lower 12 cm of the large plant was in shade. Redrawn from Waggoner and Shaw (1953).

raise it. Because vapor pressure deficit, relative humidity, and temperature are interrelated (Figure 17), it is impossible to alter one without changing the other.

The simplest method of decreasing the vapor pressure deficit is to introduce water as a vapor from evaporators or by fogging. In an actively growing crop, it can also be reduced by closing the ventilators, allowing transpired vapor to accumulate in the air until the deficit reaches a limiting value and stops transpiration by closing the stomata. Even then, water vapor may continue to enter the air, probably as the result of root pressure and

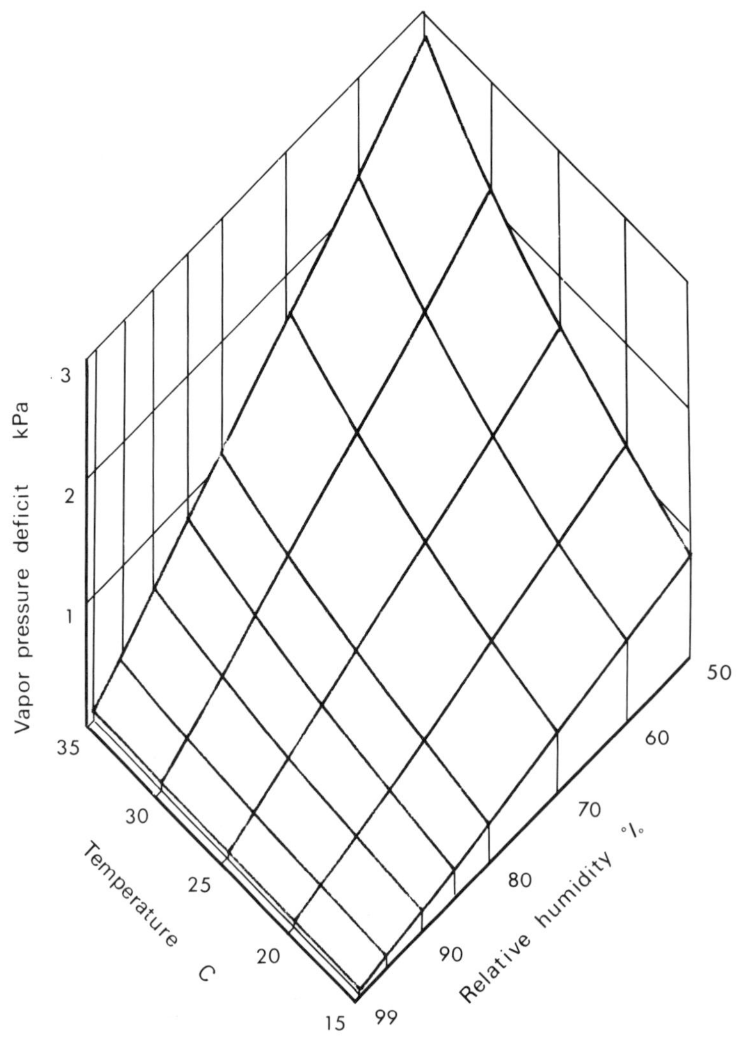

Figure 17. The relationships between water vapor pressure deficit, relative humidity, and temperature.

evaporation through thin cuticles (Bradfield and Guttridge 1984; O. F. Curtis 1936; K. E. Cockshull, personal communication). This also occurs when there is evaporation from soil or compost or fibrous organic mulch.

Conversely, the vapor pressure deficit can be raised most simply by allowing moist air to escape through the ventilators or exhausting it by fan. If a heat sink is present, such as a cold roof above a thermal screen, the screen can be opened to allow water vapor to condense on the underside of the roof. In this case, care should be taken to prevent drips from falling onto the crop. Plastic film can be treated so that water does not bead

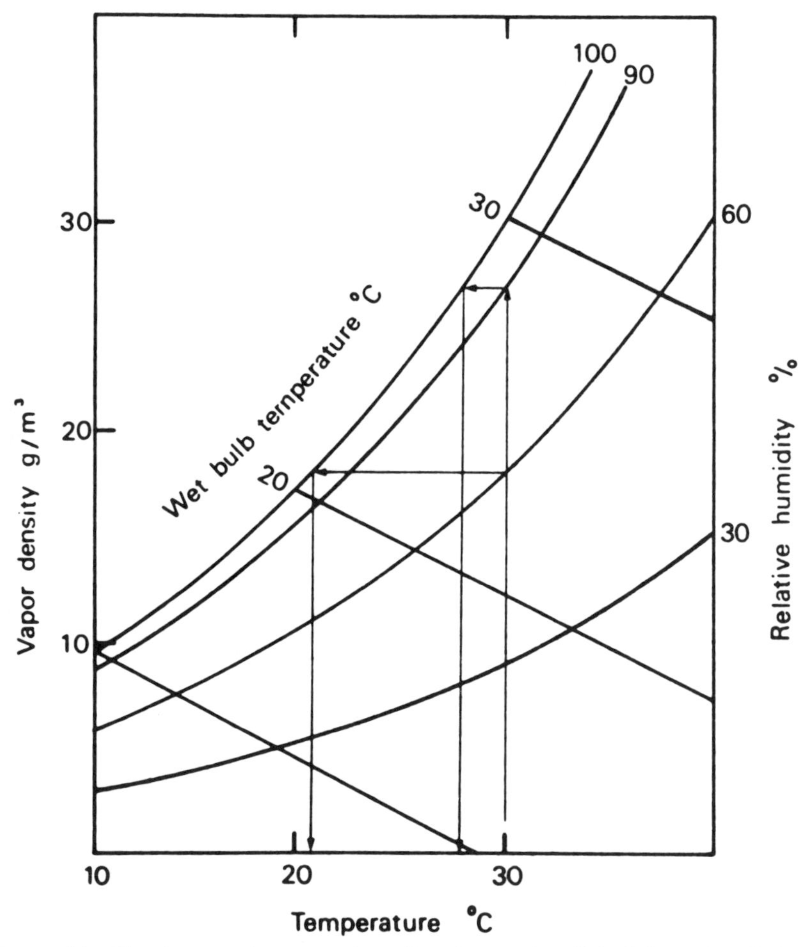

Figure 18. The temperature reaches the dew point if it falls only slightly at high relative humidity (low vapor pressure deficit). If the air temperature is 30°C, a temperature drop of only 1.8°C causes it to reach the dew point at 90% RH, but a drop of 9.2°C is needed for it to reach the dew point at 60% RH. Reprinted, by permission, from Jarvis (1989).

on it, the excess water running down to the ground in a continuous film. At the cost of some considerable expenditure of energy, air can be circulated over other heat sinks, such as chilled pipes.

The dew point is the temperature at which the vapor pressure of the air is equal to the saturation vapor pressure (that is, the vapor pressure deficit is zero). From the phytopathological point of view, the deposition of dew is one of the most important physical events in the greenhouse. It frequently occurs on clear, cool nights following warm days when humid air has been allowed to accumulate in the greenhouse. Nighttime radiation cools the structure, glass, and crop quickly. The air mass cools rather more slowly, except in the boundary layers of the rapidly cooling crop and structure. Cooling saturated air disturbs the equilibrium of saturation, and water molecules must leave the vapor phase and enter the liquid phase. This involves the loss of the latent heat of condensation (2.453×10^3 $J \cdot g^{-1}$ at 20° C) to heat sinks, which are the greenhouse structure and the crop, and so water condenses on the glass and on leaves and fruit. The temperature change needed to bring this about is very small when the relative humidity is in the range of 90–100% and rather large when the relative humidity is lower (Figure 18).

Schein (1964) pointed out the importance of these very small temperature changes at high relative humidities: as Figure 18 illustrates for air at 30° C, a temperature drop of 9.2° C is required for dew to be deposited when the relative humidity is 60%, whereas a fall of only 1.8° C is needed when the relative humidity is 90%. In practical terms, that small temperature change is less than the sensitivity of most greenhouse thermostats, thermometers, and hygrometers (Monteith 1972), almost certainly smaller than the temperature gradient in a controlled environment chamber, and far smaller than the gradient in a radiating greenhouse crop. Reports of spores germinating without free water at low vapor pressure deficits, therefore, must be treated with great circumspection. Indeed, it seems likely that spores, particularly hygroscopic spores, lying on a leaf may act as nuclei for condensing dewdrops, and the water could be retained in a mucilaginous sheath by, for example, conidia of *Botrytis cinerea* (Figure 19) (Blackman and Welsford 1916; Yarwood 1950).

Adjustments of the water vapor pressure deficit are made in response to signals from a psychrometer (wet-and-dry thermistors in an aspirated box) or from a radiometer or are made by a clock (Figure 14). The measurement of very low deficits is very difficult, requiring precise calibration of the thermistors, yet it is at low values that accurate responses are most needed so as to avoid the risk of reaching the dew point by a relatively small fall in temperature (Okada and Sameshima 1986; Schein 1964).

It is also important to recognize that leaf or fruit temperature may well fall below the ambient temperature under certain conditions. Therefore, a safety margin may have to be built into the computer to switch on heaters at a temperature a few degrees higher than the dew point at a given water vapor pressure deficit, expel humid air when the rate of temperature fall reaches a prearranged threshold level, or dehumidify the air over a heat sink. Another possibility, not used commercially, is to monitor leaf or fruit temperatures by infrared thermometry, and another is to use a synthetic fruit containing a thermistor and having thermal properties similar to those of a natural fruit. The computer's treatment of fruit temperature should be corrected for the color changes that determine the different thermal properties of the fruit as it ripens.

Ventilation

Ventilation controls are operated mainly to regulate air temperature in the greenhouse. They can be operated manually but are better linked to sensors for automatic operation (Goedhart et al 1984; Nederhoff 1984). Electric motors switched from the microprocessor open the roof or side ventilators, usually by a rack-and-pinion mechanism. Sensors detecting wind speed and direction and detecting rain and snow determine whether the ventilators are opened on the lee or the windward side of the greenhouse and the degree to which they are opened. Rheostats sense the degree of opening for display on the visual display unit and for record keeping. Whether the ventilators are opened or not is determined by the required day and night temperatures, maximum and minimum humidity settings, or incoming radiation (Figure 14). There is usually an overriding storm closure response, to avoid damage to the crop and the structure.

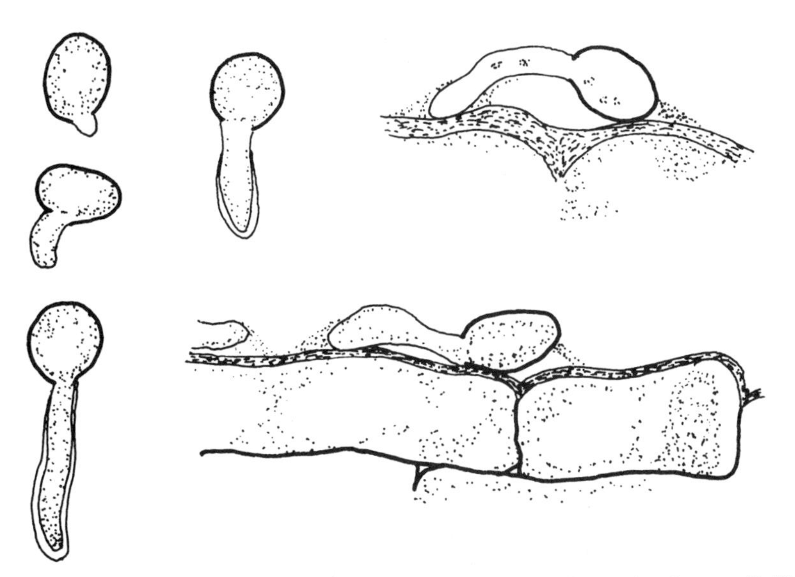

Figure 19. Stages in the germination of a conidium of *Botrytis cinerea* (left). The mucilaginous sheath is negatively stained. The conidium and germ tube (right) adhere to the host cuticle (in section) by means of the mucilage. Germination must occur in a drop of water. Redrawn from Blackman and Welsford (1916).

If ventilation is used to expel warm, humid air at sundown, then ventilator action can be determined from psychrometers, greenhouse temperature, radiation, or a clock (Figure 14).

Thermal Screens and Shading

Thermal screens are used to conserve energy under sunless conditions when the outgoing radiation exceeds the incoming radiation. The screen is opened and closed by an electric motor switched on by signals from a radiometer or a clock or by a combination of both; for example, the screen can be set not to close before a given time, or conversely, radiation input can override the clock. The degree of closure can be preset or determined by one of the other sensors, such as a humidistat in order to reduce humidity (Figure 14).

Similarly, sun-shading screens are usually controlled by a radiometer or can be controlled by time. Again, the degree of shading can be varied to suit the radiation and the crop. The thermal screen can also be used to prevent excessive temperature buildup in the crop. In both cases, a cheaper photoelectric cell can be used instead of a radiometer, but the radiometer allows accumulated radiant energy to be used as a parameter in assessing many aspects of crop productivity.

Irrigation

In soil and composts, a tensiometer can be used to determine the water potential of the medium. When linked to a computer that integrates monitoring signals from the environment, it can assess the crop's irrigation needs. These needs can be determined by environmental factors such as humidity, temperature, radiation, and light, as well as by astronomical time—the time of day as well as the time of year and the stage of crop development. Water is then precisely dispensed by a pump with an in-line impulse controlling pump frequency and stroke.

Nutrition: Fertigation

In most commercial systems, the total soluble salts in irrigation water are measured by electrical conductivity (EC), expressed in siemens per meter ($S \cdot m^{-1}$), as are the soluble salts in the water source—municipal supply, well, or creek—so that an appropriate correction can be made by the microprocessor (Hearn et al 1981).

Fertilizers in concentrated solution are metered into the irrigation system by a proportioning system controlled by solenoid valves (Figure 20). In an advanced system, individual salt components, each in its own stock solution of known molarity, can be metered by individual dosimetric pumps to deliver an appropriate volume to a dilution-mixing tank. Doses are monitored by computer, so that solutions of any chemical composition and molarity can be assembled for a crop at different stages of growth during the season or for different crops in different greenhouse ranges. Individual checks of EC are fed back to the computer. Analytical sensors for individual ions can also be used (Figure 20) (Haggett 1989; Papadopoulos and Liburdi 1989). Corrections for the salts already in the water supply are made by the computer.

Nutrition: pH

Just as the fertilizer input can be controlled, the pH of irrigation water can be determined continuously by redox electrodes and maintained at

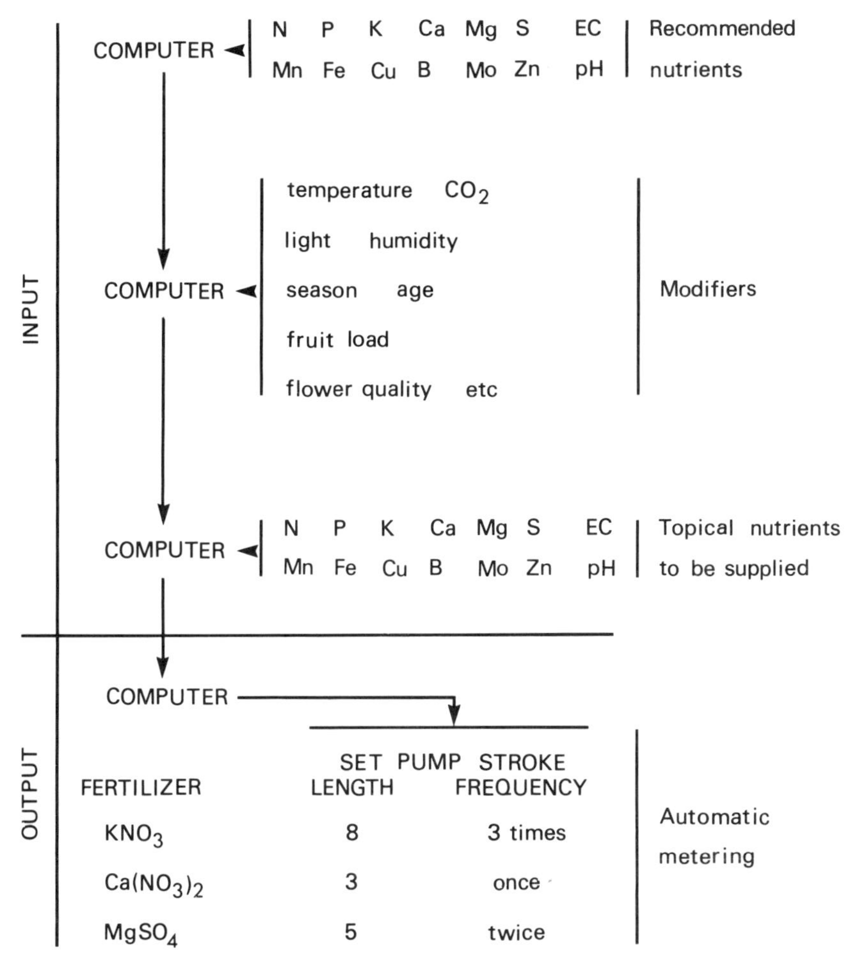

Figure 20. Schematic diagram of a computer-controlled fertigation manager, which automatically adjusts the supply of water and nutrients according to information stored in memory and inputs from sensors in the environment. Redrawn from Papadopoulos and Liburdi (1989).

a selected value by pumping in an acid, usually nitric or phosphoric acid. Appropriate corrections to nitrogen or phosphate in the fertilizer input are made by the microprocessor (Figure 14).

Carbon Dioxide Enrichment

Air outside the greenhouse contains carbon dioxide at about 320 ppm. Inside the greenhouse, even in a ventilated greenhouse, carbon dioxide is rapidly depleted by an actively growing crop (Figure 21) (Porter and Grodzinski 1985) and must be replaced from artificial sources—from burning natural gas, propane, fuel oil, sawdust, or coal dust or from supplies of liquefied carbon dioxide—if high levels of productivity are to be maintained. An oxygen inlet is required for efficient burning without the production of too much of the nitrogen oxides, ethylene, carbon monoxide, and other pollutant gases, as well as for protecting the health of workers. Sensors measuring flame color detect abnormal burning.

In order to monitor carbon dioxide levels in the crop, sensors (Monteith 1972) are sited at crop level. These can also be linked to a clock or to a radiometer (Figure 13), and carbon dioxide can be monitored in several areas sequentially by means of a manifold and scanner.

Other devices measure carbon dioxide by the differential absorption of infrared radiation (Monteith 1972).

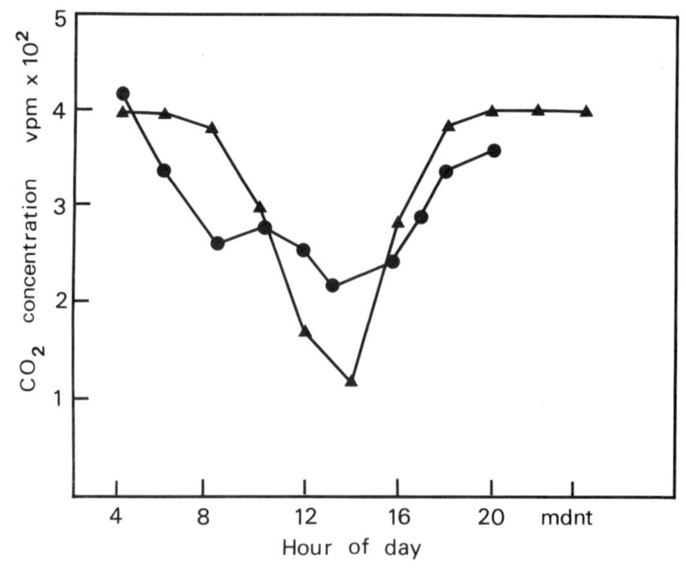

Figure 21. Diurnal changes in carbon dioxide concentration in a glasshouse growing lettuce and tomatoes (●) and in a tightly closed plastic house growing lettuce, cucumbers, and peppers (▲). Redrawn from Porter and Grodzinski (1985).

Astronomical Clock

Because plant growth is largely governed by circadian rhythms of radiation, light, and day length, greenhouse control systems invariably incorporate a clock. Biological events proceed without artificial time modification, such as time zones or summer time, and so the clock is programmed to run on astronomical time, with adjustments for day lengths imposed by the latitude and the season.

Photoperiod controls moving shade and blackout screens can be accurately regulated by the clock. Other environmental adjustments, such as changing day and night temperatures, can be made stepwise or gradually by cams on the thermostats over a selected period. This type of adjustment can be made in conjunction with other inputs, such as radiation, outside temperature, humidity, or carbon dioxide concentration, so that, for example, large, abrupt changes in temperature do not allow the dew point to be reached, nor are high concentrations of carbon dioxide pumped into the greenhouse when the crop cannot use it.

A simple concept of measuring the net environment for crop production is that of day-degrees, or heat units expressing the arithmetic accumulation of elapsed days multiplied by the average number of degrees by which a base temperature is exceeded each day. This concept can be applied to

SUMMARY

The Greenhouse Environment and Crop Protection

In the latter half of the 20th century, greenhouse technology, particularly microprocessor control of the physical and chemical control of the greenhouse environment, has made enormous strides. Whereas there are numerous algorithms for optimum crop productivity and the means to fine-focus them by feeding back the grower's experience as well as environmental and physiological information from the plant itself, there are virtually no algorithms that promote disease escape in any crop. Factors that affect infection processes at the plant shoot and root surfaces can be monitored very accurately by a wide variety of sensors if they are correctly placed in the crop. Temperature depends on radiation received at the plant surface, the plant color, rates of transpiration and evaporation, and wind speed. The water relations of the plant depend on rhizosphere water and the atmospheric vapor pressure deficit. Given adequate knowledge of the autecological requirements, it is possible to construct algorithms for disease escape, which, however, must be reconciled with algorithms for optimum and economical productivity.

vegetable crops (Kitchen 1953), but it has not been widely applied to greenhouse crops. Mauromicale et al (1988), however, were able to predict the flowering and harvest dates of beans (*Phaseolus vulgaris* L.) from a temperature base of 4.5° C in a cool greenhouse.

The day-degree concept has been used to model potato early blight, caused by *Alternaria solani* Sorauer, in Colorado potato fields (Franc et al 1988) and has been proposed for predicting insect activity thresholds (Preuss 1983). It has not, however, been applied to a greenhouse crop disease.

Overall stress in crops can be monitored in an early-warning system by electronic monitoring of chlorophyll fluorescence (Methy and Salager 1989), but again, this method has not been applied to practical problems.

Environmental Stress and Predisposition to Disease

CHAPTER 4

Schoeneweiss (1975) defined *stress* as any factor capable of producing a potentially injurious strain in a plant. Most strains are elastic (that is, they are reversible), but severe stress may induce a plastic strain, which is irreversible. Drought that induces permanent wilting, for example, is a plastic strain.

The physical stresses of the environment are measured in terms of force per unit of area, with dimensions of newtons per square meter ($N \cdot m^{-2}$) or pascals (Pa) (Eastin and Sullivan 1984; Levitt 1980). Strains, which are the reactions of plants, are measured by changes in the shape and size of cells and tissues or by changes in metabolism. Thus, the short-term stress of a very dry environment results in a reduction of the stomatal aperture, whereas prolonged stress of that type might result in loss of photosynthetic activity, loss of cell function (such as cytoplasmic streaming), or changes in the cell wall that inhibit cell expansion.

Stresses are applied to plants through several agencies: water (overabundance or drought), temperature, light intensity, nutrition (including deficiencies and toxicities of trace elements), salinity, microorganisms, pests, pesticides, and environmental pollutants (Eastin et al 1969; Hsaio 1973; Levitt 1980; Schoeneweiss 1975).

In horticulture, the routine operations of pricking out seedlings, transplanting, deleafing, training, and pruning can also induce strain in plants (Schoeneweiss 1975).

Most horticultural crops are grown under environmental conditions that differ from the optimum conditions for the production of biomass as determined by plant physiologists. With a powerful economic premium on greenhouse space, productivity per unit of area of the floor or bench is determined from the number of blooms cut, the weight of fruit, or the number of potted plants produced in as small an area per plant as possible. There is thus a stress imposed by competition for light and, for plants sharing a groundbed or pot, for water and nutrients as well. The premium on space also requires that many plants or crops be grown per unit of time and that market deadlines be met rigidly. There is essentially no market for poinsettias after Christmas or for Easter lilies after Easter. Growers compete against imports and have become adept at retarding or advancing cropping to realize top prices in the market. This can be done by manipulating temperature, irrigation, nutrients (especially nitrogen and phosphorus), the electrical conductivity of the soil solution, or day length, or it can be done by using growth regulators. Different market requirements impose different stresses. In Europe tomatoes are preferred no larger than about 50 mm in diameter, whereas in Ohio and Ontario they are preferred much larger, 75 mm or more, and so flower clusters are commonly thinned to three, four, or five flowers. Interfruit competition is therefore different. There are similar differences in the sizes of cucumbers and many other products around the world.

The reaction of plants to stress is frequently a predisposition to disease (Schoeneweiss 1975; Yarwood 1959b), that is, physical and metabolic strains readily exploited by disease organisms in the environment. Generalizing, Schoeneweiss (1975) said that environmental conditions favorable for growth usually favor infection by obligate parasites but may confer resistance to facultative parasites. The corollary is the reverse; stress induces resistance to obligate parasites and susceptibility to facultative ones. Strain can override genetic resistance in some cases (Grainger 1979).

Temperature Stress

The important temperatures that regulate pathogen activity and host defense activity are air and plant temperatures, which are not necessarily the same, and rhizosphere temperatures. Temperature is also important at the site of origin of inoculum, since it affects such things as survival, sporulation, and spore dispersal.

Greenhouse design usually aims to provide the optimum temperatures for crop productivity, which, as was pointed out earlier, may not be the optimum for vegetative growth. Temperature extremes are obviously stressful, but suboptimum temperatures (i.e., those varying from the optimum for vegetative growth) are also stressful to a degree, depending perhaps on the extent of the temperature variation. Optimum temperatures for production vary for different crops and different stages of cropping; a number of them were summarized by Hanan et al (1978) and Mastalerz (1977).

Temperature optima for plant growth processes may often coincide with optima for pathogen activity, but defense reactions in the host–parasite complex may be quite unrelated. This runs contrary to Schoeneweiss's generalization (Schoeneweiss 1975). For example, the optimum temperature overall for vegetative growth of tomato is about 25° C, and the optimum root-zone temperature for production is 16–17° C in winter and 18–20° C in spring (Geissler 1979). The optimum temperature for the growth of *Fusarium oxysporum* Schlechtend.:Fr. f. sp. *radicis-lycopersici* W. R. Jarvis & Shoemaker in culture is 27–28° C (Yamamoto et al 1974), whereas pathogenicity is best expressed at around 15–18° C (Jarvis and Shoemaker 1978). By contrast, the optimum temperature for pathogenicity by *F. oxysporum* f. sp. *lycopersici* (Sacc.) W. C. Snyder & H. N. Hans. is 27° C, which happens to be the same as the optimum temperature for the pathogen's growth in culture (Clayton 1923a).

L. D. Leach (1947) related optimum temperatures for seed decay and seedling damping-off diseases to the relative growth rates of the pathogen and the host. The growth rate of the host was measured by Kotowski's formula for the coefficient of velocity of emergence (CVE):

$$\text{CVE} = \frac{(\text{Total emergence at end of period}) \times 100}{\Sigma\,[(\text{Each daily emergence increase}) \times (\text{days since planting})]}$$

The growth rate of the fungus was measured in culture and expressed in millimeters of radial growth per day on an agar medium or milligrams of dry-weight yield per day in a liquid medium. Leach calculated the ratio of seedling growth rate (CVE) to pathogen growth rate over a range of temperatures in a study of damping-off of spinach (*Spinacia oleracea* L.), pea (*Pisum sativum* L.), and sugar beet (*Beta vulgaris* L.) by *Pythium ultimum* Trow and *Rhizoctonia solani* Kühn (Figures 22–24). In spinach, a low-temperature crop, Pythium damping-off was most severe (i.e., the ratio was low) between 12 and 20° C (Figure 22), and disease escape occurred at 4° C (the ratio was high when the host grew faster than the fungus). Rhizoctonia damping-off of spinach was moderate at 16° C and severe at 20° C, with differences between two seed lots, and there was no damping-off below 12° C (Figure 23). By contrast, there was no damping-off of seedlings of a high-temperature crop, watermelon, at 35° C, but the disease was severe at lower temperatures, at which host growth was retarded. Pythium damping-off and seed decay of pea was severe between 12 and 20° C and less severe at lower and higher temperatures, at which the growth rate of the host exceeded that of the pathogen (Figure 24). In sugar beet seedlings, damping-off caused by *P. ultimum* was severe over the temperature range 12–20° C, whereas damping-off caused by *R. solani* was more severe between 16 and 30° C. The seedborne sugar beet pathogen *Phoma betae* A. B. Frank also caused the most damage to emerging seedlings at low temperatures, when the relative growth rate was lowest; higher temperatures favored the host.

Bateman and Dimock (1959), investigating three root diseases of poinsettia (*Euphorbia pulcherrima* Willd. ex Klotzsch), caused by *Thielaviopsis basicola* (Berk. & Broome) Ferraris, *R. solani*, and *P. ultimum*, could find no satisfactory relationship between the optimum temperature for the growth of poinsettia roots (26° C) and the optimum temperature for the growth of the three pathogens in culture: for *T. basicola*, 24° C (radial growth rate on agar) and 21° C (in liquid medium); for *R. solani*, 30° C; and for *P. ultimum*, 27° C. The root rot caused by *T. basicola* was severe between 13 and 26° C (the optimum was 17° C), and *P. ultimum* also caused the most rot at 17° C, whereas *R. solani* caused more root rot as the temperature increased from 13 to 30° C.

Evidently the results of Bateman and Dimock (1959) do not lend themselves readily to the interpretation of Leach (1947). Superficially Leach's

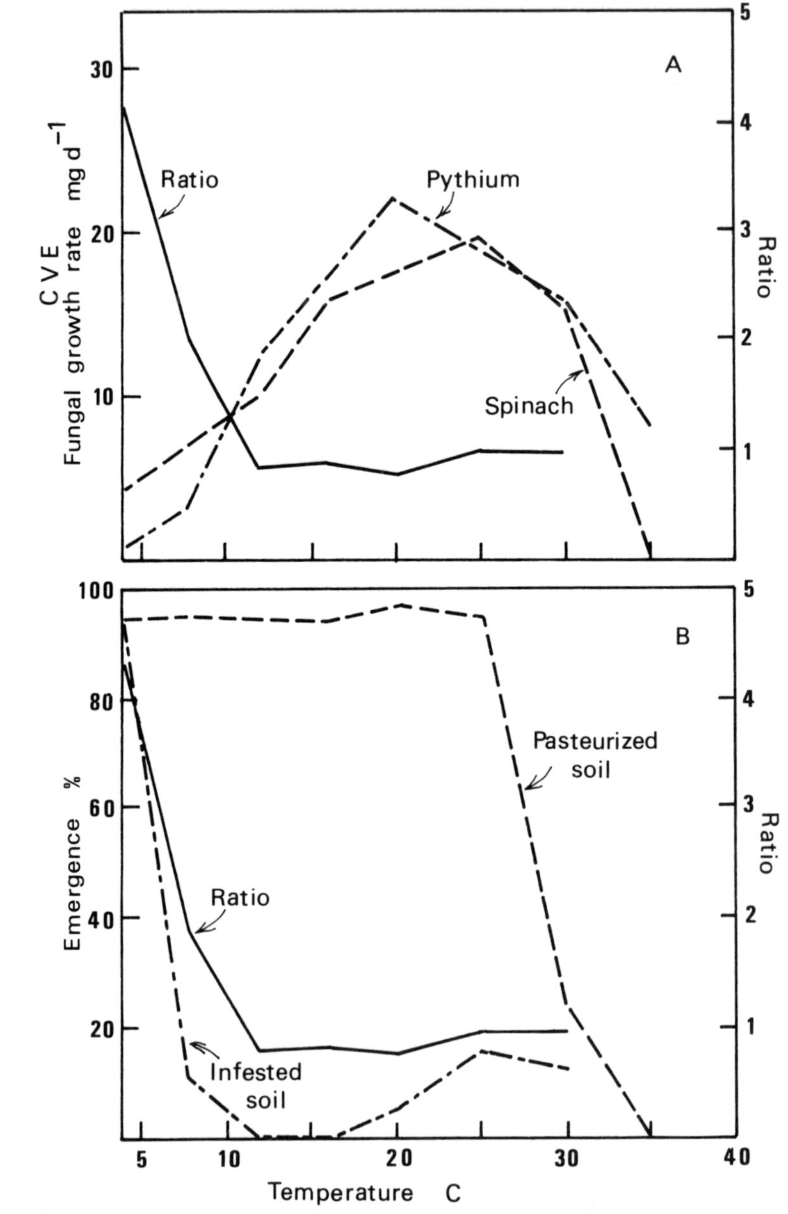

Figure 22. A, Coefficient of velocity of emergence (CVE) of spinach in pasteurized soil, growth rate of *Pythium ultimum* in a liquid medium, and ratio of emergence rate of spinach to growth rate of *P. ultimum*. **B,** Percent emergence of spinach in pasteurized soil and in soil infested with *P. ultimum* and ratio of growth rate of the host to growth rate of the pathogen. Redrawn from L. D. Leach (1947).

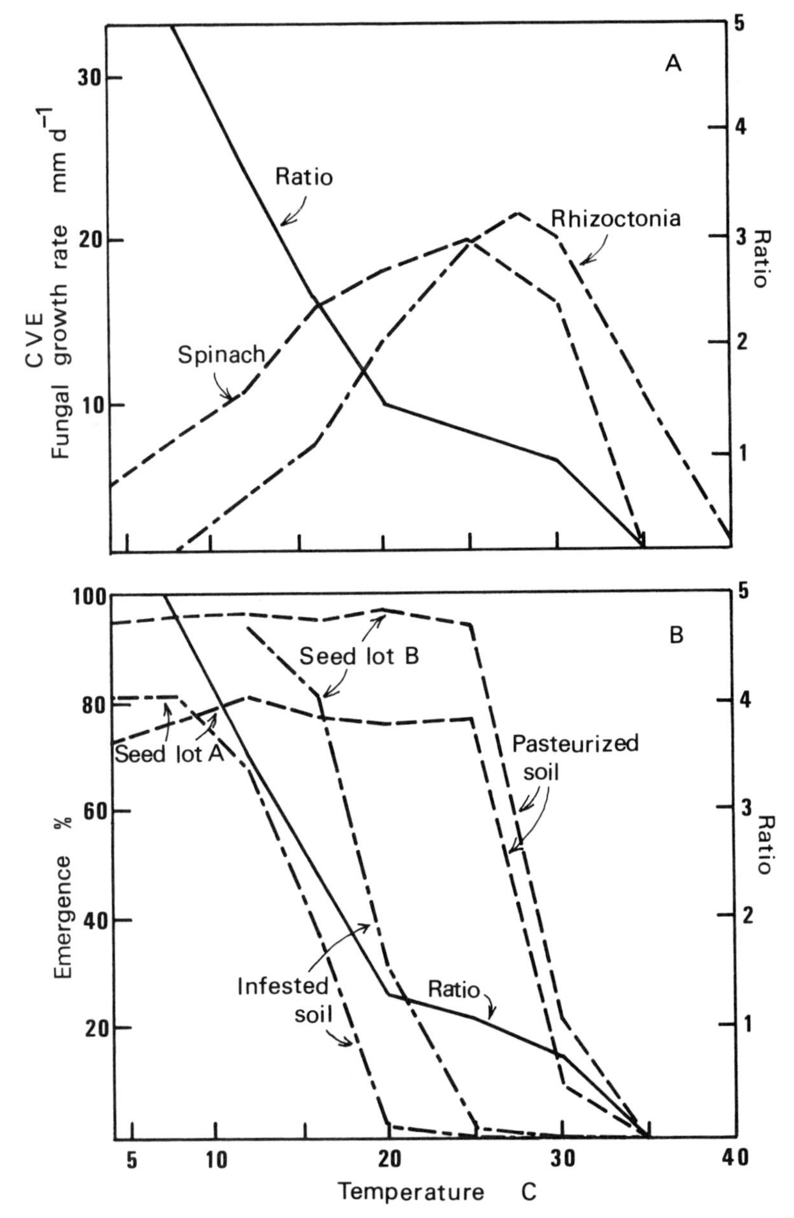

Figure 23. A, Coefficient of velocity of emergence (CVE) of spinach, growth rate of *Rhizoctonia solani* on agar, and ratio of emergence rate of spinach to growth rate of *R. solani*. **B,** Percent emergence of spinach in pasteurized soil and in soil infested with *R. solani* and ratio of growth rate of the host to growth rate of the pathogen. Redrawn from L. D. Leach (1947).

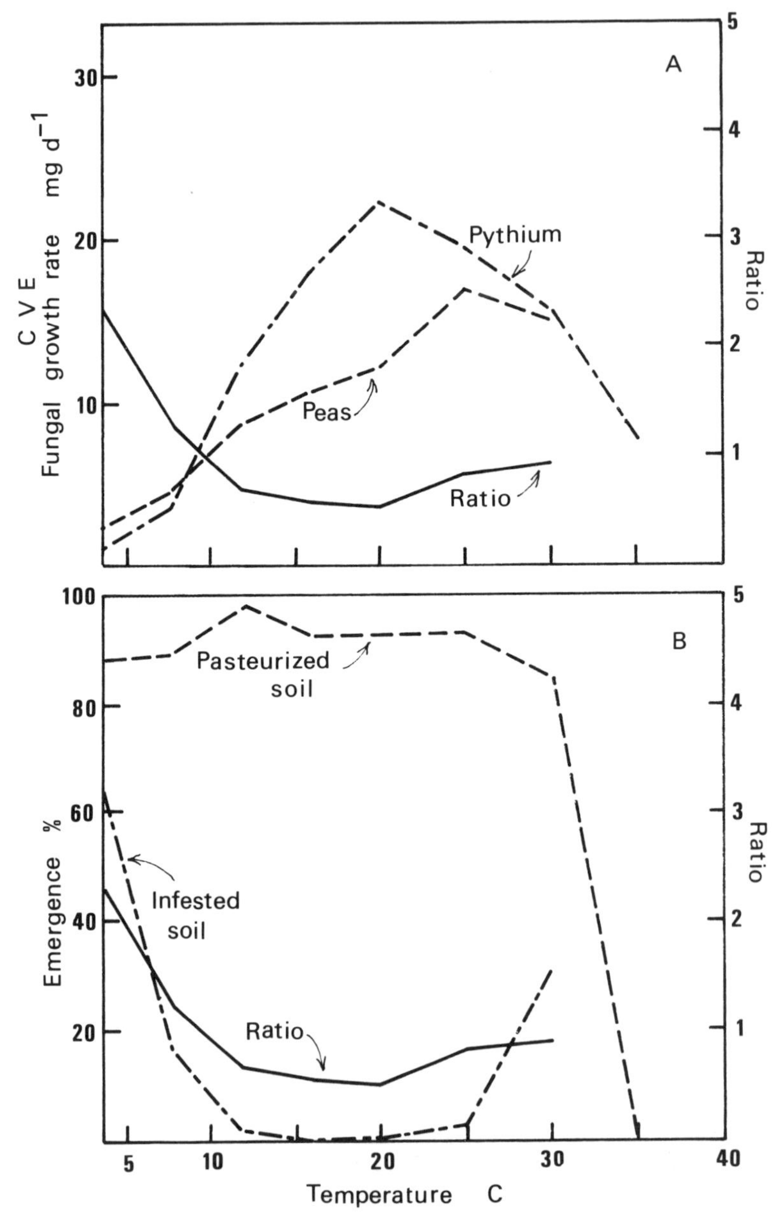

Figure 24. A, Coefficient of velocity of emergence (CVE) of peas, growth rate of *Pythium ultimum* in a liquid medium, and ratio of emergence rate of peas to growth rate of *P. ultimum.* **B,** Percent emergence of peas in pasteurized soil and in soil infested with *P. ultimum* and ratio of growth rate of the host to growth rate of the pathogen. Redrawn from L. D. Leach (1947).

concept might appear to be a useful one. However, it does not take into account the rate of defense reactions. To take a rather extreme example, *Botrytis cinerea* Pers.:Fr. is not a pathogen of potato tubers at 20° C but can be induced to become one at 5° C, at which the rate of periderm formation is insufficient to keep the fungus from invading the tissue (Ramsey 1941). Nor does Leach's concept take into account many environmental factors not directly related to temperature, such as soil moisture and pH in the case of the poinsettia root rots (Bateman 1961, 1962).

Generalizations are further complicated by the wide variations in the temperature of plant tissues and of different parts of the same plant. It is evident that simple measurements of air or soil temperature do not accurately reflect the temperature of tissues, where the stress is applied and where pathogenesis occurs.

Hanan (1965, 1970a) recorded wide differences between the temperatures of red and white carnation flowers, and both were considerably above ambient air temperature. Waggoner and Shaw (1953) recorded different temperatures in roots and shoots, even in different parts of the same shoot, shaded or unshaded, of tomato plants of different sizes (Figures 15 and 16). These discrepancies must be borne in mind when considering the effects of temperature in stressing the plant, and different parts of it, and predisposing it to disease.

Local high temperatures in tissues may lead to a breakdown in normal resistance, and further, a strain-induced principle may be translocated. Jarvis (1964) noted that when one of a pair of cucumber cotyledons was heated for a few seconds at 45, 50, or 55° C and inoculated on the abaxial surface with conidia of *Sphaerotheca fuliginea* (Schlechtend.:Fr.) Pollaci, infection occurred, and the mycelium grew through the intercellular spaces, with haustoria in the mesophyll cells; eventually conidiophores formed within the intercellular spaces as well as on the opposite surface of the cotyledon. Leaves of sunflower (*Helianthus annuus* L.), normally not susceptible to *S. fuliginea*, also became susceptible to it when heat-stressed. It may be that somewhat lower temperatures, 35–40° C, held over a longer period, can account for anomalous appearances of disease in normally resistant plants. Gardiner et al (1987) enhanced symptoms of Fusarium wilt of chrysanthemum (caused by *F. oxysporum* f. sp. *chrysanthemi* G. M. Armstrong, J. K. Armstrong, & R. H. Littrell) by raising the temperature. In two cultivars, Royal Trophy and Glowing Mandalay, the disease rating increased with increased day and night temperature to a maximum at 35° C, the highest temperature treatment. In the cultivar Torch, the highest disease rating occurred at 29° C. Measurements of the growth of the pathogen in vitro did not correlate with disease ratings at corresponding temperatures. When the day temperature was kept at 35° C and night temperatures were varied, the highest disease rating occurred with a night temperature of 29° C. There were no symptoms in Torch until the night temperature was 24° C or higher (Gardiner et al 1989). As in the first experiment, all symptomless plants were nevertheless colonized by *F. oxysporum* f. sp. *chrysanthemi*.

Similar results were obtained by Harling et al (1988) for Fusarium wilt of carnation (caused by *F. oxysporum* f. sp. *dianthi* (Prill. & Delacr.)

W. C. Snyder & H. N. Hans.). At 14–15° C, there was no wilt and very little colonization of any cultivar, but at 22° C differences between cultivars were observed, and at 26° C all but the most resistant cultivar (Carrier 929) wilted rapidly. The cultivar Red Baron showed tolerance at 22° C for 20 days and then collapsed, while Carrier 929 remained mostly uncolonized and therefore was classed as truly resistant.

Freezing stress never occurs except by accident in the greenhouse, but low temperatures predispose crops to many pathogens. Irrigating with cold water is a common fault in greenhouse management, inducing low-temperature shock in seedlings, in particular, and predisposing them to pathogens such as *Pythium* spp., *Rhizoctonia solani*, and *Botrytis cinerea*. It is likely that the defense mechanism is retarded in all cases. This happens, for example, in cucumber scab, caused by *Cladosporium cucumerinum* Ellis & Arth. (Walker 1950). Below 17° C, infection occurred readily, even in the resistant cultivar Maine 2, but at 20–22° C, incipient lesions were cicatrized, and thus a barrier was formed against further disease development. Cucumber plants are also predisposed to infection by *Didymella bryoniae* (Auersw.) Rehm by low night temperatures (van Steekelenburg and van de Vooren 1980), probably partly because of the tendency of dew to form under those conditions, enhancing infection. Gray mold, caused by *B. cinerea*, is also prevalent in a temperature regime of 16° C during the day and 7° C at night (van Steekelenburg and van de Vooren 1980).

Tomatoes are also predisposed to *F. oxysporum* f. sp. *radicis-lycopersici* by cold irrigation water and by low soil temperatures (Jarvis and Shoemaker 1978; Jarvis et al 1983). Warmed irrigation water and late spring planting overcome these stresses.

There are a number of cases (Colhoun 1973, 1979) in which adverse temperatures predisposing plants to disease are well documented. This information is invaluable in deciding on disease-escape strategy, as is information on optimum temperatures for disease expression.

The experiences of Gardiner et al (1987, 1989), Harling et al (1988), Jarvis (1964), Walker (1950), and Yarwood (1963) in which abnormal temperatures greatly affected the expression of symptoms indicate that the selection of resistant germ plasm in the greenhouse has its pitfalls. Selection is better made at a range of temperatures to ensure that no anomalies occur.

Water Stress

Water stress is applied to plants as an excess or deficit in the air or in the soil, and it induces strains in pathogens as well as host plants. There is a matrical relationship between air, soil, plant, and microorganism in water potential.

Water potential, ψ, is a thermodynamic term relating to the capacity of water to function in work processes in the biological system. It is commonly expressed in terms of pressure, with dimensions of newtons per square meter, bars, atmospheres, or pascals (Eastin and Sullivan 1984).

Water potential in the plant is conveniently considered an algebraic sum:

$$\psi = \psi_s + \psi_p + \psi_m$$

where ψ_s is the water potential exerted by solute effects; ψ_p is the turgor pressure exerted against the cell wall when cell solutes cause water to osmose into the cell, thereby increasing water activity; and ψ_m is the matric potential resulting from the adsorption of water to cell wall surfaces. In the soil,

$$\psi = \psi_s + \psi_m$$

where ψ_s is solute or osmotic potential, and ψ_m is the matric potential exerted by the adhesion of water to soil particles, which is dependent on surface tension, capillarity, pore space, and particle size, among other things. The field capacity of soil is around −30 kPa, and plants wilt at about −1.5 MPa. In high atmospheric humidity plant roots may exude water into the soil (Schippers et al 1967).

Generalizing, Cook and Baker (1983) considered that pathogens able to grow well at relatively low osmotic potentials (−1.0 to −1.5 MPa) and ceasing growth only at −9 to −10 MPa, such as *Fusarium* spp., are most virulent in dry soils. Fungi such as *Pythium* spp., *Phytophthora* spp., *Rhizoctonia solani*, and *Thielaviopis basicola*, which are able to grow at high osmotic potentials, between −0.1 and −1.0 MPa, and cease growth at −5 to −6 MPa, are most virulent in moist soils. The optimum ψ_s for active growth of bacteria and phycomycetes in the soil is about −0.1 to −0.5 MPa, and that for ascomycetes is about −0.5 to −2.5 MPa (Cook and Papendick 1972; Duniway 1976, 1979; Griffin 1963a,b; Sommers et al 1970; Zentmyer 1979).

Oospores of *Pythium aphanidermatum* (Edson) Fitzp. germinated in arginine-amended soil from saturation to −1.5 MPa, and the proportion germinating decreased at lower soil moistures (Stanghellini and Burr 1973). Stanghellini and Burr thought that the germination of oospores in wet soils is enhanced by increased availability of nutrients. This had previously been noted by Kerr (1964) in the case of pea seedlings attacked by *P. ultimum*. Depending on the soil texture and hence the rate of diffusion of solutes from germinating pea seeds, infection increased with increasing soil water. More sugars were exuded into nonaggregated sand than into aggregated loams, and the incidence of damping-off of seedlings was correspondingly higher.

Similarly, Cook and Flentje (1967) found the germination of chlamydospores of *F. solani* (Mart.) Sacc. f. sp. *pisi* (F. R. Jones) W. C. Snyder & H. N. Hans. to be affected by the seedling exudate and water potential. In a soil with a permanent wilting point at a water content of 3.6% and a field capacity of 10.7%, maximum germination occurred at a water content of 8.7% and declined in drier soil. Germ tubes grew only in soil contiguous with germinating pea seeds when the water content was 8.7% or lower; in wetter soil, they lysed. Germination was correlated with exudation as measured by loss in seed weight over 20 h, and germ tube lysis was directly correlated with that weight loss. The amendment of the soil with sucrose and ammonium sulfate resulted in germ tube lysis at all water levels. Overall, the virulence of *F. solani* f. sp. *pisi* was greatest at a water content of about 8%.

Soil structure is important in determining pore space, and hence ψ_m, as well as providing for water films on soil and roots, permitting the growth of hyphae and the movement of bacteria, nematodes, and zoospores. It also determines the resistance of the soil to root growth and the supply of oxygen to the roots and microorganisms (Drew and Lynch 1980; Stolzy and Van Gundy 1968).

Damping-off has long been associated with poorly aggregated and wet soils (J. Johnson 1914). Bateman (1961) compared the effects of soil moisture on three root diseases of poinsettia (*Euphorbia pulcherrima*). *Pythium ultimum* caused serious damage in wet soils, above 70% of moisture-holding capacity (mhc, which was 56%), and slight damage at 30–40% of mhc. On the other hand, *Rhizoctonia solani* caused severe damage at soil moistures less than 40% of mhc and decreasing damage with increasing moisture up to 80% of mhc. *Thielaviopsis basicola* was pathogenic at 36% of mhc and caused infections of increasing severity as the moisture content increased to 70% of mhc.

Phytophthora root rots are generally more severe in wet soils, and plants may also be predisposed by prior periods of drought. Blaker and MacDonald (1981) induced susceptibility to *Phytophthora cinnamomi* Rands in rhododendron cultivars by flooding the soil for 48 h or by imposing drought stress to lower their leaf water potential to −16 kPa. The cultivar Caroline was generally not very susceptible to Phytophthora root rot unless so stressed, but the cultivar Purple Splendour was always very susceptible. Duniway (1977) similarly stressed safflower (*Carthamus tinctorius* L.) by withholding water or by overwatering to leaf water potentials of −0.4 to −0.6 MPa, to induce marked susceptibility to *P. cryptogea* Pethybr. & Lafferty.

Fusarium and Verticillium wilt pathogens generally grow well in soil at −1 to −2 MPa but can also grow at −10 to −12 MPa—a response enabling their survival in drying, dead host tissue (Cook and Baker 1983). They are most pathogenic at soil moistures generally favorable to the growth of the host, possibly because active transpiration ensures abundant colonization of the host xylem (Cook and Baker 1983). Clayton (1923b) observed the most severe wilt caused by *Fusarium oxysporum* f. sp. *lycopersici* in tomatoes growing in only moderately dry soil of 26–27% moisture (mhc = 35%) and no disease in saturated soil. The time between inoculation and the appearance of symptoms declined as soil moisture increased to 32%. Foster and Walker (1947) also found that tomatoes were predisposed to this pathogen by low soil moisture, and Strong (1946) found wilt to be more severe in soil at 60% than at 85% moisture.

Soil moisture can also affect the severity of diseases of the shoot. Kendrick and Walker (1948) found tomato plants to be more severely affected by bacterial canker (caused by *Clavibacter michiganensis* subsp. *michiganensis* (Smith) Davis et al) in soil at the optimum water potential for tomato growth. By contrast, the fruit of drought-stressed cucumber plants was markedly more susceptible to internal rot caused by *Didymella bryoniae*, although wilted flowers had 60% less infection than fresh flowers (van Steekelenburg 1986). Excessive soil moisture in combination with excessive atmospheric humidity frequently results in wet leaf surfaces and invites infection from water-dependent pathogens. Increased root pressure was simulated by J. Johnson (1936), and it predisposed tomato leaves to infection by *Bacterium angulatum* and *B. tabacina* (*Pseudomonas syringae* van Hall pv. *tabaci* (Wolf & Foster) Young et al). Guttation, the exudation of drops of water and solutes from hydathodes and stomata, damages leaf tissue when the drops evaporate, because salts are left behind in toxic amounts (L. C. Curtis 1943). Toxins may also be formed in the drops by the action of microorganisms. The tissue damaged by the toxins becomes very susceptible to a number of necrotrophic pathogens (Yarwood 1959a,b), such as *Botrytis cinerea*, whose conidia are also stimulated to germinate by exosmosed nutrients in the drops (W. Brown 1922).

Water on the surface of a plant interferes with the deposition of wax (E. A. Baker 1974) and so deprives the plant of one defense mechanism. This is also commonly experienced in tissue culture in vitro. Darrow and Waldo (1932) thought that the cuticle of strawberry fruit softened in wet conditions, which would make it susceptible to *B. cinerea*.

The movement of bacteria is greatly enhanced on the surface of wet plants (Leben 1965b). Bacteria on cucumber seeds did not migrate to aerial parts of seedlings under low atmospheric humidity (at vapor pressure deficits greater than 2 kPa and 30–40% RH), but they did migrate at vapor pressure deficits less than about 0.3 kPa (above 90% RH). The bacteria tended to congregate where water films persisted, at the depressions over cell junctions and at the bases of hairs.

As guttation drops are resorbed into leaves when active transpiration resumes, bacteria in the drops may be taken into stomatal cavities and there initiate infection. Waterlogged leaves have long been known to be very susceptible to bacterial infection (Clayton 1936; Diachun et al 1942; Lelliott 1988c).

Water congestion was found to cause changes in cell permeability in cucumber challenged by *Pseudomonas lachrymans* (*P. syringae* pv. *lachrymans* (Smith & Bryan) Young et al) (Williams and Keen 1967). Tritium-labeled water accumulated in the congested areas, and inorganic ions, sugars, and amino acids were lost rapidly 24 h after inoculation. These changes paralleled the development of the characteristic vein-limited, angular lesions.

Stomatal exudation, with accompanying stimulation of bacteria and sooty molds, is responsible for a leaf spot of *Philodendron hastatum* K. Koch & H. Sello (Munnecke and Chandler 1957). The exuded drops contained 28–93% sugars, used as nutrients by several microorganisms.

An unusual example of the role of exudation was described by Wilson (1963). If tomatoes are deleafed in a humid environment, drops of water, presumably with solutes, are exuded from the petiole stubs for several hours. If conidia of *Botrytis cinerea* are present in the last drop before active transpiration resumes, they are resorbed into the xylem and become lodged in clumps some millimeters in from the cut surface. They remain quiescent there for up to 12 wk, when physiological changes in the host are believed to trigger an aggressive state. The conidia then germinate simultaneously and establish a rapid rot of the surrounding parenchyma (Figure 25). This mechanism of infection has not been described for other diseases, but it may be more common in pruned greenhouse crops than is recognized.

Very many fungi and bacteria infect plants through the agency of water. Yarwood (1939b, 1956) divided them into groups according to their humidity requirements (Table 6).

Notwithstanding Yarwood's classification of the powdery mildew fungi as a group not requiring water on the surface of the plant at any stage, many investigators have said that conidia of the Erysiphales require a water film in which to germinate, and they frequently apply inoculum of these fungi in aqueous suspensions (Spencer 1978). Yarwood (1936, 1950) thought that the powdery mildews were remarkably tolerant of low humidity and that the conidia carried enough internal water (52–72% of their fresh weight) to be able to germinate independently of external water. In addition they

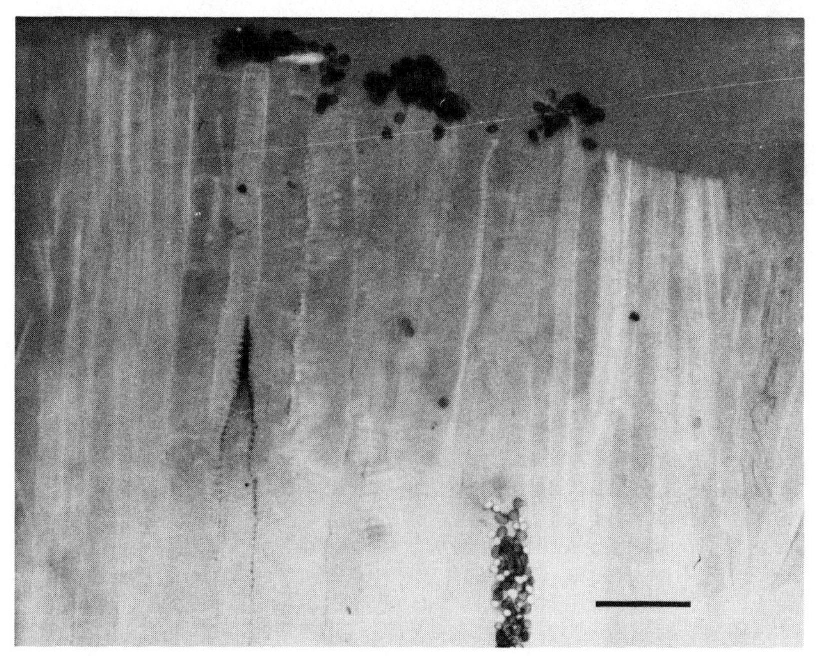

Figure 25. Conidia of *Botrytis cinerea* wedged against the open ends of the xylem vessels and in a vessel of a tomato petiole stub in transverse section. Conidia of *B. cinerea* enter the vascular system through the petiole after deleafing, carried in by the transpiration stream. They may remain quiescent, lodged against remnants of vessel cross walls, for 10–12 wk. The conidia in this photograph are stained with a fluorescent brightener. Bar = 100 μm.

are able to obtain 2.5–14% of their fresh weight as hygroscopic water. Many experiments (Delp 1954; Longrée 1939; Pathak and Chorin 1968; T. V. Price 1970; Ragazzi 1980; Rogers 1959; Weinhold 1961; Wheeler 1978) have been done on conidia of the Erysiphales at high humidities in enclosed containers in which condensation is almost certain to have occurred (Schein 1964), and so it is difficult to interpret the results with confidence. Rogers (1957, 1959) used a heat sink in a controlled environment chamber to study the water relations of the rose powdery mildew fungus, *Sphaerotheca pannosa* (Wallr.:Fr.) Lév. var. *rosae* Woronichin, on rose leaves. Humidities cycling between 55 and 100% (with condensation on the leaves) had no differential effect on the development of powdery mildew, but intermittent misting after inoculation inhibited spore germination and colony development. Longrée (1939) gave some very precise figures on the effects of relative humidity on *S. pannosa* var. *rosae* on glass:

RH (%) at 25.17°C	99.8	99.0	98.0	96.9	94.9	75.0
Germination (%)	25.8	66.2	63.1	67.7	2.0	0.0

It is doubtful that free water was not present in at least some of the treatments with high relative humidity. On rose shoots, germination declined with relative humidity at a constant temperature, and Longrée suspected that the humidity at the leaf surface was very high, even in a dry atmosphere. This guess differed substantially from the opinion of Yarwood and Hazen (1944), whose measurements indicated no more than about 50–70% RH at the leaf surface. Dimock and Tammen (1969) and Tammen (1973) reported an optimum temperature of 22°C and optimum relative humidity of 23 to over 99% for conidial germination of *S. pannosa* var. *rosae*, but conidia were short-lived at 70% RH.

Recent work on the powdery mildews has altered the picture somewhat, at least for rose powdery mildew. Powell (1990) determined that the development of a powdery mildew epidemic in rose was dependent on dew deposition and continual diurnal cycling of temperature and vapor pressure deficit. A dew period of 12 h at night, normally occurring 8 h after spores were released and alighted on leaves, ensured maximum infection, although the infection may take up to 40 h to be completed. If the inoculum was kept dry for 32 h, most of the conidia died, and infection was prevented. Shortening the dew period proportionately decreased the level of infection.

Table 6. Humidity requirements of certain groups of plant pathogens[a,b]

	Inoculation	Incubation	Infection	Sporulation	Dispersal
Anthracnose fungi	+	+	−	±	+
Bacteria	+	+	+	. . .	+
Downy mildew fungi	−	+	−	+	+, −
Powdery mildew fungi	−	−	−	−	−
Rust fungi	−	+	−	±	−
Viruses	−	−	−	. . .	+, −

[a] Adapted from Yarwood (1956).
[b] + = High; ± = intermediate; − = low.

Dry periods after successful infection permitted spore maturation and release, whereas further dew deposition permitted mycelial development and spore maturation. In contrast to minute droplets of dew, which permit spore germination, larger water drops applied as sprays during normally dry periods burst the spores. This seems to explain the different results of Yarwood (1939a, 1978), who found that immersion of mildew-infected cucumber leaves enhanced sporulation, whereas water sprays applied at about 4.5 kPa reduced the colony development of six powdery mildew fungi, including that of *S. pannosa* var. *rosae.*

Epidemics of rose powdery mildew in a Colorado greenhouse were associated with rising dew points outside by Cobb et al (1978). The disease seldom occurred in the winter, but epidemics began as dew points outside rose in early spring, and they continued into early fall. The spring and fall months were also the times when greenhouse humidity and temperature were most difficult to control. Evaporative pad cooling maintained low vapor pressure deficits, and the dew point outside rose. Powdery mildew was reduced if the evaporative pad cooling was turned off just before nightfall, allowing the vapor pressure deficit to rise. Misting during summer days had relatively little effect.

There are two powdery mildew fungi on cucurbits, *Erysiphe cichoracearum* DC. and *Sphaerotheca fuliginea.* Both are common in Europe and coexist, even in the same greenhouse, but *E. cichoracearum* seems to be rare in North America, although many early records use that name (Yarwood and Gardner 1964). Nagy (1976) found that saturated air and rain inhibit the sporulation of *E. cichoracearum*, whereas they are tolerated by *S. fuliginea.*

In all diseases except the powdery mildews, then, excessive water remaining for too long on leaves, flowers, and fruits greatly increases the risk of infection. The role of free water in infection by the Erysiphales remains uncertain, largely because of the difficulties of working at very low vapor pressure deficits with precise enough temperature control.

Osmotic Stress

The osmotic component ψ_s of the water potential equation

$$\psi = \psi_s + \psi_p + \psi_m$$

is the component determined by the concentration of solutes in the soil and the plant and in the cells of microorganisms. It plays a relatively small part in the flow of water through soil (Papendick and Campbell 1975), compared with the matric potential, but in turgid cells it plays a major role. Vacuolar sap is then largely a true solution, and $\psi_s + \psi_p = \psi_m$.

Fungal and bacterial cells must maintain a minimum hydrostatic pressure within the plasma membrane in order to grow or divide. This is maintained by osmoregulation, so that the osmotic potential ψ_s of the cell is less than that outside the cell, and water passes into the cell until equilibrium is attained. Most fungi have an osmotic optimum of around −1 to −2 MPa, but some *Fusarium* spp. can withstand −10 to −12 MPa, and bacteria −2 to −2.5 MPa (Cook and Baker 1983). The cosmopolitan greenhouse pathogen *Botrytis cinerea* was found to have an osmotic potential of −0.31 MPa, compared with −8.4 MPa for celery, one of its hosts (Thatcher 1939, 1942). The permeability of celery cell membranes increased at some distance from those already killed, in advance of hyphae of *B. cinerea*. Thatcher (1942) postulated that pectinase activity disrupted cell walls, membranes, and protoplasm of the host, increasing the availability of solutes and water to the fungus.

Osmoregulation in plants is affected by the tissue type and its age, and it is mediated mostly by concentrations of sugars, amino acids, and potassium (J. M. Morgan 1984).

Schnathorst and Weinhold (1957) advanced a hypothesis to explain the observed susceptibility to *Erysiphe cichoracearum* in normally resistant lettuce leaf tissue when the lettuce was inoculated in detached leaf culture or etiolated for short periods. Young leaves of field-susceptible lettuce cultivars were resistant to *E. cichoracearum*; in contrast, young peach leaves were susceptible to *Sphaerotheca pannosa*, but old leaves were resistant. Schnathorst and Weinhold found that sucrose solutions of $\psi_s = 1.7$ and 1.8 MPa (equivalent to the osmotic pressure of resistant lettuce and peach leaves, respectively) inhibited water uptake by conidia of the respective fungi, as indicated by their failure to germinate.

Schnathorst (1959b) examined the several responses of lettuce cultivars to *E. cichoracearum* and was able to explain the following observations with reference to the difference in osmotic pressure between the host and the parasite: the resistance of young seedlings in the field; gradients of powdery mildew in the field; the increased susceptibility of plants infected with lettuce mosaic virus; the relative resistance of lettuce cultivars in the cooler Blanco–Castroville area of the Salinas Valley, California; the

Table 7. Effects of mineral nutrition on the growth of two lettuce cultivars and on their susceptibility to *Erysiphe cichoracearum*[a]

Nutrient solution	Vigor class[b]	Disease rating[c,d]		Osmotic value of weakly susceptible GL leaves (10^5 Pa)	Osmotic value of first resistant leaves[e] (10^5 Pa)	
		GL	BB		GL	BB
Medium N, medium K	4	4	2	11.6	13.2	14.8
High N, low K	3	2	0	10.0	17.3	13.2
Low N, low K	5	3+	2	14.8	20.3	18.3
High N, high K	2	4	1	11.6	18.3	18.3
Low N, high K	5	3	1	14.8	20.3	17.3
Control (soil)	1	2	6	11.6	18.3	17.3

[a] Adapted from Schnathorst (1959b).
[b] 1 = Most vigorous; 5 = least vigorous.
[c] 0 = No visible signs of the fungus; 4 = coalescing colonies.
[d] GL = cultivar Great Lakes; BB = cultivar Big Boston.
[e] The first resistant leaf is that immediately above the uppermost slightly susceptible leaf.

increased susceptibility of shaded plants; the alteration of resistance by adjustments in mineral nutrition (Table 7); the change from resistance to susceptibility in leaf disks floating on water; the failure of *E. cichoracearum* to infect guard cells; and the field resistance of the cultivars Big Boston and Arctic King.

Schnathorst (1959b) altered the osmotic values of leaf cells of two lettuce cultivars, Big Boston and Great Lakes, both genetically susceptible, by applying nitrogen and potassium fertilizers at different rates, with these nutrients in different ratios. Plants of the cultivar Great Lakes were most resistant when grown in untreated soil and in a medium treated with a high-N–low-K nutrient solution; these plants were, respectively, the most and the third most vigorous, compared with those in the other treatments. Plants of this cultivar treated with a high-N–high-K nutrient solution were the second most vigorous, yet the most susceptible. The effects of manipulating the potassium–nitrogen ratio in the nutrient supply to lettuce are summarized in Table 7.

The effect of salinity stress by Cl^- on inoculum densities of *Pythium oligandrum* Drechs. and *Pythium* suppressiveness in soils of the San Joaquin Valley of California was investigated by F. N. Martin and Hancock (1982). Soil Cl^- levels and inoculum densities of the mycoparasite *P. oligandrum* were correlated with soil suppressiveness of diseases caused by *P. ultimum*, but Cl^- levels alone did not account for high and low levels of *P. ultimum* in soils. The addition of Cl^- or *P. oligandrum* to soil containing a high level of *P. ultimum* significantly reduced the saprophytic activity of *P. ultimum*. Martin and Hancock concluded that *P. oligandrum* is more tolerant of Cl^- and can therefore dominate *P. ultimum* in soils containing the latter at a low level. *P. oligandrum* competes successfully with *P. ultimum* for organic substrates only when its population exceeds that of *P. ultimum*.

MacDonald (1982) applied osmotic stress to rooted cuttings of chrysanthemum by exposing the roots to 0.1 or 0.2 M NaCl for 24 h before inoculating them with zoospores of *Phytophthora cryptogea*. Disease severity was three to four times that of unstressed inoculated roots. The number of zoospores on the roots was directly related to the degree of stress, and lesions spread faster and more extensively on stressed roots. The exudation of solutes was greater from the stressed roots.

Fertilizers and pH

Inorganic fertilizers, their form, and their relative composition have profound effects on the soil microflora and microfauna and on host physiology (Engelhard 1989; Henis and Katan 1975; Smiley 1975).

Henis and Katan (1975), Huber and Watson (1974), and Smiley (1975) reviewed the effects of nitrogen fertilizers. Host resistance, growth and its chemical constituents, exudates from the plant, soil and rhizosphere pH, and soil microbiology are each affected, depending on the amount and form of nitrogen.

The form in which nitrogen is supplied affects the balance of cation–anion uptake by the plant (Kirkby 1969). In order to maintain an electrochemical equilibrium, the uptake of a relatively large mass of cations as ammonium nitrogen (NH_4^+) is balanced by the return of hydrogen ions to the medium:

$$NH_4^+ \rightarrow NH_3 + H^+$$

so that the medium and the cell sap become acidic, and there is a tendency for the plant to accumulate organic anions, because of the general depression of organic acid dissociation. Cation absorption occurs at about five times the rate of anion absorption.

Fertilizing with urea also results in an acid reaction, although it is slower than that involving NH_4^+. Urea is hydrolyzed by microbial urease in two stages:

$$CO(NH_2)_2 + H_2O \xrightarrow{\text{urease}} \underset{\substack{\text{Ammonium}\\ \text{carbonate}}}{H_2NCOONH_4} \rightarrow 2NH_3 + CO_2$$

By contrast, in a plant supplied with nitrate nitrogen (NO_3^-), the cell sap and the medium become more alkaline:

$$NO_3^- + 8H \rightarrow NH_3 + 2H_2O + OH^-$$

Kirkby (1969) found that in recirculating nutrient solutions originally at pH 5.5, the absorption of NH_4^+ by tomato roots resulted in a pH of 4.0; the absorption of urea nitrogen, pH 4.5; and the absorption of NO_3^-, pH 6.2.

The changes in pH in the medium also affect the availability and uptake of the metallic cations K^+, Na^+, Ca^{2+}, Mg^{2+}, and Mn^{2+}. Smiley (1975) considered the form of nitrogen to be the single greatest influence on rhizosphere pH.

Effect on the Host

Some plants, such as peas and beans, assimilate NO_3^- mostly in the roots. Others, such as beets and clover, assimilate it mostly in the shoots, and in these plants, organic ion accumulation and cation uptake are important in maintaining the electrochemical equilibrium. Therefore, in different plants, NO_3^- has different effects on the physiology of the host and its resistance mechanisms in the roots and in the shoots. Verhoeff (1965, 1968) noted that the growth of *Botrytis cinerea* and the incidence and severity of gray mold lesions on tomato stems inoculated at deleafing wounds were negatively correlated with soil nitrogen content within the range 20–305 $g \cdot kg^{-1}$ in various soils. These results were somewhat surprising, since gray mold is generally associated with overfertilization with nitrogen (Jarvis 1977a), but Verhoeff's explanation is that senescence of the host tissue is delayed (Verhoeff 1968, 1974).

Nitrate nitrogen is one essential component of a system devised in Florida for the control of Fusarium wilts in vegetable crops and a similar system (involving lime, nitrate nitrogen, and chemotherapy) for ornamental crops (J. P. Jones et al 1989). The other essential component of both systems is hydrated lime, $Ca(OH)_2$. Fusarium wilts of the following plants were listed as amenable to control by the Florida system (Jones et al 1989): tomato (*Lycopersicon esculentum* Mill.), florist's chrysanthemum (*Dendranthema grandiflora* Tzvelev), carnation (*Dianthus caryophyllus* L.), cucumber (*Cucumis sativus* L.), muskmelon (*C. melo* L.), aster (*Callistephus chinensis* (L.) Nees), and strawberry (*Fragaria* × *ananassa* Duchesne), among several others. High soil pH resulting from the incorporation of hydrated lime severely limits the availability of essential micronutrients, such as copper, iron, manganese, molybdenum, and zinc, to formae speciales of *Fusarium oxysporum*. Jones and Woltz (1969) did not consider calcium a major factor in controlling Fusarium wilt of tomato. $Ca(OH)_2$ increased the calcium content of tomato tissue, increased soil pH, and reduced the incidence and severity of the disease in *Fusarium*-infested soil; calcium sulfate ($CaSO_4$), however, also increased tissue calcium and soil pH but did not reduce the incidence and severity of the disease.

J. P. Jones et al (1989) noted that actinomycete and bacterial populations, some of which are antagonistic to and competitive with *F. oxysporum*, are also favored by high soil pH. In the system combining NO_3^- and lime for wilt control, Jones and co-workers thought that the effect of nitrate nitrogen may be twofold: it increases soil pH, and it affects the potassium–chloride ratio in the host tissue. *F. oxysporum* f. sp. *lycopersici* is more virulent when grown on a source containing nitrate nitrogen; tomato seedlings are preconditioned and more resistant to wilt than seedlings grown with ammonium nitrogen. Ammonium nitrogen suppresses the uptake of NO_3^- and potassium, and it stimulates the uptake of Cl^-, leading to greater susceptibility (Schneider 1985). All of these factors reinforce the effects of hydrated lime. For the control of Fusarium wilts of aster and chrysanthemum, Woltz and Engelhard (1973) included a benomyl treatment with the lime–nitrate combination and obtained outstanding control.

Whereas Jones and Woltz (1969) did not consider high tissue calcium to be a primary factor in the resistance of tomato to Fusarium wilt in the lime–nitrate system, other workers have considered the role of calcium (in cell wall structure, specifically) to be important in tissue resistance. Divalent calcium bonds are responsible, in part, for pectin insolubility and resistance to enzymic degradation by pectinases. Hondelmann and Richter (1973) attributed the relative resistance of fruits of some strawberry cultivars to *Botrytis cinerea*, which secretes a battery of pectinases into the host, to a high ratio of insoluble to soluble pectin. Edgington et al (1961) also believed that the calcium status of the cell wall determines the resistance of tomato to *F. oxysporum* f. sp. *lycopersici*, whereas Corden (1965) thought that calcium inhibits the pathogen's polygalacturonase.

In lettuce, the resistance imparted to gray mold by calcium was overcome by high levels of nitrogen, and Krauss (1969, 1971) suggested that calcium uptake had been relatively decreased. Krauss also noted that phosphorus-

and nitrogen-deficient plants were resistant, probably as a result of their high calcium content and smaller cells. In plants with adequate calcium, the incidence of gray mold was less affected by anion concentrations, the pectin was less soluble, and cell membranes were less permeable.

Liming was found by Stall (1963) to reduce the incidence of gray mold in tomatoes, especially with a low rate of phosphorus application (Stall et al 1965).

The cytological roles of calcium can be summarized as an apoplastic one in the integrity of the host cell wall and a symplastic one in stabilizing membrane permeability and membrane and cytoplasm integrity (Clarkson and Hanson 1980).

Trace elements generally have little direct effect on pathogenesis (except for iron in the suppressive soil syndrome), but excesses or deficiencies stress the plant and may predispose it to infection. Chase and Poole (1984a) found chelates of copper, manganese, iron, and zinc, as well as some fungicides, to cause leaf spots on *Chrysalidocarpus lutescens* H. Wendl. almost indistinguishable from those caused by *Exserohilum rostratum* (Drechs.) K. J. Leonard & E. G. Suggs. An increase in the incidence and severity of gray mold (caused by *B. cinerea*) in strawberry following sprays of nabam–zinc sulfate and zineb was noted by Cox and Winfree (1957). Zinc accumulated in plant tissues to a level 10 times that in unsprayed plants. Stall (1963) noted that zinc or other trace elements did not increase gray mold in tomatoes in all parts of Florida, particularly in alkaline, calcareous soils or on farms with a history of regular moderate liming. The effect of zinc in increasing gray mold occurred if the soil calcium and pH were too low in as well as if they were too high. Evidently host calcium and zinc have some interaction with the pathogen.

Silicon, added to a hydroponic solution as potassium silicate at 100 $mg \cdot L^{-1}$, was found by Adatia and Besford (1986) to increase the resistance of cucumber plants to powdery mildew (caused by *Sphaerotheca fuliginea*).

Effect on Pathogens in the Rhizosphere

Inorganic ions and the pH of the soil, natural and nitrogen-induced, directly affect spore germination, the chemotaxis of zoospores and bacteria, mycelial growth, infection, and the formation and survival of resting structures of pathogens, as well as antagonists, saprophytic competitors, and hyperparasites (Henis and Chet 1975).

Like host plants, microorganisms are affected differently by the two main forms of nitrogen, NO_3^- and NH_4^+. As we have seen, the pH of the rhizosphere is reduced by ammonium nitrogen and increased by nitrate nitrogen. The growth, reproduction, and survival of pathogens is therefore influenced by the form of nitrogen as a nutrient, since they are affected by the pH of their environment. Huber and Watson (1974) listed some of the effects of the form of nitrogen on overall fungal growth, reproduction, survival, and ability to infect plants as "preferences," which are sometimes anomalous. Those benefiting from ammonium nitrogen included *Pythium aphanidermatum* and *Rhizoctonia solani*. Those benefiting from nitrate nitrogen included *Colletotrichum coccodes* (Wallr.) S. J. Hughes, *Fusarium solani* f. sp. *cucurbitae* W. C. Snyder & H. N. Hans., *P. aphanidermatum, R. solani, Verticillium albo-atrum* Reinke & Berthier, and *V. dahliae* Kleb. Those with no preference included *Botrytis cinerea, F. oxysporum* f. sp. *lycopersici, P. ultimum*, and *Sclerotinia sclerotiorum* (Lib.) de Bary. Some of these organisms occur in both lists because of the different hosts, different criteria for growth, reproduction, survival, and infectivity, and different experimental conditions used by different authors. In general, adequate but not excessive levels of nitrogen fertilizers enhance resistance to Verticillium wilt. Nitrate nitrogen appears to increase susceptibility to *Verticillium* spp., but the effect of ammonium nitrogen varies with the host species (Pennypacker 1989). Smiley (1975) put forward the generalization that pathogens most active in acid soils are favored by ammonium nitrogen, whereas those active in alkaline or neutral soils are favored by nitrate nitrogen.

Not only are there direct effects of nutrients on the activity of pathogen forms in the soil, but there are also effects on the associated microflora, sometimes contrary to the effects on pathogens. Nutrients may stimulate the germination of resting propagules in the absence of a suitable host, and then the propagules undergo lysis, a form of biological control (Papavizas and Lumsden 1980). This occurs in *F. solani* and *Thielaviopsis basicola*, among other fungi.

Soil Compaction and Oxygen Stress

Soil compaction, reviewed by Allmaras et al (1988), is not normally a problem in composts and soilless media used in greenhouse production (Bunt 1976; de Boodt and Verdonck 1972), but in soils in groundbed greenhouses in some areas, such as the very sandy groundbed soils of southwestern Ontario, pans impede root growth and make sterilization very difficult. Groundbeds with row crops requiring daily attention also get tread-compacted between rows. Compaction leads to poor soil structure, impeded drainage, ethylene evolution, and local oxygen deficiency in the rhizosphere (Allmaras et al 1988; Bunt 1961; Stolzy et al 1975; Whalen 1988). Oxygen diffuses only extremely slowly through water, at about 1×10^{-5} the rate in air (Griffin 1972). Oxygen deficiency results in chlorosis, wilt, and dieback symptoms (Bergman 1959), but the oxygen concentration has to fall to about 1×10^{-6} or 1×10^{-7} M before the cytochrome system begins to fail (Griffin 1968).

Root penetration is mechanically impeded in compacted soil at a penetrometer resistance of about 2 MPa (Allmaras et al 1988). Huisman (1982) considered the relative growth rates of roots and pathogens when roots are able to grow unrestrictedly: if the root tip grows at $0.4\ mm \cdot h^{-1}$ in well-aerated, friable soil, and the rhizosphere influence on a pathogen spore operates up to 1 mm, the approaching root tip may be recognized by the pathogen only 2–3 h before its arrival. This is too late to achieve infection in the susceptible root-tip area if germination takes 6–10 h.

Table 8. Predisposition of crops to diseases caused by various pathogens under low-intensity light

Crop	Pathogen	Reference
African violet	*Botrytis cinerea*	Beck and Vaughn (1949)
Cucumber	*Didymella bryoniae*	Svedelius and Unestam (1978)
Geranium	*Botrytis cinerea*	Melchers (1926)
Tomato	*Botrytis cinerea*	Bewley (1923)
	Clavibacter michiganensis subsp. *michiganensis*	Kendrick and Walker (1948)
	Didymella lycopersici	Verhoeff (1963)
	Fusarium oxysporum	Foster and Walker (1947)
	Verticillium albo-atrum	J. P. Jones et al (1975)

Germination then occurs in the vicinity of a mature part of the root behind the tip, with less root exudate. If infection is not then achieved, the germ tube may lyse. Alternatively, germination may not be sufficiently stimulated to begin. This constitutes a form of disease escape.

This situation is similar to that described by Miller and Burke (1975) in the case of Fusarium root rot of beans in the field. Root penetration was reduced by poor aeration, and the roots were predisposed to *Fusarium oxysporum* f. sp. *pisi* (J. C. Hall) W. C. Snyder and H. N. Hans. Under low-oxygen conditions (an oxygen content of 4%), exudates from pea seed stimulated the growth of *Pythium* and enhanced seed rot (G. E. Brown and Kennedy 1966). Here the situation was worsened by bacteria favoring sugar production in the pea seed exudates.

In hydroponic culture, a crop under stress requires about 10 times more oxygen than one not under stress, and some oxygen may be transferred from the shoot to the roots of a stressed plant (Schwarz 1989). Local oxygen deficiency at the root surface has been advanced as one of the possible causes of root death in hydroponic culture (Daughtrey and Schippers 1980).

Low-Light Stress and the Partition of Assimilates

Crops grown in poor light, in overshaded, dirty, or multilayered plastic greenhouses, are often predisposed to diseases (Table 8). Thermal screens may add to the risk. Although threshold light levels have not been established for any disease, some generalizations have been made.

It is interesting to speculate that the availability of photosynthates to the pathogen and their distribution in the plant (Daie 1985) may be a determining factor in the susceptibility of different tissues to disease. Grainger (1962, 1968) devised the ratio C_p/R_s as an empirical measure of the carbohydrate available to an invading pathogen in excess of the plant's immediate needs—the "plunderable carbohydrate," in Grainger's words (Figure 26). C_p/R_s is the ratio of the total carbohydrate content of the plant (C_p) to the residual dry weight of the shoot (R_s). Susceptibility occurs when the ratio exceeds 0.5. Very young seedlings have a very high ratio, making them hypersensitive; slightly older plants have a low ratio and are very resistant; and maturing plants have an increasing ratio and a correspondingly increasing susceptibility.

Grainger's hypothesis was supported by data from several host–parasite combinations (Figure 26). He explained the difference in susceptibility to gray mold in strawberry fruits at different positions on the cymose inflorescence by a corresponding difference in the C_p/R_s ratio. The stress applied to a fruiting crop of tomatoes or cucumbers by the number of fruits per plant may possibly be explained in a similar way: stems of both plants may be temporarily deprived of photosynthates at times of peak fruiting, when the fruits constitute a major sink for them. Tomato stems become increasingly susceptible to *Botrytis cinerea* as the fruit load diminishes (Wilson 1963). On the other hand, cucumber stems are more susceptible to *Didymella bryoniae* and some supposedly less virulent pathogens, such as *Penicillium* sp. (Jarvis and Barrie 1988) and *Aspergillus* sp. (S. D. Barrie, unpublished observation), at times of peak fruiting.

A similar situation was observed by Jarvis and Shoemaker (1978) and by W. R. Jarvis and H. J. Thorpe (unpublished observation) in Fusarium crown and root rot of tomato, caused by *F. oxysporum* f. sp. *radicis-lycopersici*. The infected plant almost always collapses at the time of maximum fruit stress, just before the first cluster is to be picked. Some growers were persuaded to prune out the maturing first and second clusters, whereupon the plants usually recovered to yield for the remainder of the

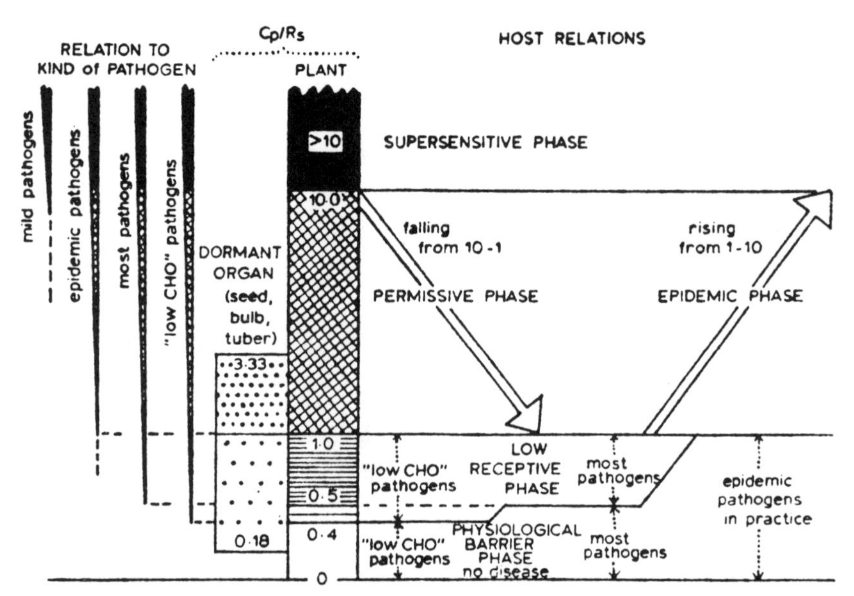

Figure 26. "Plunderable carbohydrate": C_p/R_s is the ratio of the total carbohydrate (CHO) of the plant to the residual dry weight of the shoot. This ratio is related to the susceptibility of the crop to disease. Reprinted, by permission, from Grainger (1968).

season. Some yield was sacrificed to save the plants. This phenomenon, too, may be explained by a diversion of photosynthates from developing fruits to the basal stem tissues, which the fungus had invaded.

Horsfall and Dimond (1957) divided pathogens into two broad classes: the high-sugar pathogens, attacking tissues with high levels of sugar (particularly reducing sugars), and the low-sugar pathogens. High-sugar pathogens include *Botrytis cinerea*, and low-sugar pathogens include *Alternaria* spp. and *Fusarium* spp. Like Grainger's hypothesis, that of Horsfall and Dimond may be built on a simplistic base—the concentrations of sugars in tissues—but this might represent the summation of several metabolic processes, some of which may be more directly involved in susceptibility than sugars. It is interesting, however, to note that low light intensity predisposed tomato to *F. oxysporum* f. sp. *lycopersici* (Foster and Walker 1947). It may well be that photosynthates accumulate in the first two or three clusters of developing fruits of tomato at the expense of basal stem tissues, accounting for the marked susceptibility of those tissues to *F. oxysporum* f. sp. *radicis-lycopersici* at first fruiting (Jarvis and Shoemaker 1978; W. R. Jarvis and H. J. Thorpe, unpublished observation).

Empirically, growers are often advised to prune tomato clusters and to remove fruits from the main stem of indeterminate greenhouse cucumbers to a height of about 1 m, but there is a remarkable dearth of scientific information on the levels of carbohydrates in the stem when fruit load is removed.

Ethylenebisdithiocarbamate fungicides and zinc increase the sugar status of tomato tissue and paradoxically make it more susceptible to *Botrytis cinerea* (Lockhart and Forsyth 1964). A similar increase in gray mold was observed by Cox and Winfree (1957) in strawberry following sprays of nabam–zinc sulfate and zineb, but it was not attributed to increases in tissue sugar.

The effects of day length on predisposition are seldom reported, but it can be assumed that the length of the photosynthetic period may also affect the partition of assimilates. At least four tomato diseases are increased in severity in short days: Didymella stem rot (Verhoeff 1963), Fusarium wilt (Foster and Walker 1947), Verticillium wilt (J. P. Jones et al 1975), and bacterial canker (Kendrick and Walker 1948). The promotion of adventitious rooting by long-day cultivation of *Begonia semperflorens* Link & Otto compensated for the loss of main roots attacked by *Botrytis cinerea* (Sironval 1951). This is essentially a disease-escape mechanism.

Diseases in Hydroponic Systems

The intensive production of crops, mostly vegetable crops, in a recirculating or a nonrecirculating but common supply of nutrient solution might be considered a recipe for disastrous disease epidemics. The premium on space crowds plants together, and the conditions for the multiplication of pathogens and for their plant-to-plant spread would seem to be ideal.

As Zinnen (1988) observed, the ability to grow vigorous crops in the presence of low populations of a pathogen is central to disease management in hydroponic culture. Skadow et al (1984) and Vanachter et al (1983) considered the greatest risk to be from *Pythium, Phytophthora*, and *Fusarium* spp., with *Didymella lycopersici* Kleb. a risk for tomatoes. Davies

Table 9. Pathogens of crops in hydroponic systems

Host	Pathogen	Reference
African violet	*Phytophthora parasitica*	Thinggaard and Middelboe (1989)
Anthurium	*Phytophthora parasitica*	Thinggaard and Middelboe (1989)
Carnation	*Fusarium avenaceum*	T. V. Price and Maxwell (1980)
	F. culmorum	Davies (1980)
	F. oxysporum f. sp. *dianthi*	T. V. Price and Maxwell (1980)
Cress	*Pythium* spp.	Thinggaard and Middelboe (1989)
Cucumber	*Fusarium oxysporum* f. sp. *cucumerinum*	S. F. Jenkins and Averre (1983)
	Pythium aphanidermatum	Favrin et al (1988)
	P. coloratum	Favrin et al (1988)
	P. intermedium	Gold and Stanghellini (1985) S. F. Jenkins and Averre (1983) Stanghellini et al (1988)
	P. irregulare	Favrin et al (1988)
	Pythium sp. group G	Favrin et al (1988)
	Melon necrotic spot virus	Tomlinson and Thomas (1986)
Cyclamen	*Fusarium oxysporum* f. sp. *cyclaminis*	Rattink (1990)
Gerbera	*Phytophthora parasitica*	Thinggaard and Middelboe (1989)
	Pythium spp.	Thinggaard and Middelboe (1989)
Lettuce	*Pythium aphanidermatum*	S. F. Jenkins and Averre (1983)
	P. debaryanum	S. F. Jenkins and Averre (1983)
	P. myriotylum	S. F. Jenkins and Averre (1983)
	P. ultimum	S. F. Jenkins and Averre (1983)
	Lettuce infectious yellows virus	J. K. Brown and Stanghellini (1988)
Spinach	*Pythium aphanidermatum*	Stanghellini et al (1984)
Tomato	*Clavibacter michiganensis* subsp. *michiganensis*	Davies (1980) S. F. Jenkins and Averre (1983)
	Erwinia spp.	S. F. Jenkins and Averre (1983)
	Pseudomonas solanacearum	S. F. Jenkins and Averre (1983)
	Colletotrichum coccodes	S. F. Jenkins and Averre (1983) Schneider et al (1978)
	Didymella lycopersici	Davies (1980)
	Fusarium oxysporum f. sp. *radicis-lycopersici*	S. F. Jenkins and Averre (1983)
	Phytophthora erythroseptica	T. V. Price and Nolan (1984)
	P. nicotianae	T. V. Price and Nolan (1984)
	P. nicotianae var. *nicotianae*	Skadow et al (1984)
	Pythium aphanidermatum	S. F. Jenkins and Averre (1983)
	P. debaryanum	S. F. Jenkins and Averre (1983)
	P. myriotylum	S. F. Jenkins and Averre (1983)
	P. sylvaticum	Vanachter et al (1983)

(1980), S. F. Jenkins and Averre (1983), T. V. Price and Maxwell (1980), and Schneider et al (1978) have indeed recorded disasters (Table 9). On the other hand, some experimenters (Davies 1980; Funck-Jensen and Hockenhull 1983b; Staunton and Cormican 1980) have found it difficult to induce disease in systems using the nutrient film technique (NFT). In general, excellent commercial hydroponic crops are grown year after year with little trouble from diseases.

A rigorous hygiene schedule is essential to exclude pathogens from all stages of plant production. A common ingredient for the initiation of an epidemic seems to be the inclusion of a soil or soil-mix starter block in the growing regime. Pathogens contaminating these blocks multiply quickly, especially in sterilized blocks, and then spread rapidly into the nutrient system, where plant-to-plant spread can be very fast, as Daughtrey and Schippers (1980) and Schneider et al (1978) found with Colletotrichum root rot of tomatoes (caused by *Colletotrichum coccodes*). On the other hand, Rattink (1990) did not observe any noticeable spread of disease in potted cyclamen on ebb-and-flow benches when the pathogen *F. oxysporum* Schlectend. f. sp. *cyclaminis* Gerlach was introduced either into the pots or into the nutrient solution. Within 24 h of the introduction of the pathogen into the system, the number of colony-forming units in the top half of the pots and in the nutrient effluent decreased sharply, and it was very low after 5–7 days. The bottom half of the pots contained a high number of colony-forming units, but the propagules were evidently not transported.

The nutrient film system is not entirely a biological vacuum in the sense that there are no antagonists present. A number of potential antagonists have been isolated from tomato roots (D. Price 1980; D. Price and Bateson 1976), including *Chaetomium olivaceum* Cooke & Ellis, *F. oxysporum, Penicillium brevicompactum* Dierckx, *P. nigricans* (Bainier) Thom, and *Trichoderma koningii* Oudem., all of which have been implicated in some form of biological control.

Funck-Jensen and Hockenhull (1983a,b) suggested some physiological reasons why *Pythium* spp. are less of a problem than might be supposed. The concentrations of sugars, presumably exuded from roots and released from sloughed-off root cells, appear to be an important factor in determining the inoculum potential of *Pythium* spp. in lettuce. The incidence of root rot was greatly increased if the foliage was sprayed with 1% sucrose 5 or 3 h before the introduction of a *Pythium* sp. to the roots. These treatments resulted in the death of 78% of the roots. If the roots were previously dipped in 1% sucrose for 1 h and then washed, 62% of the roots rotted, compared with 26% in a check treatment. When sucrose was added to the nutrient supply in the rinsing tank immediately after inoculation or to the incubation tank after rinsing, 76 and 100% of the roots rotted, respectively (Funck-Jensen and Hockenhull 1983b).

Funck-Jensen and Hockenhull (1983a) found that Pythium root rot in an NFT system occurred mostly under rock wool starting blocks and along the gully side, where the roots were matted together. Maximum infection occurred in narrow gullies, but roots in wide gullies with a very low nutrient flow rate were very little affected, no more than the uninoculated check treatment was. An adequate flow rate was believed to wash zoospores away from the roots. Plants grown under low light were heavily infected. *Pythium* was isolated from the nutrient solution and from healthy-looking roots, but when the crop was well grown, it was very resistant.

Aeration of the nutrient solution is undoubtedly an important factor in ensuring root health. M. B. Jackson (1980) found that slow rates of gas diffusion into and out of the solution deprived the roots of adequate oxygen and allowed damaging accumulations of carbon dioxide and ethylene. Epinasty occurred, and adventitious roots were produced. In whalehide pots with poor gas exchange there was a 95% depletion of oxygen and microbial denitrification of nitrate nitrogen. Toxins may also be produced by anaerobic microbial activity.

In tomatoes and especially in cucumbers there is frequently a very sudden

SUMMARY

Environmental Stress and Predisposition to Disease

Environmental stresses induce elastic (reversible) or plastic (irreversible) strains. The latter result in the death of tissue or whole plants, but elastic strains often result in increased susceptibility to pathogens. Stresses are applied through several agencies, such as temperature, water, light intensity and duration, nutrition (including deficiencies and toxicities), microorganisms, pests, pesticides, and environmental pollutants. Commercial crops are seldom grown under optimum conditions for the production of biomass, but rather under conditions intended to induce the maximum production of fruit, the highest flower quality, or the most desirable foliage form. This growth for productivity is often achieved under suboptimum conditions, that is, under stress.

In some cases, temperature stress is related to the relative growth rates of the host and the pathogen, but there are many examples in which defense mechanisms are impeded at optimum temperatures for production. Avoidance of stressful temperatures in greenhouse management is part of the strategy for disease escape.

Similarly, water-imposed stresses, usually overwatering or underwatering, make plants more susceptible to diseases, but no generalizations can be made. Each host–parasite combination has to be considered alone, as in the case of temperature stress.

Osmotic stress imposed by too little or too much salinity is related to stresses imposed by inappropriate levels of applied fertilizers and pH-modifying materials. The form of nitrogen is important in determining soil and tissue pH, and the balance of nitrogen with other elements, particularly potassium, is also important in determining tissue susceptibility. Calcium is an important element because of its role in the integrity and susceptibility of the host cell walls to wall-degrading

and severe condition known as root death, in which the roots appear to autolyze, becoming brown and slimy, with the cortex completely disintegrated into dead cells, which float away into the nutrient solution. No pathogens have been consistently isolated, and it is variously thought (Daughtrey and Schippers 1980; van der Vlugt 1989) that there is oxygen deprivation, an imbalance in pH, an imbalance in cytokinins, an overly cool nutrient solution, an imbalance between manganese and iron, or a physiological imbalance between the roots and the developing fruit as alternative sinks for assimilates. In more recent years better crop management, better aeration, and perhaps improved nutrient solutions have overcome this problem.

In tomato, root death, complicated by the presence of a *Pythium* sp., was prevented by using a thicker plastic liner in the NFT gullies and

pathogens. The balance of calcium with potassium and other cations is also important.

Along with water stress is the problem of soil compaction and oxygen stress, which decrease the ability of root systems to resist infection or to regenerate roots to compensate for those lost to root-rot pathogens.

Low light and shading impede the photosynthetic activity of plants. Pathogens are arbitrarily divided into two groups: those that mostly attack tissues depleted of carbohydrates, which include the wilt pathogens *Fusarium* and *Verticillium* spp.; and those, such as *Botrytis cinerea*, that mostly attack carbohydrate-rich tissues. Gradients of photosynthates exist in plants and change with cropping practices. Thus, the removal of a fruit load shifts the photosynthate sink to the stem and roots, increasing their susceptibility to certain pathogens and reducing their susceptibility to others.

These stress factors must all be considered in designing cropping practices for disease escape.

In hydroponic systems, the buffering effects of soil in moderating or delaying temperature, water, and fertilizer stresses are not present, so that accidents in greenhouse operations or breakdowns in equipment have rapid and serious effects on crops. Where nutrient solutions are recirculated, the opportunity for pathogen populations to reach damaging levels is enhanced, and plant-to-plant spread in dense plantings can be very fast. Notwithstanding these hazards, with a high standard of hygiene, strict attention to the avoidance of malfunctions in automated systems, and adequate aeration of root systems, excellent crops can be grown in hydroponic culture.

maintaining a nutrient temperature of 20–22° C (Davies 1980).

Evidently, there are very delicate physiological balances in NFT systems, easily upset, with none of the buffering that can occur in soil and, to a lesser extent, in rock wool to moderate or delay temperature, water, and fertilizer stresses. Often, subclinical infection of roots by *Pythium* spp. may occur (Stanghellini and Kronland 1986; Stanghellini et al 1988). Stanghellini and Kronland (1986) found that *P. dissotocum* Drechs. caused yield losses of up to 50% in lettuce without macroscopic root systems. The fungus attacked the tiny feeder roots to impair nutrient and water uptake.

When the temperature of an NFT system in which spinach was grown fell below 23° C, the dominant root rot pathogen was *P. dissotocum*; above 27° C, it was *P. aphanidermatum* (Bates and Stanghellini 1984; Gold and Stanghellini 1985).

PART TWO

Strategy for Disease Control

Plant diseases are the result of interactions between pathogens, hosts, and the environment. In the greenhouse, severe stresses may be imposed on a host because of the premium on space and the ability of the grower to push the crop to new limits of productivity by means of advanced horticultural technology. For the same reasons, environments geared to maximum productivity are not necessarily inimical to the saprophytic and parasitic activities of pathogens. Nor are they necessarily good environments for the successful operation, witting or unwitting, of biological control.

Most pathogens have complex systems of interlocking life cycles, which are affected by host susceptibility, density, and habit on the one hand and by interacting environmental factors on the other. By way of example, the many facets of saprophytism and parasitism in the ubiquitous gray mold fungus, *Botrytis cinerea* Pers.:Fr., are illustrated in Figure 27. Similar diagrams can be constructed for many other pathogens. Control of gray mold depends on interrupting as many as possible of the pathways depicted in Figure 27.

This can be accomplished by direct action against the fungus, such as eliminating sclerotia from seed lots, inhibiting spore germination by prophylactic fungicides, or inhibiting sporulation by antisporulants. It can also be accomplished by manipulating the environment, so that despite a prolific airborne spore inoculum, for example, infection does not occur, simply because plant surfaces are kept dry. The host, too, can be manipulated to be less susceptible—for example, in the case of gray mold, by ensuring an adequate supply of calcium. Major gene resistance to *Botrytis* spp. is elusive, but genetic resistance is or should be the primary and preferred means of controlling most vegetable diseases. This is not so for ornamentals, which are bred primarily for novelty of form and color. Enhancing the activity of biological control organisms is another form of manipulating the environment and is being increasingly put into practice.

Controlling diseases in the greenhouse, then, is a very complex undertaking, which requires a thorough knowledge of the autecology of pathogens and their commensal microorganisms. Implicit in autecology are environmental stresses on both the host and the pathogen, as well as stresses on and stresses imposed by other microorganisms. Greenhouse technology imposes its own set of stresses on all three components of the host–pathogen–environment synecology.

Strategy for control can be arbitrarily divided, for discussion, into eliminating inoculum, limiting disease spread, manipulating ecology to prevent disease escape, exploiting intrinsic resistance, and enhancing biological control.

Finally, all disease control measures must be integrated into a flexible system that is compatible with insect control, with crop production systems, and not least with economics.

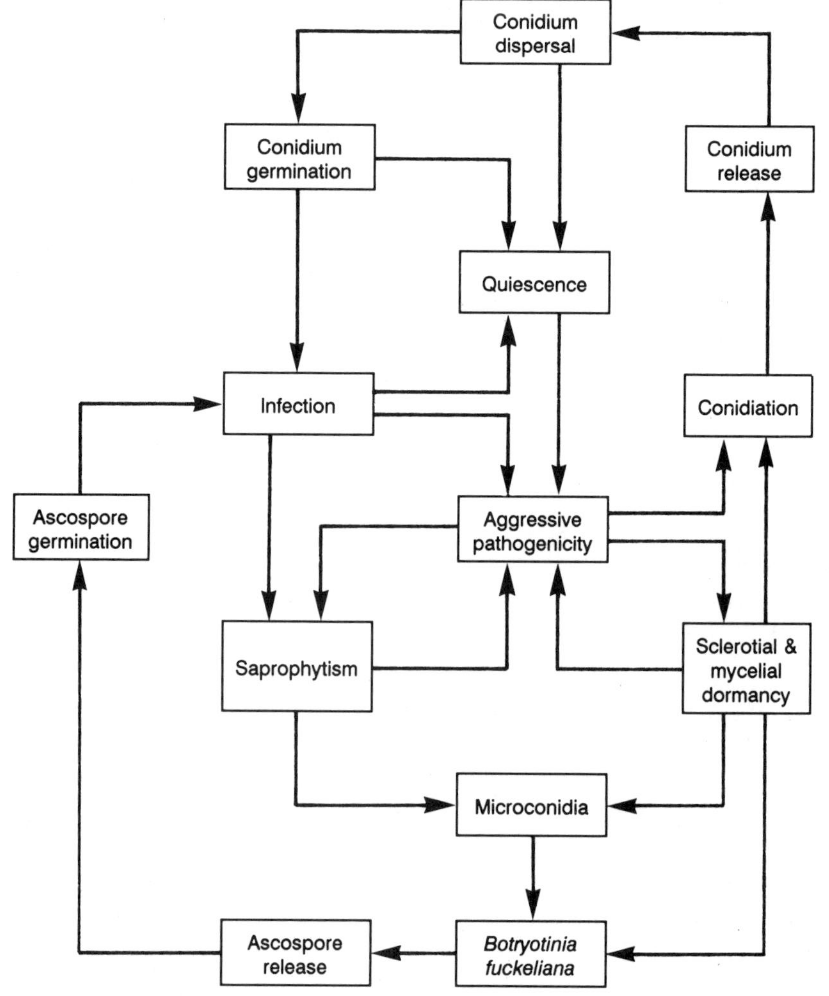

Figure 27. The life cycle of *Botryotinia fuckeliana* (anamorph *Botrytis cinerea*). Controlling gray mold involves interrupting as many pathways of the life cycle as possible.

Eliminating Inoculum

CHAPTER 5

Sterilizing Media

The sterilization of soil and soil substitutes is a major cost in greenhouse production. Steam sterilization, even with air–steam mixtures (K. F. Baker and Roistacher 1957a,b), is expensive in energy consumption and labor. In experiments on the value of steam sterilization of potting media for controlling *Pythium splendens* H. Braun on *Peperomia obtusifolia* (L.) A. Dietr. and *Cylindrocladium spathiphylli* Schoulties, El-Gholl, & Alfieri on a *Spathiphyllum* sp., Chase and Poole (1985b) concluded that the expense outweighed the benefits. Disease levels were no greater in unsterilized media than in steam-sterilized media. Fumigation with volatile chemicals is also expensive and, in the case of methyl bromide, leaves unacceptable bromine residues, which contaminate leafy vegetables and the environment. Only solarization is not expensive in the consumption of fossil fuels, but like all forms of sterilization, it leaves greenhouse space nonproductive during the treatment period and for some time afterwards, until the soil temperature can be tolerated by plants again. Nevertheless, the usual monocrop system of greenhouse production means that populations of soilborne pathogens tend to build up to damaging levels unless they are periodically reduced (Ebben 1971; Ebben and Last 1975; Last and Ebben 1966).

Preparation of Soil and Composts for Sterilization

Microorganisms are more susceptible to heat and fumigants when they are in an active metabolic state, but many plant pathogens form resistant structures—oospores, chlamydospores, resting mycelium, or sclerotia—in dry, cool soils. It is therefore advantageous to sterilize soils, either by heat or by chemicals, when they are warm and moist. This is most convenient immediately after a crop is removed. Early sterilization also allows more time for toxic residues, such as methyl bromide, manganese, and ammonia, to disperse and for a saprophytic microflora to become established before the next crop is planted. A highly conducive soil is thus permitted to become suppressive.

It is generally recommended that before sterilization or pasteurization all plant residues should be removed, particularly thick roots containing fungal resting structures (Rowe and Farley 1978; Rowe et al 1977) and viruses (Broadbent and Fletcher 1966; Broadbent et al 1965), which might escape inactivation or death and constitute a source of recontamination of otherwise sterile soil (Bollen 1969a,b; Hege and Ross 1972). The soil should be well tilled, with the clods and pans (which are surprisingly common in greenhouse soils) broken up, and it should be porous, to allow the steam or fumigant to percolate freely (McKeen 1954; Munnecke et al 1971).

Loose-structured soils are more effectively fumigated than peaty and organic or heavy, wet clay soils (Hanson and Nex 1953; Hemwall 1960; Siebering and Leistra 1979), but steam is more effective in a well-aerated clay soil with cracks providing access to the deeper levels (Nederpel 1979; Newhall 1955).

Soil temperature is important for both steaming and fumigation. Steam condenses quickly in a cold soil, increasing the moisture content of the top 10 cm of soil by as much as 10–11%, and the water impedes further steam penetration (Hege and Ross 1972; Heijna 1966). Temperature is even more important for fumigation; too cold a soil restricts volatility and permits unwanted side effects to occur, whereas too warm a soil loses the fumigant too quickly. The diffusion of methyl bromide is satisfactory at 10°C, but microorganisms are not very active metabolically at that temperature and are less susceptible. The rate of diffusion of D-D in a mineral soil was four times faster at 24°C than at 6.5°C (McClellan et al 1949b), and it took only 1 day to kill nematodes (*Meloidogyne incognita* (Kofoid & White) Chitwood) at 27.5°C and 3 days at 16.5°C, whereas nematodes still survived after 13 days at 6.5°C.

Soil moisture is important in permitting a free flow of steam through a soil (Hege and Ross 1972; Nederpel 1979). A wet soil rarely attains a sterilizing temperature, and the more steam is pumped in, the wetter the soil becomes. This is even more important in a water-absorbing peat or peat mix; Runia (1986) considered that peat should be steamed only when almost dry.

Van Berkum and Hoestra (1979) considered soil moisture the most important factor in fumigation. Methyl isothiocyanate releasers require a soil just moist enough for the microorganisms to be in a susceptible state and for the fumigant to be released (Leistra et al 1974). On the other hand, methyl bromide diffuses too fast in a dry soil, which must be brought to an optimum moisture level (Goring 1962). As a rule of thumb, the soil should be at seedbed moisture for 2 wk.

The overall biological efficacy of fumigants and their rate of disappearance from soil can be expressed as the product of concentration (dose) and exposure time. These effects can be simulated by computer models for sets of environmental factors (Siebering and Leistra 1979).

Sterilization by Steam

K. F. Baker and Roistacher (1957a) advocated heating soil to 82°C for 30 min to sterilize it. Most plant pathogens are killed by temperatures

Table 10. Thermal inactivation of selected plant pathogens and pests

	Temperature (°C)	Exposure time (min)	Reference[a]
Most bacteria	60–70	10	1
Thermotolerant bacteria	90	30	2
Botrytis cinerea	55	15	1
Cylindrocarpon destructans	50	30	2
Didymella lycopersici	50	30	2
Fusarium oxysporum			
f. sp. *dianthi*	60	30	2
f. sp. *gladioli*	57	30	1
Phialophora cinerescens	50	30	2
Phytophthora cryptogea	50	30	2
Pythium sp.	53	30	2
P. irregulare	53	30	2
P. ultimum	46	20–40	1
	50	30	2
Rhizoctonia sp.	52	30	1
	53	30	2
R. solani	53	30	2
Sclerotinia sclerotiorum	50	5	1
Sclerotium rolfsii	50	30	2
Thielaviopsis basicola	48	30	2
Verticillium albo-atrum	53	30	2
V. dahliae	58	30	2
Most pathogenic fungi	60	30	1
Most actinomycetes	90	30	1
Foliar nematodes	49	15	1
Anguillulina dipsaci	56	11	1
Heterodera marioni	48	15	1
Meloidogyne incognita	48	10	1
Pratylenchus penetrans	49	10	1
Most viruses	100	15	1
Insects and mites	60–70	30	1
Worms, slugs, centipedes	60	30	1
Most weed seeds	70–80	15	1

[a] References: 1, K. F. Baker and Roistacher (1957a); 2, Bollen (1969b).

around 70° C. Some viruses, however, such as tobacco mosaic virus and cucumber green mottle mosaic virus, are difficult to inactivate even at 100° C (Runia 1986). Table 10 gives some examples of inactivation temperatures. There are exceptions to the generalizations in Table 10; some *Pythium* spp. and some isolates of *Fusarium oxysporum* Schlechtend.:Fr. are more thermotolerant than others (Bollen 1969b).

The techniques of soil sterilization by the injection of steam have been outlined by K. F. Baker and Roistacher (1957b) and Nederpel (1979). Various hollow-tined injection devices, such as the Hoddesdon grid (Plate 18), are available for introducing steam at depth into groundbeds, but they are laborious to use. Most growers rely on balloon-sheet steaming, in which steam is introduced under a plastic sheet (strengthened by a nylon net and anchored at the edges) spread on the soil surface, so that the sheet balloons under pressure (Plate 19) (Grainger and Kennedy 1964; Hege and Ross 1972; Heijna 1966; Nederpel 1979). Temperatures of 50°C or more can be achieved at a depth of 30 cm in the soil and 70°C at about 23 cm. Grainger and Kennedy (1964) thought that most of the heating at that depth was by condensed hot water.

A common problem is for growers to attempt to sterilize too large a groundbed area with a single steam inlet; in general an area no longer than 20 m and no wider than 5 m can be sterilized with one inlet (Nederpel 1972).

The efficacy of groundbed steaming is improved by passing the steam through perforated pipes permanently buried about 45–60 cm deep in the soil (Nederpel 1979). In all cases a condensate collection system has to be used to permit free steam flow.

Steam sterilization of small batches of soils, composts, and rock wool slabs is easy to accomplish in a variety of containers (Aldrich and Nichols 1969; K. F. Baker and Roistacher 1957b; Runia 1986) and under plastic sheets on a concrete floor. Steaming, however, creates a biological vacuum, which could be filled equally well by pathogens as by saprophytes. Soil that has been newly and completely sterilized by steam is highly susceptible to reinfestation by airborne pathogens (Rowe et al 1977) or organisms in windblown dust, splashed water, or dirty tools. It is the saprophytes that we wish to encourage to fill the vacuum, in order to set up suppressiveness in the soil with populations of potential biological control agents.

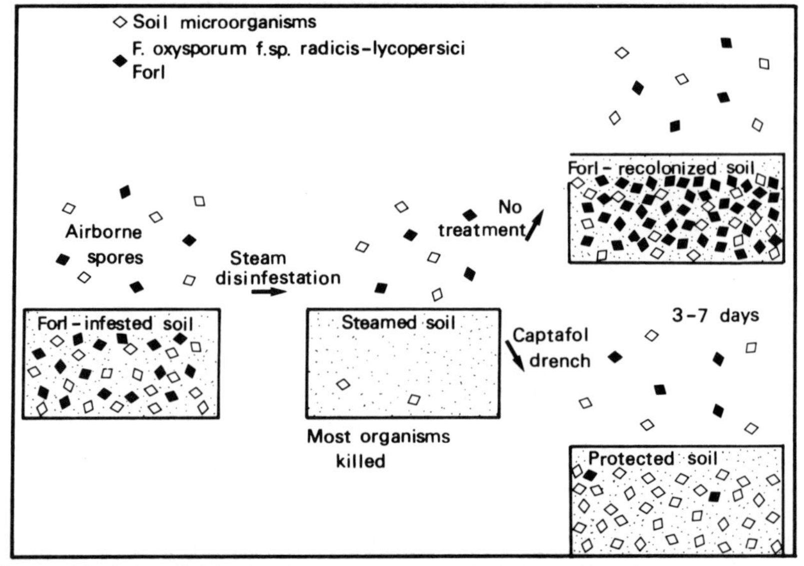

Figure 28. A fungicide sieve to prevent reinfestation of newly sterilized soil by airborne microconidia of *Fusarium oxysporum* f. sp. *radicis-lycopersici* (Forl), the cause of Fusarium crown and root rot of tomato. Redrawn from Rowe and Farley (1978).

Table 11. Active manganese contents of six soils and exchangeable manganese contents after steam sterilization[a]

Soil type	Active Mn[b] ($\mu g \cdot g^{-1}$)	Exchangeable Mn ($\mu g \cdot g^{-1}$)	
		After steaming	Control
Sand	45	30	10
Loamy sand	62	45	15
Clay loam	240	180	28
	345	140	25
Clay peat	198	160	20
	195	165	18

[a] Adapted from Sonneveld (1979).
[b] Active Mn includes exchangeable Mn.

Rowe and Farley (1978) devised a fungicide sieve for newly sterilized tomato groundbeds, which are colonized rapidly by airborne microconidia of *F. oxysporum* f. sp. *radicis-lycopersici* W. R. Jarvis & Shoemaker (Rowe et al 1977). Immediately after steaming, and while the soil was still hot, they sprayed captafol over the groundbed from an overhead irrigation system (Figure 28). Captafol is active against *F. oxysporum* f. sp. *radicis-lycopersici* but permits the growth of many saprophytic organisms, which exert biological control over the pathogen (Marois and Mitchell 1981a,b; Marois et al 1981).

Chase and Poole (1985b), however, investigating the benefits of steam sterilizing (at 88°C for 1.5 h) potting media for *Peperomia* and *Spathiphyllum*, thought that suppressive microorganisms might not be as important as had been suggested previously. When *Peperomia* was inoculated with high populations of *Pythium splendens* and *Spathiphyllum* with *Cylindrocladium spathiphylli*, disease severity was never greater in the steam-treated medium than in the comparable untreated medium. Suppressivity may depend largely on the soil type and the microorganisms available to return quickly.

A major disadvantage of steam treatment at temperatures greater than 80°C is that manganese becomes available to plants in toxic amounts (in soluble form at levels greater than 12 $\mu g \cdot g^{-1}$ or in exchangeable form at levels greater than 50 $\mu g \cdot g^{-1}$), particularly in acid soils, and unless this free manganese is leached from the soil, it may remain toxic for 60 days or more (Sonneveld 1979) and contribute to iron deficiency. The soil type is important in this respect (Table 11). Manganese becomes available in the soil as bivalent ions in the soil solution or bound to clay or humus (exchangeable manganese) or in a readily reducible form (active manganese).

The nitrogen in steamed soil also undergoes marked changes. In nonsterile soil, nitrifying bacteria convert applied ammonium fertilizers into nitrates very quickly. Steaming destroys these bacteria, and nitrites are formed, together with phytotoxic amounts of ammonia (more than 40 $\mu g \cdot g^{-1}$) (Sonneveld 1969). These toxic materials, together with the condensate left after steaming, all have to be leached from the soil by water applied at 25–50 $L \cdot m^{-2}$. This water has to be drained off, and the groundbed is then out of production for 2–3 wk. Buried pipes used to supply the steam also serve to drain the condensate from the groundbed, if they are properly sloped. Drainage of the leachates creates environmental pollution problems (Runia et al 1988; E. A. van Os et al 1988). In the Netherlands, where the concern is greatest, research concentrates on recycling greenhouse drainage water (van Os et al 1988).

The chemical changes in soilless media appear not to have been investigated to any appreciable extent. Runia (1986, 1988) studied the disinfestation of a number of soilless substrates by steam and other means. She advocated that the substrate be as dry as possible before steaming and recommended a temperature of 70° C for all pathogens except viruses. Inactivation of cucumber green mottle mosaic and tobacco mosaic viruses required 100° C.

For peat, Runia (1986) recommended at least 3 h of steam treatment. The dry peat is stacked over a perforated inlet tube in a ridge no higher than 60 cm and 100 cm wide.

Pasteurization

Heating soil to 70° C kills most plant pathogens without leaving a biological vacuum (K. F. Baker 1962a, 1970; K. F. Baker and Olsen 1960) and leaves the soil suppressive, with its saprophytic microflora. At the same time, problems of excessive exchangeable manganese, nitrites, and ammonia are obviated (Sonneveld 1979; Sonneveld and Voogt 1973, 1975). Nederpel (1979), however, pointed out disadvantages in labor and energy requirements.

A uniform temperature of 70° C in a mass of soil or other medium is almost impossible to achieve in practice with steam alone, but it becomes possible when air is mixed with steam immediately before it enters the soil (K. F. Baker 1962a; K. F. Baker and Olsen 1960; K. F. Baker and Roistacher 1957b; Brazelton 1968; Bunt 1976; Dawson et al 1965, 1967; Morris 1957; Nederpel 1979). The aerated steam system has three phases (Brazelton 1968), the warm-up, cook, and cool-down phases. The last uses air only, to terminate the cook phase (30 min at 60–80° C) so as not to destroy potential biocontrol microorganisms. The system requires relatively little energy. A small, easily controlled steam flow is injected into a large-volume, low-pressure airflow (Figure 29). For potting mixes, a number

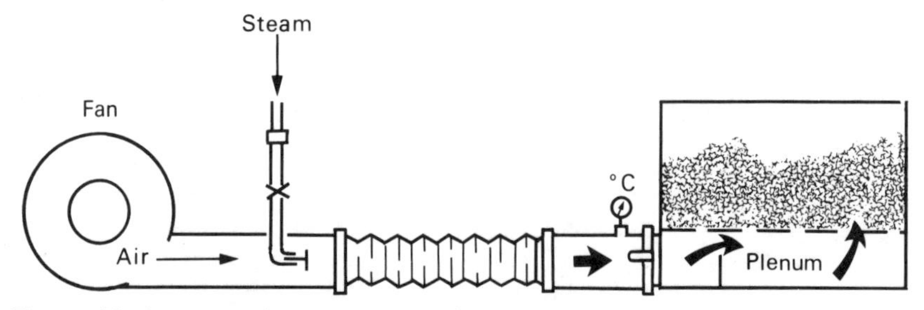

Figure 29. An aerated steam system for pasteurizing soil and potting mixes. Redrawn from Brazelton (1968).

of portable bins can be used in rotation with a single stationary air–steam source, and the escaping air–steam mixture can be recirculated. Thermostats control the steam valve and airflow gate.

Dawson et al (1965) found that heating soil with an air–steam mixture for 10–30 min at 60° C eliminated attacks on lettuce and tomato seedlings by *Rhizoctonia solani* Kühn and improved their growth. In soils heated to over 80° C, however, growth was poorer than in soils heated to 60–75° C; this effect was attributed to relatively high soil concentrations of soluble and exchangeable manganese and nitrite. The release of ammonia was greatly increased at 52° C. When reintroduced in preheated soils, *R. solani* caused more damage to lettuce in soil preheated to 60° C than in unheated soil, but less than in soil preheated to 100° C.

After heating soils in situ with air–steam mixtures, Dawson et al (1967) obtained an increase in yield in subsequent cucumber crops from 8.25 kg (check) to 10.75 kg per plant in soil heated to 88° C and, in a second experiment, from 23.75 kg (check) to 47.75 kg per plant. They attributed this to the control of a sclerotial fungus, later identified as *Phomopsis sclerotioides* van Kesteren. In these experiments, concentrations of soluble and exchangeable manganese increased from 4 $mg \cdot kg^{-1}$ in untreated soil to 9, 43, 87, and 108 $mg \cdot kg^{-1}$ in soils heated to 66, 77, 88, and 100° C, respectively.

Negative-Pressure Steam Sterilization

A significant advance in soil steaming was made by Heijna (1966) and Runia (1983). A fan is used to set up a negative pressure within buried pipes, so that steam supplied at the soil surface is actively moved downwards. This arrangement provides more uniform and deeper distribution of heat than sheet steaming (Table 12).

Negative-pressure steaming requires less fuel to generate steam than conventional balloon-sheet steaming. Runia (1983) supplied 200 $MJ \cdot m^{-2}$ for sheet steaming, 170 $MJ \cdot m^{-2}$ for conventional buried-duct steaming, and only 115 $MJ \cdot m^{-2}$ for negative-pressure steaming. The period of steaming can be reduced to 4 or 5 h, and radiation losses are lower than those from the ballooning sheet and from the soil surface after the sheet has been removed.

Table 12. Average temperatures (° C) achieved in 8 h at various soil depths by sheet steaming and negative-pressure steaming[a]

	Clay soil depth (cm)			Sandy soil depth (cm)				Loam soil depth (cm)		
Steaming system	45	25	10	45	35	25	10	50	30	10
Sheet	44	94	100	33	38	63	98	65	79	87
Negative pressure	67	90	98	50	88	92	99	58	94	100

[a] Adapted from Runia (1983).

Steam Sterilization of Rock Wool

Runia (1986) found that rock wool can be sterilized by steam if the plastic covers are removed from the slabs and the slabs are stacked not more than 1.3 m high on a pallet. The slabs are more readily sterilized dry than wet, since higher temperatures are attained more quickly. Slabs with the plastic covers still on and stacked 1.5 m high on pallets took 5 h to reach 100°C; with the covers removed, the effective time was only 2 h. In single layers in situ in the greenhouse, slabs in their bags took 2 h to reach 91–100°C; free of their covers, they took 1 h to reach 83–100°C (Runia 1986). If the rock wool is contaminated by cucumber green mottle mosaic or tomato mosaic virus, Runia (1986) recommended that 100°C should be maintained for 10 min.

Composting

Composting is a complex process, combining thermal eradication of a portion of the microflora of organic waste materials with certain physical and chemical properties of composts that are valuable for disease control (Hoitink and Fahy 1986).

The thermal processes of composting resemble those of pasteurization, and temperatures in large compost piles may be sufficient to kill some plant pathogens, aided by the antibiotic action of thermostable microorganisms (Table 13). Hoitink and Fahy (1986) recognized three phases in composting: an initial phase of 1–2 days during which temperatures rise and readily degradable compounds are decomposed; a thermophilic phase, possibly lasting for months, during which mostly cellulose is degraded; and a curing or stabilization period, when temperatures decline, decomposition rates decline, and moderately thermotolerant organisms recolonize the compost. For maximum decomposition rates, temperatures should lie between 38 and 55°C.

Bollen (1985) reviewed the eradication of plant pathogens during composting and suggested three inactivation mechanisms: exposure to high

Table 13. Some greenhouse crop pathogens killed by composting

Pathogen	Reference
Erwinia carotovora subsp. *carotovora*	Lopez-Real and Foster (1985)
E. chrysanthemi	Hoitink et al (1976)
Armillaria mellea	Yuen and Raabe (1984)
Botrytis cinerea	Hoitink et al (1976); Lopez-Real and Foster (1985)
Didymella lycopersici	Phillips (1959)
Phytophthora cinnamomi	Hoitink et al (1976)
Pythium irregulare	Hoitink et al (1976)
Rhizoctonia solani	Wijnen et al (1983); Yuen and Raabe (1984)
Sclerotium rolfsii	Yuen and Raabe (1984)
Verticillium dahliae	Yuen and Raabe (1984)
Tomato necrosis virus	Wijnen et al (1983); Lopez-Real and Foster (1985)

temperatures over long periods, the toxicity of metabolites during heating, and microbial antagonism in the cooler outer layers or during the curing period. Many pathogens are killed by a 30-min exposure to 55°C or by longer exposures to lower temperatures, as they are in solarization (Katan 1981). Some, including *Verticillium albo-atrum* Reinke & Berthier, may survive (Hoitink et al 1976). Tobacco mosaic virus in infested tobacco residues was not inactivated by composting for 6 wk at 50–70°C (Hoitink et al 1982). However, Avgelis and Manios (1989) could not detect tomato mosaic virus in infected tomato residues composted at temperatures below 47°C, and they concluded that it was degraded by microbial action rather than by heat.

Composts vary widely in their nutritional and physical effects on plant growth, both stimulatory and inhibitory (Hoitink and Fahy 1986). They are often conducive to diseases caused by *Fusarium, Pythium*, and *Rhizoctonia* spp. if they have been overheated to 55 or 60°C, whereas material taken from the cooler outer layers of composts is suppressive (Hadar and Mandelbaum 1986; Kuter and Hoitink 1985; Nelson and Hoitink 1982).

Hardwood bark composts are generally, but not invariably, suppressive to *Fusarium, Phytophthora, Pythium*, and *Rhizoctonia* spp.; softwood bark composts to *Phytophthora* and *Pythium* spp.; and composted municipal sludge only to *Pythium* spp. (Hoitink and Fahy 1986).

Composting of bark must proceed to completion before it is used it on plants, because of the phytotoxicity of intermediate low-molecular-weight aliphatic acids (Still et al 1976).

Properly processed composts should not be confused with trash piles and loose debris, which are sources of numerous pathogens, such as tomato mosaic virus (Broadbent 1965a; Broadbent et al 1965), *Phytophthora infestans* (Mont.) de Bary (Bonde and Schultz 1943), *F. oxysporum* f. sp. *radicis-lycopersici* (Jarvis et al 1983), *Didymella lycopersici* Kleb. (Phillips 1959), *Microdochium panattonianum* (Berl.) Sutton, Galea, & Price in Galea, Price, & Sutton (Galea and Price 1988), and *Alternaria* spp. They are also sources of insect pests (van Lenteren and Woets 1988).

Solarization

Solarization is a form of soil pasteurization whereby solar energy is entrapped beneath plastic sheets spread on the soil surface. The process was first described by J. Katan et al (1976) and has been comprehensively reviewed by Katan (1979, 1980, 1981). It is a cheap and effective means of pasteurizing soil, provided that at least 30 days is available during a period of high solar radiation. The depth of heat penetration and hence the efficacy are improved by prolonging the period of solarization (Garibaldi and Tamietti 1984; Kassaby 1985; Katan 1979, 1980, 1981; Katan et al 1976; Mahrer and Katan 1981; Pullman et al 1981). Mahrer and Katan provided a theoretical model that predicts soil temperatures in different types of soil in various weather and under different widths of polyethylene mulch (Figure 30).

Pullman et al (1981) derived a logarithmic relationship between temperature and the time required to kill the soilborne pathogens *Pythium ultimum* Trow, *Rhizoctonia solani, Thielaviopsis basicola* (Berk. & Broome) Ferraris, and *Verticillium dahliae* Kleb. (Figure 31).

As with other forms of sterilization and pasteurization, the medium to be solarized should be moist, but not over-wet, so that the microorganisms are in an active metabolic state. Clods should be broken up and thick root debris removed.

Solarization has long been used by Canadian greenhouse growers in midsummer in attempts to inactivate pests and pathogens in empty, closed greenhouses but with no plastic cover directly on the soil (H. J. Thorpe, personal communication). With modern tendencies to keep the greenhouse in production all through the summer and into autumn, this practice has fallen into disuse. It is, however, practiced in Mediterranean countries (Garibaldi and Tamietti 1984; Goisque et al 1984; Malathrakis et al 1983; Tamietti and Garibaldi 1980; Tjamos 1983; Tjamos and Faridis 1980) and as a preplant treatment for strawberry beds in Japan (Kodama and Fukai 1982).

Garibaldi and Tamietti (1984) found that polyvinyl chloride film was superior to polyethylene in solarizing soils in northern Italy (Table 14). They also found a glass-covered house to be much more effectively solarized than a plastic-covered house (Table 15). This result can be attributed to

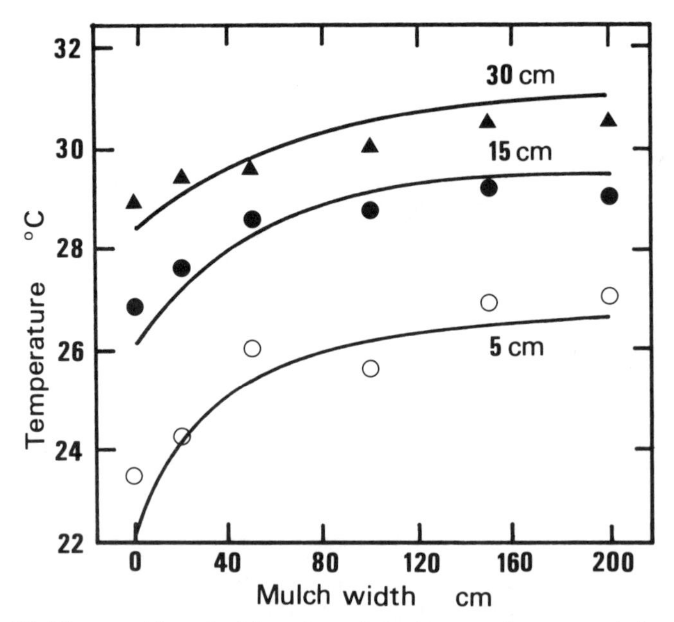

Figure 30. Observed (symbols) and predicted (curves) average daily minimum soil temperatures at depths of 5, 15, and 30 cm at the center of a polyethylene film covering the soil, as a function of cover width. Redrawn from Mahrer and Katan (1981).

Table 14. Soil solarization under polyethylene (PE) and polyvinyl chloride (PVC)[a]

Soil depth (cm)	Hours of temperature above 45° C (per day)		Mean temperature (° C)		Accumulated hours of temperature above 40° C	
	PE	PVC	PE	PVC	PE	PVC
7	3.55	5.50	46.08	50.10	371	410
15	0.92	0.93	42.48	49.73	308	347
20	0.00	0.00	37.56	38.85	36	156

[a] Adapted from Garibaldi and Tamietti (1984).

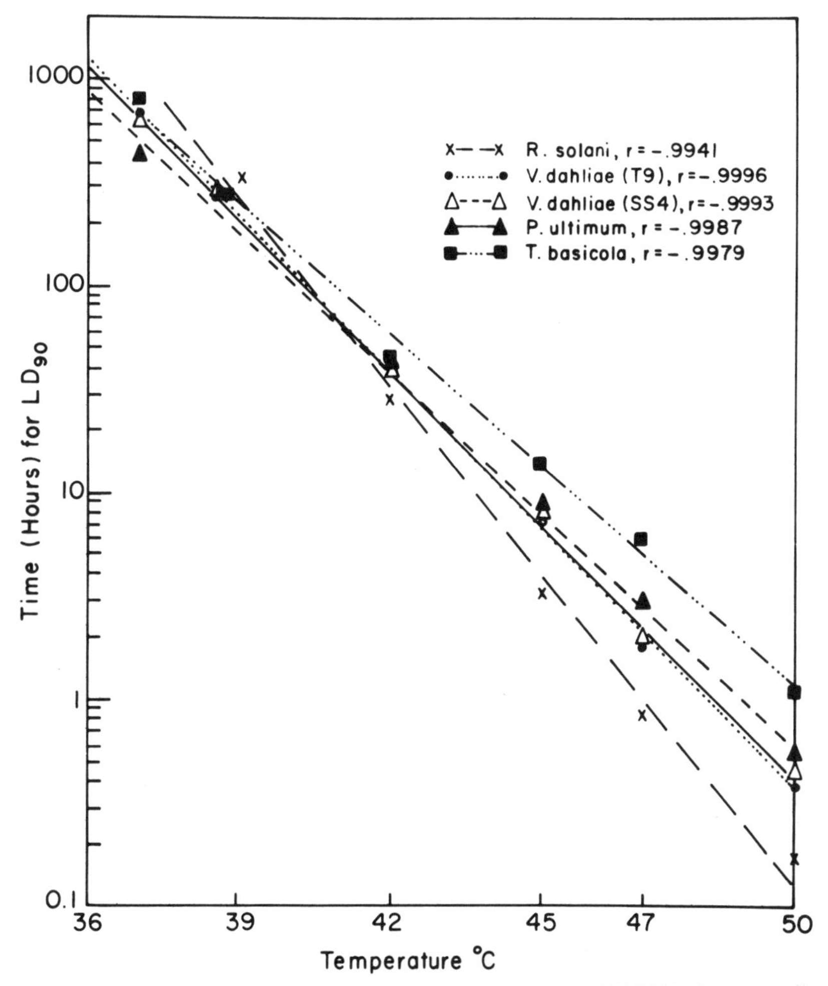

Figure 31. Time and temperature exposures required to kill 90% of propagules of *Rhizoctonia solani, Verticillium dahliae* (two isolates), *Pythium ultimum*, and *Thielaviopsis basicola* on potato dextrose agar. Reprinted, by permission, from Pullman et al (1981).

Table 15. Soil solarization at three soil depths in a glass-covered and a plastic-covered greenhouse in northern Italy[a]

Greenhouse covering	Accumulated hours of temperature above 40° C		
	5-cm depth	12-cm depth	20-cm depth
Glass	577	560	375
Plastic	287	191	0

[a] Adapted from Garibaldi and Tamietti (1984).

Table 16. Severity of corky root rot of tomato after fumigation of soil with methyl bromide and subsequent solarization[a]

Treatment[b]	Disease index[c]	
	Experiment 1	Experiment 2
Solarization and MeBr (68 $\mu g \cdot g^{-1}$)	0.48	1.26
Solarization and MeBr (34 $\mu g \cdot g^{-1}$)	0.47	1.46
Solarization	0.91	1.58
No treatment	2.12	2.27

[a] Data from Tjamos (1983).
[b] Methyl bromide (MeBr) was applied in autumn. The experiments were conducted in Greece.
[c] Scale of 0–5, from least to greatest severity.

the relative opacity of glass to infrared radiation (Trickett and Goulden 1958).

Solarization would seem to be a very useful practice where mobile or temporary plastic tunnels are used, if the area to be planted is solarized before the tunnel is erected over the site.

It would also seem to have considerable potential in the sterilization of potting composts stored outside, as well as pots, flats, and recycled peat bags and rock wool slabs. Experimental data, however, appear to be lacking. Besri (1982) almost completely eradicated *Didymella lycopersici* from tomato stakes by solarization. He achieved temperatures greater than 50° C between April and September in California, and this reduced the incidence of cankers in the next staked crop from 20% in the check plants to 2% in plants with disinfested stakes.

Tjamos (1983) obtained improved control of tomato corky root rot (caused by *Pyrenochaeta lycopersici* R. Schnieder & Gerlach) by treating soil with a low dose of methyl bromide before solarization (Table 16). This helps to reduce the levels of bromine residues in soils and crops. Garibaldi and Tamietti (1984), however, found that a half-rate treatment with methyl isothiocyanate followed by a 30-day solarization provided no advantage in controlling the same disease.

The efficacy of solarization is enhanced by combining it with allelopathy. Ramirez-Villapudua and Munnecke (1987) found that cabbage yellows (caused by *Fusarium oxysporum* f. sp. *conglutinans* (Wollenweb.) W. C. Snyder & H. N. Hans.) was eliminated in plots in which cabbage residues were incorporated into the soil before solarizing for 4 or 6 wk in California. They suggested that fungitoxic gases, perhaps sulfur-containing volatiles

(Lewis and Papavizas 1970, 1971, 1974), were trapped beneath the plastic sheet and contributed to the reduction in populations of fungal propagules. It would therefore seem worthwhile, subject to economic consideration, to advocate growing a crucifer green manure crop for incorporation into a soil to be solarized.

Katan (1989) considered it likely that high soil temperatures affect the number of antagonists of pathogens, their antagonistic capacity, or both. Presumably, high-temperature but sublethal stress could induce strains in pathogens that make them more susceptible to antibiosis.

Dielectric Heating

Dielectric heating raises the temperature of material by the energy from electric fields generated by microwaves on the order of 27 MHz (Diprose and Evans 1988) or 2,450 MHz (van Wambeke et al 1983). It is ideally suited for heating materials containing water. Composts are best treated at a water content of about 20–30% (Diprose and Evans 1988; van Wambeke et al 1983). The load is heated evenly to 100° C in 3–4 min, and the heating period can be prolonged by thermally insulating the apparatus.

Compost is continuously fed onto a conveyor belt, in a layer about 10 cm thick and 25 cm wide, passing through a microwave generator. A 50-kW machine can heat as much as 2 t of compost per hour to 100° C. Although the capital cost is high, the running costs for total sterilization of compost are relatively low.

Dielectric heating can also be used to sterilize rock wool slabs for recycling (M. F. Diprose, personal communication). Runia (1986), however, found microwaves to be poorly distributed in rock wool in a microwave oven, and they failed to inactivate cucumber green mottle mosaic virus.

The same method can be used to sterilize mushroom casing (Diprose and Evans 1988).

Gamma Irradiation

A dose of 500 krad inactivates artificially introduced tomato mosaic virus and cucumber green mottle mosaic virus in rock wool slabs, but more than 1 Mrad is required for inactivating natural contamination (Runia 1986). In potting soil, 100 krad eradicated *Phytophthora nicotianae* Breda de Haan var. *nicotianae*, and 250 krad eradicated *Fusarium* spp. (Rattink 1982).

Fumigation

Sterilization of soil and soilless media by fumigation depends on infiltrating the medium with the vapor of a volatile chemical. Some of these chemicals have a broad spectrum of activity against the soil microflora, fauna, and weed seeds, whereas others have a more limited spectrum (Table 17) (Vanachter 1979; van Assche 1979; van Berkum and Hoestra 1979). Fumigants behave differently in different soils and under different conditions, and these effects were discussed at length by Siebering and Leistra (1979).

A frequent response to fumigation is enhanced crop growth in addition to reduction in disease. This effect, known as the increased growth response (IGR) (Altman 1970; Wilhelm 1965; Wilhelm and Nelson 1970), is the result of the control of organisms that cause inconspicuous damage to the fine feeder roots but nevertheless impede the uptake of nutrients and water. *Pythium* spp., nematodes, and several other microorganisms are responsible and have been termed *root nibblers* (Altman 1970; Wilhelm 1965; Wilhelm and Nelson 1970).

In general, plant-pathogenic bacteria are not well controlled by fumigation. Munnecke and Ferguson (1960) found that methyl bromide failed to control *Clavibacter michiganensis* subsp. *michiganensis* (Smith) Davis et al, *Xanthomonas campestris* (Pammel) Dowson pv. *pelargonii* (Brown) Dye, and *Agrobacterium tumefaciens* (Smith & Townsend) Conn in U.C. mix. Control of *C. michiganensis* subsp. *michiganensis* proved even more difficult in the presence of farmyard manure (Slusarski 1983). Unfortunately, nitrifying bacteria are readily killed, and so fumigation is usually followed by inhibition of nitrification, which may last for several weeks (Good and Carter 1965; Marks et al 1972; Williamson 1953; Wolcott et al 1960), particularly if the soil is cold and wet. This often necessitates an adjustment of nitrogen fertilization (Tu 1972).

Like pasteurization with steam, fumigation usually leaves a fumigant-tolerant microflora, particularly at depth and at the edges of the treated volume of soil. Some fungi are quite tolerant (Kreutzer 1963). *Trichoderma* spp. were shown to be tolerant of methyl bromide (Munnecke et al 1973; Ohr et al 1973). Ebben (1971) found that *Pyrenochaeta lycopersici*, causing corky root rot of tomatoes, was difficult to control with methyl bromide when a high population of the pathogen was present in the soil. Investigating the control of cucumber black root rot (caused by *Phomopsis sclerotioides*), White and Hague (1971) obtained an increase in cucumber yield with a

Table 17. Some common fumigants and their spectra of activity[a]

Fumigant	Activity against:				
	Bacteria	Fungi	Nematodes	Insects	Weeds
Chloropicrin	+	±	+	+	+
Dazomet (MIT[b] releaser)	...	+	+	+	±
Dichloropropane–dichloropropene mixture (D-D)	±	±	+	±	±
1,3-Dichloropropene (1,3-D)	...	−	+	...	±
Ethylene dibromide (EDB)	...	−	+	...	...
Metham sodium (MIT releaser)	−	±	±	+	+
Methyl bromide	−	±	+	+	+
Methyl bromide–chloropicrin mixture	+	+	+	+	+
Methyl isothiocyanate (MIT)	+	±	+	+	±

[a] + = Strong activity; ± = activity against some organisms or at high rates; − = little or no activity.

[b] Methyl isothiocyanate.

high concentration of methyl bromide and a long exposure time, although there was still a substantial level of root damage. A low dose of methyl bromide was followed by an increase in the number of roots infected. White and Hague were unable to explain this, but it seems possible that the fungus remaining unsterilized by being too deep or within thick roots may have constituted an inoculum in a biological near-vacuum. This is the so-called boomerang effect. It is not a universal problem; Florida growers have regularly fumigated soils for many years without it. Altman (1970) discussed the boomerang effect, in which rapid reinvaders or fumigant-resistant organisms may cause increased levels of disease. Examples include an increased incidence of crown gall caused by resistant *A. tumefaciens* and Verticillium wilt after treatment with ethylene dibromide.

Fumigants are available in a number of formulations: as a gas (chloropicrin); a gas compressed to a liquid (methyl bromide); as liquids alone or in various mixtures (dibromochloropropane [DBCP], D-D, ethylene dibromide [EDB], methyl isothiocyanate [MIT]); or as granules (dazomet). Not all of these are legal pesticides in many countries.

Methyl bromide is the most widely used fumigant in the greenhouse, but since it is more reactive than the other bromine-containing fumigants, it leaves residues of bromide in the soil, which have several side effects (Hoffmann and Malkomes 1974). Methyl bromide reacts with amines and thiols in organic matter in the soil to form methylammonium bromides and thiol complexes, which react further to release inorganic bromine (Hoffmann and Malkomes 1979), sometimes in concentrations of 30–50 $\mu g \cdot g^{-1}$ or more after standard fumigation procedures, depending on the soil type, moisture, and temperature (Malkomes 1971). Fumigated peat contains substantial amounts. These concentrations are phytotoxic to many crops, especially aralia (van Assche 1979), carnations (Kempton and Maw 1974; Sciaroni et al 1972), chrysanthemums (Williamson 1953), melon, spinach, garlic, pansy (J. P. Martin 1966), and onion (Wilson and Norris 1966). Sometimes this effect may be complicated by the general release of other salts after fumigation (Hoffmann and Malkomes 1974). Cultivars may differ in susceptibility to bromine. Woltz and Waters (1975) found methyl residues to be more toxic to chives than bromine residues. Leafy vegetables may take up bromine at levels exceeding national tolerance levels established for daily dietary intake (Hoffmann and Malkomes 1974, 1979; Kempton and Maw 1972; Malkomes 1972). Lettuce, parsley, and spinach are particularly at risk (Kempton and Maw 1972; Malkomes 1972). Fruits such as cucumber (Malathrakis and Saris 1983; Malkomes 1972) and tomato (Kempton and Maw 1973; Malkomes 1972) do not accumulate so much bromine. A comprehensive list of residues found in fruits and vegetables was given by Hoffmann and Malkomes (1979).

In order to reduce levels of bromine and other released salts, it is usually necessary to leach soils with water at the rate of 25–50 $L \cdot m^{-2}$. This presents considerable environmental problems, and some countries ban the use of methyl bromide for this reason, as well as out of concern for human dietary health (van Leeuwen and Sangster 1987).

Chloropicrin (2%, v/v) is usually added to methyl bromide to improve efficacy as well as to add a lachrymatory and odorous hazard warning. Van Wambeke (1983) found enhanced efficacy in a mixture of methyl bromide and methyl chloride, so that a lower dose could be used.

Methyl bromide and chloropicrin are applied from pressurized cans through perforated plastic tubes lying on the soil surface beneath a plastic sheet. A film of water condensed on the underside of the sheet enables it to retain the gas. This method can be used for seed trays, flats, and heaps of potting mixes. On a larger scale, groundbeds are treated by injecting the gas under pressure through chisels moving through the soil. Plastic covers are laid down immediately after injection, and van Wambeke (1983) emphasized the importance of the quality of the plastic.

Liquid fumigants can also be applied by chisel injection, the chisels spaced 15–20 cm apart and 15–20 cm deep, or by hand injector at 15–20 cm intervals (van Berkum and Hoestra 1979). The more volatile fumigants are retained in the soil by a wet plastic cover, and the less volatile ones are retained by watering or simply by compacting the surface with a roller.

Granular dazomet is row-drilled or broadcast, worked into the soil to a depth of 20 cm, and sealed in with water or by the use of a roller.

Most fumigants require 2–3 days to kill organisms in the soil, depending on the soil type and temperature. After 7–14 days the soil is worked to allow gas residues to escape, and leaching is also required for methyl bromide. Depending on the fumigant, a waiting period is necessary before planting or seeding, to ensure that all of the fumigant has disappeared. This may be checked by attempting to germinate seeds of cress (*Lepidium sativum* L.) or lettuce (*Lactuca sativa* L.) in a sealed jar containing a soil sample.

In most countries, the application of methyl bromide is permitted only if it is done by licensed operators under strictly enforced conditions, to protect the operator, the environment, and food quality (van Leeuwen and Sangster 1987).

Disinfesting Water and Hydroponic Solutions

Water supplies from open reservoirs, streams, and wells are liable to be contaminated by, among other things, *Pythium* spp. (Hendrix and Campbell 1973), *Phytophthora* spp. (Erwin et al 1983), *Erwinia* spp. (Pérombelon and Kelman 1980), cucumber green mottle mosaic virus (Lecoq 1988), melon necrotic spot virus (Bos et al 1984), tomato mosaic virus (Broadbent 1976; Broadbent and Fletcher 1963), and chrysanthemum stunt viroid (Monsion and Dunez 1971). In addition, pathogen vectors in the Chytridiales may be expected (Tomlinson and Thomas 1986). It is uneconomical to apply standard sterilization techniques, such as steam sterilization or dielectric heating, to the large volumes of water needed for irrigation, and so bulk water purification by gravel and sand filtration beds have to be used if contamination is a recurring problem (Holden 1970).

Not infrequently, open reservoirs, streams, and wells contain runoff fertilizers and pesticides. The former can be taken into account in calculating nutrient levels for greenhouse crops, but pesticides, particularly herbicides and inappropriate materials such as aldicarb, present major problems, which require growers either to await biological detoxification or to seek alternative water sources.

The disposal of water and nutrients draining from rock wool and similar nonrecirculating systems also presents major problems, particularly in the Netherlands, where the soil water table is very high. The waste amounts to about 18 $m^3 \cdot ha^{-1} \cdot day^{-1}$, 10–50% of what is supplied to the crop (E. A. van Os et al 1988). Inactivation of pathogens is necessary before the drainage water can be allowed to run into the external water table or before it can be recirculated.

Chlorine and other halogens. Chlorination cannot be used to sterilize hydroponic solutions; the effective concentrations of free chlorine cause phytotoxicity, which generally occurs at 3–10 $mg \cdot L^{-1}$. Chlorine supplied at low levels as sodium hypochlorite has been tried as a continuous disinfestant, but without great success (Ewart and Chrimes 1980; Runia 1988). Runia obtained a moderate but incomplete kill of *Fusarium oxysporum* f. sp. *melongenae* Matuo & Ishigami with hypochlorite supplying chlorine at 1–5 $mg \cdot L^{-1}$, but this failed to inactivate cucumber green mottle mosaic virus. On a laboratory scale, phytotoxicity was overcome by adsorbing residual Cl^- on activated charcoal.

Runia (1988) also investigated iodine as a disinfestant. On a laboratory scale, the nutrient solution was shaken with an iodine-charged ion-exchange resin for up to 30 min. Iodine failed to inactivate cucumber green mottle mosaic virus and tomato mosaic virus.

Tomlinson and Faithfull (1979) found that two bromine-containing cationic surfactants, didecyldimethylammonium bromide and cetyltrimethylammonium bromide, inactivated zoospores of *Olpidium brassicae* (Woronin) P. A. Dang., the vector of lettuce big vein virus.

Ozone. Ozone at 1.5 $mg \cdot L^{-1}$, bubbled with air through a column of nutrient solution at 30 $L \cdot min^{-1}$, killed *Phytophthora nicotianae, Verticillium albo-atrum, V. dahliae, Fusarium oxysporum* f. sp. *lycopersici* (Sacc.) W. C. Snyder & H. N. Hans., and *F. oxysporum* f. sp. *melongenae* within 20 min, and it partially inactivated tomato mosaic virus. Cucumber green mottle mosaic virus required a 60-min exposure (Runia 1988). Ozone slightly reduced the concentration of iron-chelated diethylenetriaminepentaacetic acid (Fe-DTPA) over a 2-h exposure.

Ultrafiltration. Ultrafiltration through pores of about 10 μm at 300 kPa eliminated *Fusarium oxysporum* f. sp. *lycopersici, Verticillium albo-atrum*, and cucumber green mottle mosaic virus, but not tomato mosaic virus (Runia 1988).

Ultraviolet irradiation. Better success has been obtained with ultraviolet irradiation in small-scale trials, but the flow rates required for large commercial operations have not been realized. P. M. Smith (1977) and P. M. Smith and Ousley (1984) killed 98–100% of zoospores of *Phytophthora cinnamomi* Rands at dosages of 54–96 $mW \cdot cm^{-2} \cdot s^{-1}$ when the zoospores (600 spores per milliliter) were circulated in a jacket around a UV source. The dosage required to kill fragments of mycelium was higher, 130–160 $mW \cdot cm^{-2} \cdot s^{-1}$. Adams and Robinson (1979) killed 99% of a number of pectolytic bacteria in river water with an array of UV sources emitting light at 253.7 nm. They obtained a flow rate of 11,250 $L \cdot h^{-1}$ and used 16 $W \cdot cm^{-2} \cdot s^{-1}$. These results were essentially repeated by Ewart and Chrimes (1980), who noted, however, significant iron chlorosis when a nutrient solution for tomatoes was irradiated. Iron is precipitated from solution by UV irradiation. Stanghellini et al (1984) controlled *Pythium aphanidermatum* (Edson) Fitzp. in a nutrient solution applied as an aeroponic spray to the roots of spinach. They used a UV source emitting 30 $mW \cdot cm^{-2} \cdot s^{-1}$ at 253.7 nm, but they also had to compensate for precipitated iron, which reduced the iron content of the solution from 4.5 to 0.1 $\mu g \cdot mL^{-1}$ within 24 h. Daughtrey and Schippers (1980) failed to demonstrate control of root death of tomatoes with UV irradiation, and Benoit and Ceustermans (1989) failed to control lettuce root diseases with consistency by irradiating the nutrient solution. Runia et al (1988) similarly failed to control cucumber green mottle mosaic virus, tomato mosaic virus, or *Phytophthora nicotianae* by UV irradiation of the nutrient solution.

Heat sterilization. The demand for large quantities of nutrient solution can be reduced if it is recirculated, but it needs disinfesting to avoid the buildup of pathogens in the system (Runia et al 1988; E. A. van Os et al 1988). Runia et al and van Os et al described a system in which water

SUMMARY

Sterilizing Media

The first essential stratagem in controlling greenhouse crop diseases is the eradication of inocula. Where the same crops are grown repeatedly in soil, the population of soilborne pathogens builds up and must be reduced to relatively harmless levels by full or partial sterilization. Full sterilization of soil by steam or dry heat is very expensive and often brings problems with phytotoxic levels of manganese and ammonia. Most fumigants, however, do not control bacteria very well. Less expensive is soil pasteurization, in which soil is heated to about 80°C by an air–steam mixture. This kills most plant pathogens but leaves a rich thermotolerant microflora, which can exert considerable biological control over residual pathogens. Solarization can also accomplish this very cheaply in capital outlay, but at the cost of keeping the soil and other substrates out of production for a few weeks. The disinfestation of water and recirculating nutrient solutions presents special problems, because of the large volumes involved. Chlorine kills many pathogens but is generally phytotoxic at levels effective against microorganisms. Ultraviolet sterilization is feasible, but it precipitates iron in nutrient solutions. Heat sterilization precipitates calcium and is very expensive for large volumes.

draining from a rock wool system passes through two heat exchangers, one raising the temperature to 95° C and maintaining it in a coil for 30 s, and the other giving up that heat to preheat incoming drainage water (Figure 32). Heated in drainage water for 10 s at 94° C, some old spores of *Fusarium oxysporum* f. sp. *melongenae* survived, but those of *Verticillium dahliae* in concentrations of 2×10^5 and 2×10^4 spores per milliliter were killed by 10 s of exposure at 90 and 83° C, respectively (Runia et al 1988). The results, however, were somewhat erratic from experiment to experiment, but Runia et al could not explain the variability satisfactorily. They thought that their system could cope with 36 $m^3 \cdot ha^{-1} \cdot day^{-1}$ but not the 200 $m^3 \cdot ha^{-1} \cdot day^{-1}$ needed for potted plants or the 1,000 $m^3 \cdot ha^{-1} \cdot day^{-1}$ needed in the nutrient film technique. A problem with the system was the precipitation of calcium, and acid washing was required to remove the deposits.

Disease-Free Planting Materials

Seed, cuttings, transplants, and even produce contaminated or infected by pathogens are major sources of inoculum in greenhouse crops and a serious impediment to local, national, and international trade (K. F. Baker and Chandler 1957). Voluntary or official regulation of commerce, coupled with therapeutic measures, is an essential part of disease-free production.

Concerted action on eradication has proved very successful in many cases and has been of considerable value to local industries (K. F. Baker 1952, 1956; K. F. Baker and Chandler 1956; Bald 1956; Dimock 1956, 1962; Hollings 1965; Lyle 1956; Tammen et al 1956; Wilhelm and Raabe 1956).

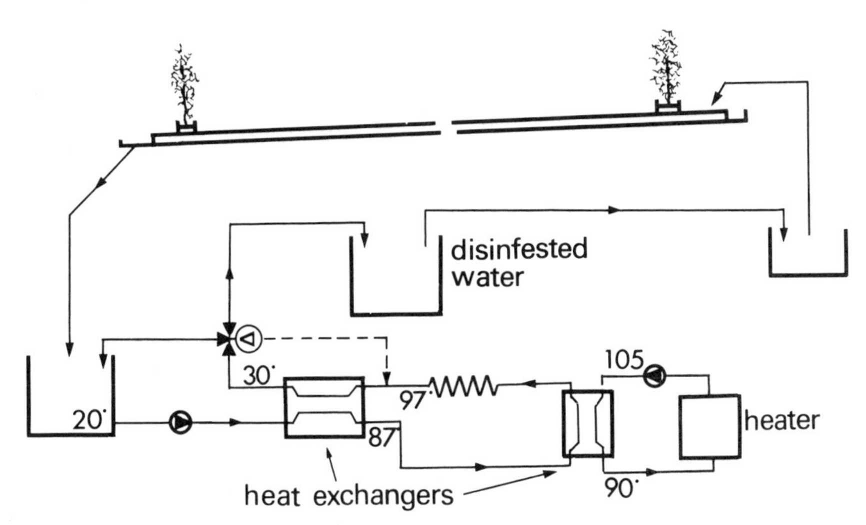

Figure 32. A system for disinfesting water and nutrient waste from greenhouses. Redrawn from E. A. van Os et al (1988).

Lelliott (1984) discussed the costs and benefits of eradication campaigns against alien pests and diseases in the United Kingdom, citing the example of *Puccinia horiana* Henn., which causes white rust of chrysanthemum (Water 1981). Although the costs of eradicant measures were very high in some years in some crops, Lelliott considered the costs of vigilance in its exclusion from imported material outweighed the costs of living with the disease as indigent over a 5-yr period. At the same time, the Netherlands, an exporting country, also realized the economic benefits of vigilance in maintaining the health of potted chrysanthemum and cut flowers (Veenebos 1984).

There are three main ways of obtaining pathogen-free planting material. The first is to ensure that the planting material comes from a disease-free area and disease-free stock. This insurance can be legislated by quarantine regulations at international and regional levels, supported by adequate inspection, certification, and pathogen detection procedures. A complete program for the production of healthy strawberry mother plants was described by Beech (1989); it is typical of the concerted action that would be successful for many other crops.

Second, on the assumption that contamination or infection may have occurred, pathogens can often be eradicated by a culture-indexing program whereby multiplicative propagation is done only from healthy plant parts, such as stem tips or meristems or even protoplasts (micropropagation). Some pathogens can also be eradicated by heat treatment (thermotherapy) or by pesticides (chemotherapy).

Third, once a healthy planting stock has been acquired, there are basic procedures whereby the grower can maintain it with minimum risk of recontamination or reinfection. This is the basis of nuclear stock schemes.

Quarantine

Seed crops are often grown in arid areas remote from the main production areas. The crops are inspected at frequent intervals, rogued if necessary, and treated with pesticides to reduce the incidence of disease and to restrict populations of pathogens. Mother plants and rootstocks are treated in a similar manner.

Weeds are often sources of pathogens for crops of this type. *Pseudomonas cichorii* (Swingle) Stapp, a seedborne pathogen of lettuce, is harbored by several weeds in lettuce fields (Ohata et al 1982). Tomlinson et al (1970) found 20 species of weeds, including the very common *Stellaria media* (L.) Vill., carrying the aphid-borne cucumber mosaic virus, which has a host range of 775 species in 86 families (Quiot et al 1979). Both pathogens are transmitted very quickly from weeds to crops (Doolittle and Walker 1926; Ohata et al 1982). Katan (1971) found a number of symptomless carriers of the tomato wilt pathogen, *Fusarium oxysporum* f. sp. *lycopersici*, in tomato fields. They included species of *Oryzopsis, Digitaria*, and *Amaranthus*.

Provided that official inspection schemes are adequately staffed by well-trained personnel, they can be quite effective in restricting the reintroduction of pathogens. A scheme called MT0 (mosaic tolerance zero) in the United States almost eliminated lettuce mosaic from seed stocks, whereas in Europe a voluntary tolerance level of 0.1% of contaminated seed permitted lettuce mosaic to remain a problem (Lot 1988). A method for the detection of bacteria in *Pelargonium* × *hortorum* L. H. Bailey in the context of the general quarantine arrangements for growing geraniums in France was described by Digat (1987).

Inspection

The value of inspection schemes is highlighted by breaches in both export and import inspections. Matteoni et al (1988) documented the importation of tomato spotted wilt virus into Canada in New Guinea impatiens, *Impatiens wallerana* J. D. Hook., together with its thrips vector *Frankliniella occidentalis* (Pergande). Allen and Matteoni (1988) recorded a similar case with cyclamen ringspot, another disease caused by tomato spotted wilt virus, and Linfield (1987) reported the importation of *Fusarium graminearum* Schwabe as the cause of a new cobweb disease of imported carnation flowers.

No formal inspection scheme for Fusarium crown and root rot of tomato, caused by *F. oxysporum* f. sp. *radicis-lycopersici*, was in place when outbreaks of the disease were attributed to contaminated imported seed in Israel (Krikun et al 1983), contaminated peat compost in France (Couteaudier et al 1985), and infected tomato transplants in the United Kingdom (Anonymous 1988b). Any occurrence of this disease in the United Kingdom has to be reported to the U.K. Plant Health and Seed Inspectorate. Affected crops are liable to be destroyed under strict quarantine conditions. Bacterial canker of tomato, caused by *Clavibacter michiganensis* subsp. *michiganensis*, is also a notifiable disease in the United Kingdom.

Certification

In many cases, voluntary or official certification of seed and mother plants is an essential part of the nursery and seed supply businesses, with a constant program of indexing to guarantee substantial freedom from pathogens.

Accreditation of crops ranges from field inspections to detailed sampling and indexing of samples of seed and planting material, and many methods have been developed for specific pathogens. A typical program, for strawberries, was described by Beech (1989).

M. J. Richardson (1979, 1981) listed about 750 fungi, 100 bacteria, 150 viruses, and 20 nematodes that are seedborne internally or externally. Richardson (1983) also outlined the standard techniques of identifying and quantifying seedborne fungi and bacteria, and Saettler et al (1989) described procedures for bacteria. These methods are equally applicable to the indexing of plant material. They are

- Direct examination for lesions, malformation, and discoloration, which are often diagnostic

- Incubation in a moist environment, with microscopic examination for sporulation of fungi
- Incubation in small volumes of water or nutrient broth, in which the presence of large numbers of bacteria soon becomes evident
- Isolation of bacteria and fungi on agar media for further diagnostic tests
- Serology for the diagnosis of bacteria, viruses, and some fungi
- Phage typing of bacteria
- Inoculation of indicator plants

Detection of Pathogens

The detection of very small populations of organisms or pathogens in very low numbers of seeds and plants presents staggering organizational problems in a laboratory. A sampling system ensuring confidence in the detection of very few of pathogens in very large populations is essential (Healy 1955; Rouse et al 1985). Isolation and identification can be very tedious, often involving ambiguous determinations on selective media (Saettler et al 1989), but recent advances in molecular biology enable the diagnostician to undertake hitherto daunting tasks. S. A. Miller and Martin (1988) reviewed some of the techniques available. They are particularly suitable for bacteria and viruses:

- Immunoassays
  - Enzyme-linked immunoabsorbent assay (ELISA) (Clark 1981)
  - Colloidal gold as an immunohistochemical marker (Benhamou et al 1985)
  - Radioimmunoassay, with radioactive tagging (Ghabrial and Shepherd 1980; Savage and Sall 1981)
- Nucleic acid hybridization for pathogen detection (Hames and Higgins 1985)
  - Nucleotide sequencing with ^{32}P
  - Dot-blot assay (Meinkoth and Wahl 1984)
  - Nonradioactive labeling (Renz and Kurz 1984)
  - Restriction fragment length polymorphism
- Nucleic acid probes using double-stranded RNA (Dodds et al 1984)
  - Synthetic oligonucleotide probes (Bar-Joseph et al 1985)

Culture Indexing

Selection of pathogen-free material simply involves rejecting all material that reacts positively in detection assays and retaining negative samples for continued observation and multiplication as a nuclear stock. These methods also provide a quantitative estimation of the incidence of contaminated material where official tolerance levels have been established.

In many cases, the selection of apparently healthy material is enough to give pathogen-free stock. K. F. Baker and Chandler (1956) selected shoots of *Fittonia verschaffeltii* (Hort. ex Lem.) Coem. var. *argyroneura* (Coem.)

Regel and *Pellionia pulchra* N. E. Brown high above the level of soil splash and took tip cuttings there. Many of these were free of *Rhizoctonia solani*; they were further subjected to a hot-water treatment at 51°C for 30 min. Similarly, cuttings of *Syngonium podophyllum* Schott taken well above the soil proved to be free of *Pythium* spp. Seed was also taken from *Aloe variegata* L. and *Haworthia alternata* Haw. high above the ground, and plants raised in pasteurized soil were further treated at 46°C for 20 min (for small plants) or 40 min (for large plants).

Dimock (1956) stressed the importance of taking cuttings of chrysanthemum well above the water-splash level to ensure freedom from *Septoria obesa* Syd., *S. chrysanthemella* Sacc., and the foliar nematode *Aphelenchoides ritzemabosi* (Schwartz) Steiner & Buhrer. He recommended trickle irrigation in all such cases. Lyle (1956) recommended no pesticide sprays on rootstock material of *Rosa multiflora* Thunb. ex J. A. Murray, to avoid infection by *Diplocarpon rosae* F. A. Wolf (Lyle 1938) on plant surfaces wetted by ineffective materials.

In order to avoid being misled by latent infections delaying symptoms, Munnecke (1956) advocated a 6-mo watch on mother plants of geranium (*Pelargonium* × *hortorum*), and he further stressed the importance of a high standard of hygiene in the handling of cuttings, with frequent sterilization of knives with methanol.

A cultured cutting technique (Dimock 1962) can combine indexing (Nyland and Milbrath 1962) with building up a pathogen-free stock. Chrysanthemum cuttings (Dimock 1956) and carnation cuttings (Gasiorkiewicz and Olsen 1956; Tammen et al 1956) can be treated in this way. Terminal cuttings are taken from healthy-appearing shoots and cut in two; the terminal piece is propagated, while the basal piece is virus-indexed (by grafting or sap inoculation of indicator plants) or is surface-sterilized and plated on agar for evidence of bacterial or fungal contamination. Culture indexing can also be used for selecting rose wood free from *Verticillium* spp. (Wilhelm and Raabe 1956) and for obtaining virus-free stock of herbaceous and woody material (Nyland and Milbrath 1962).

In addition to cuttings taken from shoot tips for propagation, meristems are often cultured to obtain virus-free stock (R. Baker and Phillips 1962; Hollings 1965). The meristem of a healthy-appearing shoot is removed by dissection, propagated on a filter paper bridge irrigated with a nutrient solution, and later indexed.

Micropropagation

Micropropagation is a technique of cloning vegetative material from limited mother plants (Aitken-Christie and Davies 1988; Styer and Chinn 1983). It is largely self-indexing, since any contamination by pathogens usually becomes apparent from the beginning. Since it often uses explants from disease-free tissue, particularly meristems, micropropagation resembles the tip-cutting technique (Dimock 1956; Gasiorkiewicz and Olsen 1956; Tammen et al 1956), but meristems are usually freer of pathogens than shoot tips.

Micropropagation requires advanced laboratory facilities as well as a quarantine house. Nevertheless, it is particularly useful for eradicating viruses from, for example, carnations (H. van Os 1964) and alstroemeria cultivars (Hakkart and Versluijs 1988), and it is widely used by commercial propagators. The regeneration of healthy plants from protoplasts requires even more elaborate facilities (Roest and Gilisson 1989).

Although it might be supposed that the clonal propagation of plants from a tiny piece of tissue or from a few cells or even protoplasts would ensure genetic uniformity in a clone, experience has proved otherwise. This is contrary to the general experience over centuries, in which whole organs, such as cuttings, runners, and tubers, retaining organizational integrity, remain genetically stable for the most part.

Pierik (1987) indicated that mutations in micropropagated material depend on the propagation system used, the growth regulator, whether the explant tissue is differentiated or not, the genotype, the number of subcultures, whether a plantlet is derived from one or more cells, whether the plant is a chimera, and the ploidy level of the plant. Pierik (1988) also suggested two explanations for the high rate of mutation in plantlets. One is that the expression of a mutation that has already occurred in the explant as a result of endomitosis or nuclear fusion, especially in polyploids, is enhanced in some way during propagation. Second, it may be that mutations are induced directly by the technique, especially in aneuploid cells. Synthetic growth regulators may be inducers.

Mutants also occur in plant cell cultures (Maliga 1984), which are sometimes used to initiate pathogen-free clones.

Plantlets in the high-humidity conditions of tissue culture frequently suffer from deranged water relations, a condition known as glassiness, or vitrification (Pierik 1988), and have poorly developed cuticular waxes (Sutter 1985). Such plantlets are difficult to acclimatize to open greenhouse conditions and are unusually susceptible to pathogens.

Explant material is often contaminated by bacteria, and commercial techniques are open to contamination from outside sources. Meristem culture reduces these risks, but if antibiotics are used they are frequently phytotoxic or increase the risk of further pathogenesis by antibiotic-resistant organisms. A frequent contaminant is a bacterium, probably *Bacillus subtilis* (Ehrenberg) Cohn, producing a condition known as white ghost (Pierik 1988).

Thermotherapy

Thermotherapy is a successful and widely used technique for the eradication of pathogens—bacteria, viruses, fungi, nematodes, and mycoplasmas—from seed and other planting material. It depends on a differential susceptibility of the host and its internal or external parasite to high temperatures (K. F. Baker 1962a,b). The greater the difference between the thermal death points of the host and the parasite (Table 10), the better the chance for success. Pathogens are generally more heat-susceptible than saprophytic microorganisms (Baker 1962a).

Seed or excised pieces of plant material are usually treated, but whole rooted plants can be treated as well, particularly to eradicate viruses (Cohen 1977; Hollings 1962, 1965), when such factors as reduced light, root temperature, and the carbon dioxide level are critical. Cohen successfully treated chrysanthemum at 39° C by day and night and carnation at 39° C by day and 34° C by night, but daphne was killed after 5 wk.

A number of other factors affect thermotherapy of seeds and excised plant pieces, including time, moisture content, and dormancy of the material.

Time. Thermotherapy is a function of both time and temperature, so that relatively low temperatures, 35–40° C, used for virus eradication can be maintained over long periods, up to several weeks, whereas higher temperatures, on the order of 50° C, can be used for 30 min or less. The temperature–time relationship is sigmoidal (Pullman et al 1981; J. H. Smith 1923) and can be expressed as a linear relationship by the general equation

$$T_{\mathrm{m}} = c + d(\ln E)$$

where T_{m} is the midpoint lethal temperature (the inflection point on the sigmoid curve), c is the intercept, d is the slope, and E is the exposure time. The relationship can be modeled (Ingram 1985; Roebroeck et al 1991; van Asten and Dorpema 1982) and used to predict either lethal temperature or exposure time. This relationship is applicable to thermotherapy, soil solarization, or damage to the host (Roebroeck et al 1991).

Moisture content. The greater the moisture content of plant tissues, the greater their susceptibility to heat damage. Thus seeds are much more tolerant of heat than herbaceous cuttings, and treatments that increase water content (for example, hot-water treatment) are more damaging than dry heat (K. F. Baker 1962a,b; K. F. Baker and Chandler 1957).

Dormancy of materials. Microorganisms that are actively growing are more susceptible to heat damage than those that are dormant, such as chlamydospores, oospores, and sclerotia. By the same token, and also associated with water content, dormant host material is more tolerant of heat. Thus, gladiolus cormels freshly dug were killed by hot-water treatment at 55° C, whereas dormant ones survived 60° C (K. F. Baker 1962b). However, host susceptibility to damage is also a function of growing conditions. Whereas gladiolus cormels grown during the summer in a subtropical area withstand 57° C for 30 min, those grown in a temperate area or during the winter in a subtropical area are damaged at temperatures over 55° C (Roebroeck et al 1991). The probability that the pathogen is also dormant in a dormant host does not seem to alter their relative heat tolerance. The age and condition of seed also determine its resistance to heat, and these characteristics are bound up with water content and food reserves. Seed is better treated within 1 yr of harvest (Bant and Storey 1952).

Other factors. Other characteristics of host tissue that affect thermotherapy include its pretreatment conditioning, freedom from surface damage, the ease of penetration of hot water through surface hairs, the size of the tissue samples, and the cultivar.

Gladiolus cormels are more successfully treated than corms, since they can tolerate higher temperatures. They are also more successfully treated if presoaked before heating and if 5% ethanol is added to the hot water (Roistacher et al 1957).

Heat treatment is not necessarily entirely successful in every plant or in all parts of the plant, and so it is necessary to take tip cuttings or meristem cuttings and to wait perhaps as long as 2–3 mo to be sure that symptoms have not merely been suppressed or that the virus is not detectable by indexing.

The practicalities of heat treatment have been described by K. F. Baker and Chandler (1957). Since the method depends on maintaining a temperature lethal to a pathogen without damage to the host, the temperature control has to be very good, preferably within 0.5°C. The shorter the treatment, the better must be the temperature control. It is emphasized that only apparently clean planting material can be successfully treated; dirty and obviously infected and rotting material cannot be rescued.

Some examples of bacteria and fungi successfully eliminated by thermotherapy are given in Table 18, and some others are given elsewhere (Anonymous 1971). Thermotherapy of stock infected with viruses has been very successful, for the most part (Hollings 1962, 1965; Kassanis 1954). The heat treatment of dormant stock depends on the ability of the host material to withstand high temperatures, generally between 35 and 54°C,

Table 18. Examples of elimination of pathogens from horticultural planting material by heat treatment

Host	Pathogen	Temperature (°C)	Treatment duration	Reference
Begonia, tuberous	*Meloidogyne incognita*	49	30 min	Gillard and van der Brande (1955)
Celery seed	*Septoria* spp.	50	25 min	Bant and Storey (1952)
Chrysanthemum	*Aphelenchoides ritzemabosi*	46	5 min	Southey (1978)
Cucumber seed	Cucumber green mottle mosaic virus	70	3 days	Lecoq (1988)
Gladiolus cormels	*Fusarium oxysporum* f. sp. *gladioli*	53–57	30 min	Roistacher et al (1957)
	Stromatinia gladioli	53	30 min	
Narcissus bulbs	*Ditylenchus dipsaci*	43.3	4 h	Southey (1978)
Pelargonium	*Puccinia pelargonii-zonalis*	38 34	48 h 4 days	I. M. Smith (1988)
Strawberry crowns	*Aphelenchoides ritzemabosi*	46	5 min	Southey (1978)
Tomato seed	*Clavibacter michiganensis*	56	30 min	Goode and Sasser (1980)
	Pseudomonas syringae pv. *tomato*	56	30 min	
	Xanthomonas campestris pv. *vesicatoria*	56	30 min	

better than viruses. Budwood can be successfully treated in hot water. Nondormant plants and tissues, however, are usually damaged by hot water. Hot-air treatment is more successful in treating these materials, with temperatures usually between 35 and 40°C, maintained for about 4 wk. Hollings (1965) listed a large number of viruses that could be eliminated from stock material by hot-air treatments. Some are given in Table 19. Plants and viruses differ widely in their tolerance of this treatment, and there are no particular groups of viruses that are more successfully treated than others. Chrysanthemum stunt virus is exceptionally heat-tolerant and can withstand boiling water (Brierley 1952).

Mycoplasmas can also be eliminated by thermotherapy. Van Slogteren et al (1976) eliminated them from gladiolus corms by keeping the corms in warm air (38°C) for 4 wk and storing them at 20°C for 6 wk, and they eliminated them from hyacinth bulbs by a regime of 30°C for 6 wk, 38°C for 2 wk, and 44°C for 3 days.

Commercial tomato seed is almost always treated for 30 min in water at 56°C, and although the germination of some cultivars is reduced by as much as 10%, the benefit in reducing the incidence of bacterial canker (caused by *Clavibacter michiganensis* subsp. *michiganensis*) outweighs the disadvantages (Goode and Sasser 1980).

Chemotherapy

Treatment of seed with a pesticide has two main purposes: one is to eradicate pathogens from the surface of the seed and, if possible, from internal tissues; the other is to protect the seed and seedling from infection by soil-inhabiting pathogens. For the latter purpose, a pesticide must be virtually water-insoluble and nonvolatile, so that protection is maintained as long as possible when the seed is sown. To eradicate microorganisms from the surface of dry seed before sowing, on the other hand, volatile pesticides are needed, to penetrate all parts of the seed coat and preferably also the tissues to a certain extent. These two types of pesticidal action

Table 19. Elimination of some viruses from greenhouse crop plants by hot-air treatment[a]

Host	Virus	Temperature (°C)	Treatment duration (wk)
Carnation	Carnation mottle	38	8
	Carnation ringspot	36	4
Chrysanthemum	Tomato aspermy	36	4
	Chrysanthemum flower distortion	35	8–20
	Chrysanthemum green flower	38	4
Hydrangea	Hydrangea ringspot	35	9–14
Pelargonium	Pelargonium leaf curl	38	4
Periwinkle	Strawberry green petal	42	3
Rose	Rose mosaic	34–36	4–10
Strawberry	Strawberry crinkle	37	2–3
	Strawberry mottle	38	2

[a] Adapted from Hollings (1965).

are hardly compatible in a single substance (H. Martin 1964; Purdy 1967; Sharvelle 1969).

Pesticides are generally effective in killing only surface organisms and some that are just below the surface, but because of the difficulties of penetrating into hairy seed coats, cracks, and the hilum, pesticides are not a guarantee of clean seed.

When systemic fungicides first appeared, soaking seed in them was widely investigated (Edgington and Peterson 1977). Maude (1964) eradicated *Septoria apiicola* Speg. from celery seed (*Apium graveolens* L. var. *dulce* (Mill.) Pers.) by soaking it in thiram, which is slightly systemic, for 24 h at 30° C. Orlikowski et al (1974) and Strider (1973) soaked seed of *Gerbera jamesonii* H. Bolus ex J. D. Hook. and statice (*Limonium* sp.), respectively, in benomyl to eradicate *Botrytis cinerea*.

Bulbs, corms, and rhizomes are less successfully treated by pesticide dip and soak treatments, partly because infections are more deep-seated than in seeds, and partly because many pathogens are resistant to the systemic fungicides, for example, *Botrytis tulipae* (Lib.) Lind in tulip bulbs (Duinveld and Beijersbergen 1975). Nevertheless, fungicides may be useful in reducing the effect of a disease and make postemergence control measures easier

SUMMARY

Disease-Free Planting Materials

Unless special precautions are taken, pathogen-infested seed, cuttings, and transplants can be sources of severe epidemics in crops and serious impediments to local and international trade. In many cases, the movement of planting material is stringently regulated by quarantine legislation. There are three main ways of obtaining healthy planting material: first, preferably, it should come from a disease-free area and disease-free stock; second, pathogens can often be eradicated by a culture-indexing program or by thermotherapy or chemotherapy; third, healthy planting stock can be maintained by a nuclear stock program aimed at minimizing recontamination or reinfection. The success of quarantine depends on the ability to detect pathogens even at very low population levels, so that official health certificates receive a high level of confidence. Techniques of starting with healthy stock include propagation of tip cuttings or meristem tissue and micropropagation from single cells or protoplasts, all of which reduce the risk of concomitant propagation of pathogens. Thermotherapy can often disinfest plant material, including seeds, cuttings, and whole plants. Sublethal heat is applied for periods of a few minutes for seeds to several days or weeks for whole plants being treated to eradicate viruses. Chemotherapy is less successful for deep-seated infection, but it is useful for surface disinfestation of seeds.

(D. Price and Briggs 1974). Fungicides have been used for the treatment of iris rhizomes (MacWithey 1967) and Easter lilies (Lenz et al 1971).

The treatment of cuttings by dipping them briefly in pesticides is not generally recommended, because of the risk of increasing bacterial contamination. There is the same risk in dipping in liquid formulations of rooting hormones; Lelliott (1988a) noted it in treating cuttings of chrysanthemum infected by *Erwinia chrysanthemi* Burkholder et al pv. *dianthicola* (Hellmers) Dickey.

Nuclear Stocks

Once plant material has been freed of pathogens, it is vital that stocks be built up for commercial use and maintained in a disease-free condition for a period exceeding the known latent period of the disease (Stout 1962). Geranium mother plants freed of *Xanthomonas campestris* pv. *pelargonii* are kept isolated from production florist crops and indexed at frequent intervals, and the stock is changed every 2 or 3 yr (Lelliott 1988c). Lelliott (1988a) stressed the importance of watching the latent phase in the case of *Erwinia chrysanthemi* pv. *dianthicola*. The pathogen may easily remain undetected when plant growth is slow.

If possible, nuclear stock is best grown in isolation in areas well away from commercial production. When it is grown near production areas, however, the distances of isolation are often fixed by law to exceed the likely distances of transmission of pathogens by wind, water, or insects or other vectors. Weeds are also eradicated.

Mother plants for cuttings should be kept in a separate greenhouse, well isolated from production houses, with well-kept lawns between them to lessen the risk of vector transmission of diseases. Weeds and all other incidental plants, such as grapevines, oleanders, and figs, which are not uncommon in production greenhouses, should be eradicated. Hanging baskets to fill up space should also be banned. Separate tools and clothing and limited access by personnel are advisable for the mother-plant house, which is, in fact, a quarantine house. Suitable facilities have been described by Beech (1989) and Stout (1962).

Pesticides

Pesticides in the greenhouse are a mixed blessing. Efficacious materials properly applied have a valid place in the integrated control of insect and mite pests and diseases, but there are some serious disadvantages in the injudicious use of pesticides in greenhouse crops. Visible pesticide residues on florist crops diminish the retail value. Pesticides are not weathered in the greenhouse as they are in field crops, and residues exceeding recommended daily dietary intakes may occur on vegetables picked every 2 or 3 days over several weeks or months. Integrated control of insects and mites has reached an advanced level in greenhouses (van Lenteren and Woets 1988), and the delicate population balances between beneficial and destructive insects and mites are easily upset by an inappropriate pesticide (Morgan and Ledieu 1979). Legal intervals between application and harvest are usually 2 or 3 days but extend to 30 days in the case of cucumbers sprayed with dinocap in Canada. Clearly this interval obviates the use of dinocap for all save the very early part of the season. The closed environment of the greenhouse also causes volatile pesticides to be a problem to workers, who are required almost daily to handle crops freshly treated with pesticides.

National and regional authorities govern the use of pesticides and establish safe limits of residues. Regulations vary widely among authorities, and there are similarly wide differences in recommended rates and schedules of application. Moreover, pesticides are more or less ephemeral in commerce, because of the continual revision of regulations and because of changing economics. Many compounds can no longer be used because of the appearance of resistant populations of insects, mites, and pathogens. It is therefore impossible to give specific recommendations, but the precautions and practices of application are common to most of the pesticides used in greenhouses.

The purposes of chemical control have been summarized by Skylakakis (1983) as sanitation, to reduce or eliminate the inoculum initiating disease in a crop; eradication, to inactivate a pathogen already in a host; and protection, to prevent healthy tissue from becoming infected or prevent infected tissue from becoming diseased and infectious.

In epidemiological terms,

$$dx/dt = QR(1 - x)$$

The proportion of infected plants (x) changes with time (t), depending on the initial inoculum (Q), the infection rate (R), and the diminishing proportion of plants remaining healthy as the disease progresses ($1 - x$) (Vanderplank 1963). When the availability of susceptible diseased tissue is limiting,

$$dx/dt = rx(1 - x)$$

where x is the disease present at time t, and r is the apparent infection rate (Vanderplank 1963).

Sanitation causes a delay Δt in reaching a given level of disease severity:

$$\Delta t = 1/r(\ln x_0/x_0')$$

where x_0 is the initial inoculum without sanitation, and x_0' is the initial inoculum with sanitation. If the inoculum can be removed completely by sanitation, $x_0' = 0$.

Chemical control can also reduce the apparent infection rate r and prolong the delay in reaching a given disease level:

$$\Delta t = (\Delta r)t/(r - \Delta r)$$

It also extends the latent period p (Vanderplank 1963) in the equation

$$dx/dt = R_c(x_{i-p} - x_{t-i-p})(1 - x)$$

where i is the period for which infected tissue remains infectious, and R_c is the corrected infection rate.

Application of Foliar Pesticides

The objective of pesticidc application systems is to deliver pesticides to the target plant areas and pests with the minimum of waste and contamination of the environment and with maximum efficacy. Each target requires an appropriate drop size and density (Heine 1980; Hislop and Baines 1980). A number of different types of applicators are available. These have been reviewed by Coffee (1980), Heine (1980), Law (1980), and Wenner (1979).

In contrast to the machinery developed specifically for certain field crops such as potatoes or cereals, few applicators have been developed for specialized use in greenhouses, where crops range from vegetable seedlings a few centimeters tall in flats on benches to lettuce in hydroponic production, massed beds of carnations, and rows of tomatoes and cucumbers 2–4 m tall or more.

Air-blast sprayers of the type used in top-fruit orchards are useful for disinfesting empty greenhouses with formalin or hypochlorite solutions. They provide wide coverage, reaching most of the superstructure, behind the heating pipes, and into other hidden areas.

Pulse fog generators produce a stream of hot gas from a gasoline-powered pulse-jet engine or from electrically heated sources (Anonymous 1976). Pesticides are introduced into the hot gas and are dispersed as particles about 5–100 μm in diameter. These generators are designed to disperse a pesticide fog throughout a greenhouse from a set position at the end, but many achieve a throw of no more than 10–15 m (Jarrett et al 1978; W. M. Morgan 1981).

Oil-based pesticide formulations can also be thermally dispersed from a smoke bomb. Again dispersal depends largely on convection currents, and coverage is not generally good.

Benn (1987) described a cold-fogging device requiring 45% less pesticide than conventional cold-dispersal applicators. Its efficacy depended on climate, an auxiliary forced flow, and an evaporation retardant.

Gravity-fed backpack or trolley-mounted high-volume sprayers, gasoline-powered and either air-assisted or pressurized, are useful only for relatively small areas. Careful operators can achieve good coverage "to runoff," but even so, runoff wastes about 95% of the spray.

Hydraulic nozzle sprays force out a liquid under pressure. There are three basic nozzle designs: the fan nozzle, with an orifice shaped to form a sheet of liquid into a fan, which breaks up into drops; the impact nozzle, in which an obstruction breaks the flow; and the cone nozzle, shaped so as to deliver a cone of spray. These sprayers deliver drops of widely varying size; that is, they have a wide droplet spectrum. The small drops drift and constitute a hazard to the operator and the environment, and the large drops are wasteful, as they bounce off or miss the target. Bals (1978a,b) calculated that 8.6% of drops were over 261 μm in diameter but constituted 47.6% of the spray volume.

Several modifications of nozzle design have been made to reduce drift and waste. The most significant development is controlled drop application, a system for delivering drops in a narrow size range, around 250 μm in diameter, with better coverage, less drift, and less waste than wide-spectrum sprays (Fraser 1958). Controlled drop application generates drops from the edges of spinning disks or cups, and the drop size is predictable from the equation

$$d\omega(DPT^{-1})^{1/2} = K$$

where d is the drop size, ω is the angular velocity of the disk, D is the disk diameter, P is the density of the pesticide formulation, T is surface tension, and K is a constant (Walton and Prewitt 1949). The angular velocity determines the centrifugal force required to overcome surface tension. The correct speed and pesticide flow rate must be selected according to the pesticide formulation, which affects surface tension.

A further development in spray technology induces an electrostatic charge on spray drops as they leave the sprayer (Cayley et al 1984; Law 1980; Matthews 1989). The drops are charged to $-10\ \text{mC}\cdot\text{kg}^{-1}$ by solid-state circuitry with inputs of 1 kV and 50 mW or less (Matthews 1989). Canopy penetration is achieved aerodynamically, and the charged drops are attracted to the crop. Up to sevenfold increases in deposition have been recorded, and the system is particularly suitable for ultra-low-volume applications, using only half the conventional doses of pesticides. Leaves with points, such as those of cucumber, set up contrary electrostatic fields and reduce the efficacy of the sprayer.

Another electrostatic system, the Electrodyn, developed by Coffee (1979, 1980), produces narrow controlled-drop spectra anywhere in the range of 40–200 μm, and the spray is largely unaffected by air movement and gravity. It works simply by adjusting the high- and low-voltage inputs; there are no moving parts. The liquid is charged by a coulombic force applied directly to the surface, and it receives a further propulsive charge at the nozzle. Turbulence and drift are considerably reduced. This ultra-low-volume system would appear to be particularly advantageous in greenhouse crops, because it does not restrict the nozzle design; the spray can be distributed from a linear source, for example, perhaps suitable for tall vertical rows of plants. Oil-based formulations of certain pesticides can be applied at rates of only $0.5\text{–}1.0\ \text{L}\cdot\text{ha}^{-1}$.

Electrostatic deposition can be combined with forced air to improve foliage penetration. Adams and Palmer (1989) evaluated the performance of an air-assisted electrostatic applicator for the control of the whitefly *Trialeurodes vaporariorum* (Westwood) in a tomato greenhouse. A permethrin formulation was applied to the apical foliage, in drops having a median diameter of 18 μm, to avoid direct contamination of developing fruit and foliage, where larval stages of the whitefly parasite *Encarsia formosa* Gahan occurred. The deposition on abaxial leaf surfaces was considered good, and the permethrin controlled first-instar larvae there. The inner leaves received less coverage than the peripheral leaves, but the efficacy of the insecticide was not impaired.

Techniques for collecting and determining the size of deposited spray drops have been outlined by Arnold (1980). There have been relatively few studies on efficacy in relation to drop size, distribution, and redistribution on leaves (Frick 1970; Hislop and Baines 1980; Lindquist and Powell 1980). Using the model system of conidia of *Botrytis fabae* Sardiña on leaves of *Vicia faba* L. and the fungicides copper oxychloride and maneb, Hislop and Baines found that the proportion of inoculum drops in which the fungus was killed could be expressed by the equation

$$\ln[-\ln(1 - P)] = 0.95 \ln N + 0.54 \ln A - 0.038\, w^{-1/2} - 0.37$$

where P is the proportion of inactive inoculum drops, N is the number of spray drops per square centimeter, w is the amount of fungicide (μg) per spray drop, and A is the area of spread (cm^2) of each inoculum drop together with any contiguous spray drops.

Hislop and Baines (1980) suggested that although coverage of a crop with fungicides is the prime factor in protection, there may be a limit to the value of increasing cover, because of an interaction between the number of fungicide drops and their size and because of the effects of inoculum drop volumes. There also seems to be a limit to the value of increasing the dose at a constant level of coverage. With copper, that limit was reached at about 1 $\mu g \cdot cm^{-2}$. Better protection is afforded when the inoculum drop volume is large rather than small, and a fungicide that is volatile (e.g., maneb) and therefore redistributed provides better protection than one that is not (e.g., copper).

Lindquist and Powell (1980) evaluated four low-volume applicators in experimental greenhouses and a commercial house with roses 2 m tall. One of the applicators was a pulse jet designed to be placed at one end of a greenhouse, and the others were portable. All of them delivered 60–70% of drops within only 10 m of the applicator and none or very few up to 30 m away. Deposition on the lower surfaces of leaves was generally poor, and more pesticide was deposited on apical leaves than on lower leaves. There was no difference between applicators in efficacy in controlling the whitefly *T. vaporariorum*. The choice of pesticide was more important. These investigators suggested that coverage of lower surfaces would be improved by keeping air circulation systems running during spraying.

Similar results were reported by Jarrett et al (1978) when they applied *Bacillus thuringiensis* Berliner as a fog, by controlled drop application, or in a high-volume spray.

Dust formulations. Because of problems in accurate placement and because of the relatively high proportion of dust escaping into the environment, dusts are virtually never used in commercial greenhouses.

Application of Soil Pesticides

Fungicides and nematicides are placed in the soil for three main purposes: to prevent the development of disease from seedborne pathogens; to protect the seed and seedlings from soil-inhabiting pathogens with a wide host spectrum, such as *Pythium* spp.; and to protect the mature, growing crop from more specialized pathogens such as *Meloidogyne incognita* and formae speciales of *Fusarium oxysporum* (de Tempe 1979). For these functions, soil pesticides require different properties (Domsch 1964) and are placed in the soil in different ways (Purdy 1967; Rodríguez-Kábana et al 1977).

For seed protection, a pesticide ideally should have some penetrant action, moving into the seed to kill pathogens in or just below the testa. It should be fairly water-soluble and have an appreciable vapor phase. The old organomercurial compounds had those properties. Seed protectants are ideally placed directly on the seed, as a dust or in a soak immediately before sowing, or incorporated into a pellet embedding the seed.

For protection of seedlings from soil inhabitants, a pesticide must be more persistent, and not water-soluble, volatile, or easily decomposed. Examples of this type are copper compounds, ethylenebisdithiocarbamates, thiram, and captan. These materials require a somewhat larger sphere of influence than that provided by pelleting or seed dressing, and they are best applied in the furrow or by drenching or by being incorporated dry into the surface layers of the soil.

For protection of the growing crop from more specialized pathogens, pesticides must be mixed homogenously with relatively large bulks of soil (Munnecke 1957). Roots have a vast surface area to protect, and in addition, decreasing the population of a pathogen in a volume not yet penetrated by roots requires homogenous and widespread mixing. The pesticide should be volatile. Fumigants are suitable for this type of general soil disinfestation.

To some extent protection of rapidly growing roots is afforded by systemic chemicals applied to the leaves. Oxamyl applied with sucrose, sprayed on the leaves of tomato plants, was translocated to the roots and significantly reduced the feeding of *Meloidogyne javanica* (Treub) Chitwood, but the dose was 30 times that applied in the root zone as a drench to prevent feeding (Atilano and van Gundy 1979).

General soil disinfestants, such as dazomet, are applied as granules to the soil surface, worked in with a rotary tiller, and, depending on volatility, sealed in by compacting the soil surface or with water or plastic film. Liquid formulations can be injected 15–25 cm into the soil by hand injectors and applied at 20- to 30-cm intervals, or they can be delivered under pressure by tractor-mounted chisels (Whitehead 1978). When granular formations are worked into groundbeds with a rotary tiller, the mixing is always less than completely homogenous and is limited in depth. Potting mixes are more successfully treated in bulk mixers, such as a concrete mixer. Quintozene specifically used to protect seedlings from *Rhizoctonia solani* can simply be raked into the top 5 cm of soil (Munnecke 1957).

Metham sodium can be applied in overhead or drip irrigation systems. It has been successfully applied by preplant drip irrigation to control *M.*

incognita, M. javanica, Pythium ultimum, and *Fusarium* spp. in carrots and tomatoes (Roberts et al 1988).

The efficacy of soil pesticides, like that of fumigants, depends on the physical and chemical characteristics of the soil and on its water relations and temperature (Munnecke 1957). Efficacy is similarly variable in soilless potting mixes. Stephens and Stebbins (1985) readily controlled the damping-off fungi *Rhizoctonia* and *P. ultimum* in peat, composted hardwood bark, and composted pine bark. Control by fungicides was generally less effective in processed pine bark. The volatile quintozene was slightly more effective than ethazole (etridiazole) with thiophanate-methyl or benomyl at low rates. There was less damping-off in untreated composted hardwood bark than in other media, and fungicides did not affect this property.

Phytotoxicity is a primary concern with soil fungicides. Powell (1988) found only benomyl applied as a drench to be safe for controlling *R. solani* in unrooted cuttings of poinsettia set in a soil–peat–perlite mix; quintozene and etridiazole impeded rooting.

Pesticides in Hydroponic Systems

Pesticides have not proved to be panaceas in the control of diseases of crops in hydroponic production. Shoots can be sprayed in the normal way for prophylactic control (Benoit and Ceustermans 1989), but the addition of pesticides to the nutrient solution is fraught with technical difficulties of solubility, settling in the tank, and blockage of emitters (Jamart et al 1988). Phytotoxicity and pesticide resistance also are common problems.

Often the target disease organisms are phycomycetes, and the fungicides etridiazole, furalaxyl, and metalaxyl have been tried frequently. All are usually phytotoxic at effective fungicidal rates (Daughtrey and Schippers 1980; Ewart and Chrimes 1980; D. Price and Dickinson 1980; T. V. Price and Fox 1986; T. V. Price and Maxwell 1980; Skadow et al 1984; Staunton and Cormican 1980; Tomlinson and Faithfull 1979, 1980). Bates and Stanghellini (1984) had success in controlling *Pythium*-incited root rot of spinach by misting the roots with metalaxyl. Benomyl, tried for its systemic properties, is phytotoxic at fungicidal concentrations in the nutrient film technique (Runia 1987).

Vanachter et al (1983) pointed out that systemic fungicide levels in edible crops could easily exceed legally permissible levels and cited a case of excessive metalaxyl residue in tomatoes in Belgium. This would be less of a problem in ornamental crops, provided that a compromise between fungitoxicity and phytotoxicity can be reached (Jamart et al 1988).

Copper fungicides, mainly copper oxinate, have occasionally been tried (D. Price and Dickinson 1980; Tomlinson and Faithfull 1979), as well as oxyquinoline sulfate (Jamart et al 1988), the latter for ornamentals. Copper materials are generally phytotoxic in the nutrient film technique.

Chlorine as a general disinfestant has had little success, although S. F. Jenkins (1981) found Cl^- to be satisfactory for cucumber and tomato culture at concentrations between 2.5 and 5.0 $mg \cdot L^{-1}$. Concentrations less than 2 $mg \cdot L^{-1}$ were ineffective, and more than 10 $mg \cdot L^{-1}$ was phytotoxic. Other workers found Cl^- to be phytotoxic at effective concentrations (Ewart and Chrimes 1980; P. M. Smith 1977). Smith found that although zoospores of *Phytophthora cinnamomi* were killed by a 2-min exposure to Cl^- at 1–2 mg · L^{-1}, a 4-h exposure at 100 mg · L^{-1} was required for killing mycelium, well beyond the phytotoxic tolerance levels of the nursery stock with which she worked.

There has been better success in controlling chytrid vectors of some viruses. Tomlinson and Faithfull (1979, 1980) tried a number of fungicides to control *Olpidium brassicae* (Woronin) P. A. Dang., the vector of lettuce big vein virus, and concluded that the surfactant in a commercial preparation of carbendazim (Bavistin) was the effective ingredient. A systematic trial of a number of anionic, cationic, and nonionic surfactants showed a formulation of alkyl phenol ethylene oxide condensate to be particularly effective. Added to the nutrient solution in amounts of 20 mg · L^{-1} every 4 days, it gave good control of big vein. Benoit and Ceustermans (1989), however, experienced occasional failures of control with the surfactant at 300 mg · L^{-1} in lettuce culture in a nutrient film system, but they offered no explanation.

Tomlinson and Thomas (1986) obtained good control of melon necrotic spot virus in cucumbers through the control of its vector *Olpidium radicale* Schwartz & Cook. Young plants were exposed to the surfactant at 30 mg · L^{-1} for 5 wk and then routinely to 20 mg · L^{-1} throughout the production season. There was no phytotoxicity.

Pesticide Resistance

Continuous use of a single pesticide in a monocrop system exerts intensive selection pressure for pesticide-resistant forms of pathogens, resulting in rapid and total loss of efficacy, which has happened with benomyl and metalaxyl (M. S. Wolfe 1981). Other fungicides have been remarkably durable: it was only about 25 years after the introduction of dichlofluanid that resistance to it appeared in *Botrytis cinerea* (Malathrakis 1989). Captan, first sold in 1952, was another durable fungicide, but Pepin and MacPherson (1982) found strains of *B. cinerea* resistant to it. Parry and Wood (1959b) obtained resistant strains by exposing *B. cinerea* to gradually increasing concentrations of captan in agar media.

There are many reports of resistance of fungi to all groups of fungicides. As long ago as 1947 *B. cinerea* was induced to form strains resistant to the then-new chlorinated nitrobenzenes when exposed to their vapor (Reavill 1954), and since then it has produced strains resistant to virtually every group of fungicides (Table 20) (Lorbeer 1980). New fungicide-resistant strains are formed as the result of selection pressure. Bolton (1976) considered fungicide resistance in *B. cinerea* to be the result of selection pressure on resistant strains already present in the wild.

When metalaxyl resistance appeared in *Bremia lactucae* Regel in the United Kingdom, all the isolates were found within a 20-km radius of the initial outbreak in 1984 (Crute et al 1987). By the end of 1984 and in 1985, resistance was found in most of the lettuce areas of the country. All the isolates were of sexual compatibility type B2, and all were of the same virulence phenotype, so that Crute et al concluded that the resistant isolates came from one clone with a single origin. The resistant pathotype was virulent on factors R1–R10 and R12–R15 but lacked virulence for R11 and R16–R18. This virulence phenotype was also the most common among metalaxyl-sensitive isolates.

Resistance, once acquired, is a heritable factor. *Botrytis cinerea* is a heterokaryon containing wild-type nuclei and nuclei with mutant, pesticide-resistant genes. Summers et al (1984) considered heterokaryosis to provide a mechanism for maintaining dicarboximide resistance even when resistance is associated with low fitness. Lorenz and Eichorn (1982) found no differences in the numbers of nuclei in conidia and hyphal cells of *B. cinerea.*

The buildup of resistance and the fitness of pesticide-resistant strains of fungi to survive and incite disease in the field are interconnected; both depend on the dose efficacy, the persistence of the pesticide, the type of disease, and the life cycle of the pathogen (Dekker 1987). Compounds with a high risk of building up resistance are the benzimidazoles and the acylalanines; the dicarboximides are considered to have a moderate or low probability of resistance, perhaps connected with the mode of action (Dekker 1976, 1987). Resistance to dinocap was never found in *Sphaerotheca fuliginea* (Schlechtend.:Fr.) Pollaci during 30 years' use (Schepers 1984).

Once resistant strains have appeared, most of them survive for long periods, and the risk of reenhancing resistant populations with the renewed use of ineffective fungicides is very high (Dekker 1987). Schepers (1984)

Table 20. Examples of fungicide resistance in pathogens of greenhouse crops

Fungicide	Pathogen	Reference
Benomyl	*Botrytis cinerea*	Watson and Koons (1973)
	Didymella bryoniae	Malathrakis and Vakalounakis (1983)
	Sphaerotheca fuliginea	Schroeder and Provvidenti (1969)
	S. pannosa var. *rosae*	Jarvis and Slingsby (1975)
Bitertanol	*S. fuliginea*	Schepers (1984)
Captan	*B. cinerea*	Pepin and MacPherson (1982)
Carbendazim	*D. bryoniae*	Malathrakis and Vakalounakis (1983)
Chloronitrobenzenes	*B. cinerea*	Reavill (1954)
Copper	*B. cinerea*	Parry and Wood (1958)
Cypendazole	*D. bryoniae*	Malathrakis and Vakalounakis (1983)
Dicarboximides	*B. cinerea*	Pommer and Lorenz (1982)
Dichlofluanid	*B. cinerea*	Malathrakis (1989)
Dicloran	*B. cinerea*	Lankow (1971)
Ethylenebisdithio-carbamates	*B. cinerea*	Parry and Wood (1959a)
Fenarimol	*S. fuliginea*	Schepers (1983)
Fosetyl-Al	*Phytophthora citrophthora*	Ali et al (1988)
Iprodione	*B. cinerea*	Northover and Matteoni (1986)
Metalaxyl	*Bremia lactucae*	Crute et al (1987)
	Pseudoperonospora cubensis	Pappas (1980)
Phosphorous acid	*P. citrophthora*	Ali et al (1988)
Procymidone	*B. cinerea*	Panayotakou and Malathrakis (1983)
Thiophanate-methyl	*B. cinerea*	Tezuka and Kiso (1970)
Triforine	*S. fuliginea*	Schepers (1983)

found the resistance of *S. fuliginea* to benzimidazoles and dimethirimol to persist for 10 yr after the withdrawal of the materials from commerce. Its resistance to pyrazophos persisted for 7 yr in cucumber greenhouses. This fungus was also resistant to bitertanol, fenarimol, imazalil, and triforine (Schepers 1983) and to benomyl (Schroeder and Provvidenti 1969).

Davis and Dennis (1981) investigated the survival and infective ability of dicarboximide-resistant strains of *B. cinerea.* Notwithstanding the generally poor production of conidia, the resistant strains survived at least 9 mo in strawberry leaf litter, were as pathogenic to strawberry as sensitive strains, and competed successfully in equal concentrations in mixed inocula. Once dicarboximide-resistant, *B. cinerea* never became fully sensitive through several subcultures (Lorenz and Eichorn 1982). T. Katan and Ovadia (1985) noted that in greenhouses in Israel the population tended to shift toward dicarboximide sensitivity during the summer, and then resistant strains rapidly reappeared after a few dicarboximide sprays in the winter.

The rate of reproduction and the frequency of appearance of fungicide-resistant mutants is dependent on the mode of fungicide use (Dekker 1987). M. S. Wolfe (1981) considered these implications in terms of a crop with a fungicide-treated (*t*) and an untreated (*u*) fraction attacked by a pathogen population with a fungicide-sensitive (*s*) and an insensitive (*i*) or resistant fraction. The highest rate of fungal reproduction occurs in the sensitive population on the untreated crop (*su*), followed in descending ranking by the insensitive population on the untreated crop (*iu*), the insensitive population on the treated crop (*it*), and the sensitive population on the treated crop (*st*), with the lowest reproductive rate. An effective fungicide, which controls the fungus almost completely, has to be used in the case of *su*. Against this fungicide, the sensitive pathogen fraction fails almost completely, so that $st = 0$, and *iu* is barely detectable, because it is so rare and unadapted. If the use of the fungicide is increased, the reproduction of *su* becomes limited as the treated crop area increases, but there is also an increase in the area in which the resistant fraction (*it*) can reproduce without competition. The resistant fraction becomes relatively more frequent even with no change in the basic reproductive rates of *i* and *s*, but the increase in $iu + it$ does raise the probability of selecting forms with a higher reproductive rate. Efficacy therefore depends on the interaction of differences in the reproductive rates of *i* and *s* on crop areas *u* and *t*.

Skylakakis (1983) reviewed the theory of selection pressure for pesticide resistance under two main headings: directional and disruptive selection. Directional selection operates when an untreated pathogen population is distributed widely and unimodally with respect to fungicide sensitivity; when it is exposed to a fungicide, the whole distribution shifts, as measured by a parameter such as ED_{50}. It is likely that this type of selection operated in the case of the resistance of *S. fuliginea* to dimethirimol (Skylakakis 1982). Disruptive selection, on the other hand, operates when the pathogen population contains a sensitive and a resistant subpopulation in a bimodal, nonoverlapping frequency distribution. Exposure to the fungicide brings about a change in the relative proportions of these subpopulations. Skylakakis (1983) considered resistance to the benzimidazoles and metalaxyl to have originated by disruptive selection. It appears to operate faster than directional selection.

Skylakakis (1982, 1983) derived a standard selection time t_s in an attempt to produce a predictive model for the emergence of fungicide resistance:

$$t_s = 1/(r_2 - r_1)$$

where r_1 and r_2 are the infection rates for the fungicide-sensitive and fungicide-resistant subpopulations, respectively. Skylakakis (1983) calculated that *Phytophthora infestans* had a standard selection time of only 4.1 days in the presence of an efficient fungicide, *Erysiphe graminis* DC. had 7.8 days, and an *Ustilago* sp. would only marginally increase its resistant frequency, which would remain undetectable for an indefinite period.

Skylakakis (1982) worked through a number of examples of observed and modeled appearances of resistant pathogen populations. Among them was the case of resistance to dimethirimol in *Erysiphe cichoracearum* DC., one of the causes of powdery mildew of cucumber in the Netherlands. Skylakakis's model predicted the appearance of resistance in 98–263 days; after two to four applications of dimethirimol to cucumber crops, resistance appeared between 112 and 224 days after treatment, in quite good agreement with the prediction.

Pesticide durability. In order to prolong their efficacy when resistant strains appeared, pesticides have been used as little as possible where the risk of increasing a resistant pathogen population was thought to exist. Programs were evolved in which a pesticide is used at a fraction of its normal rate, in a tank mix with other pesticides, in alternation with other pesticides, or in combination with a biological control.

Skylakakis (1981, 1984) reviewed six reports of the efficacy of fungicide mixtures in delaying the buildup of resistant fungus populations: only one report was judged to have provided adequate evidence of delay, one was inconclusive, three provided insufficient evidence, and in one resistance developed faster under treatment with a mixture. All of the studies had predicted delay. Skylakakis (1984) concluded that the models were not realistic or the experimental techniques were not very sensitive, because of the rapidity of the buildup of resistance or because of variations in sampling and cross-contamination between plots. Another factor influencing these results was that alternation or mixtures with other pesticides tended to control diseases better than a systemic pesticide alone when a significant proportion of the population was resistant, so that plots with these treatments tended to have lower disease severity than those treated with only the fungicide at risk.

Again using standard selection time, Skylakakis (1983) attempted to derive theoretical strategies for delaying resistance buildup. The alternation of fungicides does not appear to affect the intensity of selection pressure but decreases the period over which it operates. By contrast, mixtures of fungicides do influence the intensity of selection pressure and are best used in the case of slow pathogens. Their delaying effect is enhanced when the second fungicide is equally effective against both subpopulations.

Notwithstanding this type of model, practical experience with alternating programs and mixtures has been less than satisfactory; most of these programs evolved empirically, rarely according to model predictions. M. S. Wolfe (1981) concluded that it was impossible to predict the efficacy of heterogeneous pesticide treatments.

T. Katan and Ovadia (1985) found that the appearance of dicarboximide-resistant strains of *Botrytis cinerea* was delayed when chlorothalonil sprays were interspersed with the dicarboximide treatment in two of three experiments in greenhouse crops, but not in the third experiment.

Northover and Matteoni (1986) found iprodione-resistant strains of *B. cinerea* in Ontario greenhouses following regular use of this compound (every 7–10 days), but not after irregular use (less than one spray per month). In greenhouses where iprodione had never been used, they found no resistant strains, except for one resistant strain in stock newly imported from the United States. Interspersed sprays of captan did not prevent an increase in the resistant population, but it did control gray mold in neighboring vineyards.

Cross resistance. Georgopoulos (1977) defined cross resistance as resistance to two or more pesticides mediated by the same genetic factor. This is distinct from multiple resistance determined by different genes (de Waard 1984).

Jarvis and Slingsby (1975) isolated a benomyl-resistant strain of *Botrytis cinerea* from rose. This strain, in culture, was also resistant to captan, copper, ferbam, anilazine, and chlorothalonil, but it remained sensitive to dicloran and Dikar (a mixture of dinocap and mancozeb).

Crute et al (1987) described metalaxyl-resistant isolates of *Bremia lactucae* that were cross-resistant to other phenylamide pesticides, but isolates resistant and sensitive to phenylamides showed similar responses to the unrelated systemic materials propamocarb and fosetyl-Al.

Resistance to metalaxyl developed very quickly after its introduction (Dekker 1987). Clerjeau et al (1984) found that fosetyl-Al controlled acylalanine-resistant strains of *Plasmopara viticola* (Berk. & M. A. Curt.) Berl. & De Toni in Sacc. and *Phytophthora infestans*, and they attributed earlier reports of cross resistance to contamination of test disks in the bioassay by the volatile metalaxyl, stimulating the subpopulation resistant to it.

Cross resistance in which resistance to one pesticide is usually accompanied by resistance to others of the same chemical group or with a similar mode of action is described as being positively correlated. This is common among the benzimidazoles.

More rare is negatively correlated cross resistance, in which a mutation may give resistance to one pesticide but increased sensitivity to other pesticides with related or unrelated modes of action (de Waard 1984). Kato et al (1984) described negatively correlated cross resistance in which methyl *N*-(3,5-dichlorophenyl)carbamate (MDPC) controlled gray mold and powdery mildew of cucumber among other diseases caused by benomyl-resistant strains but not those caused by benomyl-sensitive strains. Negatively correlated cross resistance also occurs among acylalanines, carboximides, dicarboximides, and ergosterol biosynthesis inhibitors (de Waard 1984).

Mixtures of pesticides may have additive, synergistic, or antagonistic toxic action, and together with negatively correlated cross resistance they can be exploited to control resistant populations of pathogens (de Waard 1984).

Pesticides and Biological Control

In addition to failing to kill or inhibit a fungus directly because of fungicide resistance, fungicides can also fail to control diseases or even exacerbate them by killing commensal antagonistic microorganisms.

Van Dommelen and Bollen (1973) found that when benomyl failed to control *Botrytis cinerea* in cyclamen because of resistance in the pathogen, some plants remained relatively disease-free. *Penicillium brevicompactum* Dierckx and *P. stoloniferum* Thom (syn. *P. brevicompactum*) isolated from these plants were very antagonistic to *B. cinerea* in culture. They were also even more resistant to benomyl than *B. cinerea* was, and van Dommelen and Bollen concluded that they played a significant part in gray mold control under natural conditions, an effect lost unless the antagonists were also resistant to the fungicide.

In attempts to control Fusarium basal rot of narcissus (caused by *Fusarium oxysporum* f. sp. *narcissi* W. C. Snyder & H. N. Hans.), Langerak (1977) noted that treated bulbs remained protected long after thiram, organomercurials, and formalin had been reduced to nonfungicidal levels; protection against the pathogen was provided by antagonistic species of

Table 21. Effects of some greenhouse crop fungicides on the insect predators *Phytoseiulus persimilis* and *Encarsia formosa*[a]

	Effect[b] on:			
	P. persimilis		*E. formosa*	
Fungicide	Eggs	Adults	Pupae	Adults
Benomyl	H	H	...	S
Captan	...	S	...	S
Carbendazim	...	I	...	S
Chlorothalonil	...	S	S	S
Copper oxychloride	S	S	...	S
Cupric ammonium carbonate	...	S	...	I
Dichlofluanid	...	S	S	S
Imazalil	I	S	...	I
Iprodione	S	S	...	S
Maneb	...	S	...	I
Pyrazophos	H	I	...	H
Thiophanate-methyl	H	S	...	S
Thiram	...	...	...	S
Triforine	S	S	...	S
Vinclozolin	...	S	...	S
Zineb	...	S	...	S

[a] Adapted from Morgan and Ledieu (1979).
[b] H = harmful; S = safe at recommended rate; I = intermediate effect.

SUMMARY

Pesticides

Pesticides have a role in greenhouse production, but they are subject to restrictions not generally applicable to those used on field crops. In the greenhouse, pesticides are not weathered, so that residues tend to be higher on vegetables and fruit and may exceed recommended daily dietary intakes. In addition, residues are more visible, detracting from the appearance of foodstuffs and particularly of flowers and ornamental foliage plants. In the confines of a greenhouse, workers handle treated plants almost daily and are placed at greater exposure risk.

Pesticides are used to reduce inoculum, to inactivate a pathogen already in the host, to protect healthy tissue from becoming infected, and to prevent infected tissue from becoming diseased and infectious. In practice, pesticides are rarely eradicant; their principle role is in prophylaxis.

Few spray applicators are designed specifically for use in the greenhouse; most are adapted or unadapted field applicators, such as variants of orchard sprayers for whole-house sanitation. Pulse fog generators disperse pesticides in hot gases but have a limited throw and provide poor coverage in dense, tall row crops. There are few, if any, effective sprayers for large areas of potted plants on benches or seed trays on the floor. Most greenhouse spraying is done by backpack sprayers, variously powered. Spraying is arbitrarily done to runoff, which wastes about 95% of the spray. Electrostatic sprayers are rather less wasteful, although their efficacy is reduced in crops with pointed leaves, such as cucumbers, which set up contrary electrostatic fields. There is evidently considerable scope for more research on greenhouse spray technology.

Pesticides are incorporated into soil to prevent the development of diseases caused by seedborne pathogens, to protect the seed and seedlings from soil-inhabiting pathogens with a wide host spectrum (such as *Pythium* spp.), and to protect the maturing crop from specialized pathogens (such as formae speciales of *Fusarium oxysporum*). None of the current soil pesticides can perform all these functions. To protect seeds, a pesticide should be somewhat water-soluble, in order to be able to penetrate the testa; but to protect seedlings, it should not be very water-soluble or volatile or easily decomposed, and it should have a wider sphere of influence than the spermosphere. An even greater sphere of influence is required for pesticides directed at specialized pathogens, and these compounds have to be thoroughly mixed with relatively large volumes of soil. Soil pesticides are applied in ways appropriate to their function. Seed protectants are generally incorporated in pellets around the seed; seedling protectants are applied locally to the sowing furrow; and general soil disinfestants are incorporated dry in large volumes of soil or are injected as volatile liquids or gases. The efficacy of soil pesticides depends on the physical and chemical characteristics of the soil. Phytotoxicity is a primary concern; cuttings that are being rooted are particularly at risk.

Pesticides can be used normally as sprays applied to shoots but are difficult to use in nutrient solutions to control root pathogens. Many that are safe in soil are phytotoxic in nutrient solutions, and when rates are reduced because of that, efficacy declines. There are also technical problems with solubility, settling in the tank, and blockage of emitters. There are also problems with pathogens resistant to some of the common pesticides used in hydroponic production, such as metalaxyl.

It is a common experience that the use of a single pesticide repeatedly in a monocrop system soon leads to the selection of pesticide-resistant populations of pathogens and to rapid and widespread loss in efficacy. Resistance, once acquired, is a heritable factor, and resistant strains are usually persistent with good fitness for survival. Selection pressure for resistant strains is strong, with whole populations shifting to pesticide sensitivity, as measured by ED_{50}. This is termed directional selection. Disruptive selection operates when exposure to a pesticide changes the relative proportions of two subpopulations with different pesticide sensitivities. It happens faster than directional selection. From models, it is possible to predict the appearance of resistant strains, but predictions of effective programs for alternating or mixing pesticides to delay or annul resistance have proven more difficult. In practice, alternating programs and mixed pesticide applications have generally not been very successful. The problem is compounded when resistance to two or more pesticides is mediated by the same gene, producing cross resistance. Negatively correlated cross resistance occasionally occurs, in which a mutation to resistance to one pesticide confers increased sensitivity to another. Mixtures of pesticides may have additive, synergistic, or antagonistic toxic action, which along with negatively correlated cross resistance can be exploited to control pesticide-resistant pathogens.

Pesticidal and biological control are inextricably tied together. Diseases may be exacerbated by a pesticide if the pathogen is resistant but its commensal biological control microorganisms are not. It is therefore advantageous if the biological control agents are resistant to the common pesticides, and indeed biotechnology can produce such organisms. Also at risk are biological controls for insects, especially entomopathogenic fungi, but also parasitic and predatory insects.

Penicillium and *Trichoderma*. Similarly, Henis et al (1978) evolved an integrated control of damping-off of radish caused by *Rhizoctonia solani* by the combined use of quintozene and a quintozene-resistant strain of the antagonist and hyperparasite *T. harzianum* Rifai. Sundheim (1982) and Sundheim and Amundsen (1982) used the hyperparasite of powdery mildews, *Ampelomyces quisqualis* Ces., in the presence of triforine to control *Sphaerotheca fuliginea* on cucumber. Hijwegen (1986) found *Tilletiopsis minor* Nyland, an antagonist of *S. fuliginea*, to be unaffected by dimethirimol but sensitive to fenarimol. He suggested that fenarimol could be used in an integrated control program if fenarimol-resistant strains of *T. minor* could be selected on media containing the fungicide.

Locke et al (1985) isolated a benomyl-resistant strain of *Trichoderma viride* Pers.:Fr. and used it to colonize a pasteurized soil mix to prevent reinvasion of the mix by *F. oxysporum* f. sp. *chrysanthemi* G. M. Armstrong, J. K. Armstrong, & R H. Littrell. The control of Fusarium wilt of chrysanthemum obtained by this means was as good as that from the standard benomyl–fertilizer program. Papavizas and Lewis (1981) and Papavizas et al (1982) induced new biotypes of *T. harzianum* resistant to benomyl and also to captan and captafol, so that its biocontrol capabilities would be enhanced in the presence of benomyl.

In addition to the effects of pesticides on microorganisms controlling diseases, pesticides of all types have marked effects on insect predators and parasites, and so in an integrated control program these toxicities have to be evaluated. Table 21 summarizes the effects of fungicides on two insect predators (Morgan and Ledieu 1979).

Limiting Disease Spread

CHAPTER 6

Limiting Inoculum Production

Among the stratagems for controlling diseases—namely, limiting inoculum production, reducing the rate of disease increase, and reducing the time in which an epidemic might develop (Williams 1978)—limiting the inoculum is very successful if it can be reduced to almost zero and the rate of disease increase is low. These three factors, initial inoculum x_0, rate of disease increase r, and time t, are related by Vanderplank's equation (Vanderplank 1963)

$$\ln \frac{x}{1-x} = rt + \ln \frac{x_0}{1-x_0}$$

where x is the amount of disease in a crop at time t.

Reducing the inoculum through sanitation—for example, sterilizing the soil, maintaining elementary greenhouse hygiene, and using pathogen-free planting material—is the basis of all the work to be done preparatory to growing a crop. Vertical genetic resistance also gives x_0 near 0 (Vanderplank 1963).

Williams (1978) cited the example of treating celery seed, of which 1% was assumed to be infected, with a thiram soak to eradicate *Septoria apiicola* Speg. If the treatment is 99% effective and r is low, then only 0.13% of the celery plants would be expected to have leaf spot at harvest at 90 days. If the treatment is only 50% effective, however, 7% of the plants would be diseased.

This stratagem would be less successful against a disease such as late blight of tomatoes, in which r is much higher than the 0.08 units per day assumed for celery leaf spot. Vanderplank (1963) calculated $r = 0.46$ for late blight of potatoes, caused by *Phytophthora infestans* (Mont.) de Bary. If r for celery leaf spot were doubled to 0.16, all plants would be diseased at harvest, even with the seed treatment (Williams 1978).

In order to be able to limit the potential of the initial inoculum, it is important to be able to recognize it and to know the environmental conditions under which it becomes a decisive factor in epidemics in their logarithmic phase (Vanderplank 1963), in which

$$x = x_0 e^{rt}$$

Last et al (1969) found that inoculum of *Pyrenochaeta lycopersici* R. Schneider & Gerlach, the soilborne fungus causing corky root rot of tomato, built up at a compound rate over a 5-yr period. With an inoculum infecting 4% of plants at the beginning of the period, 12% of plants became infected between 5 and 17 wk later, a threefold increase; with 8% infected plants at the beginning, 40% became infected over the same period, a fivefold increase. Over a 5-yr period with successive tomato crops, the percentage of infected plants rose from 4.4% in the first season to 40.3% in the fifth season. Reducing the inoculum to a very low initial level limits crop losses within a season and over a number of seasons.

Ideally x_0 would be 0 if all sanitation procedures were successful. In practice this is rarely so, and pathogens readily appear if given the appropriate environment. Reducing x_0, then, consists of preventing those conditions. However, Plaut and Berger (1981) noted that sanitation was not necessarily as effective as had been supposed. In three diseases in a greenhouse—peanut early leaf spot, caused by *Cercospora arachidicola* S. Hori on *Arachis hypogaea* L.; bean rust, caused by *Uromyces appendiculatus* (Pers.:Pers.) Unger on *Phaseolus vulgaris* L.; and gray blight of begonia, caused by *Botrytis cinerea* Pers.:Fr. on *Begonia semperflorens* Link & Otto—the rates of disease increase were greatest when the epidemic began with the lowest of three initial disease levels. A low initial disease level was associated with an accelerated rate of disease increase (Figure 33), so that the final disease level was about the same as in an epidemic starting with

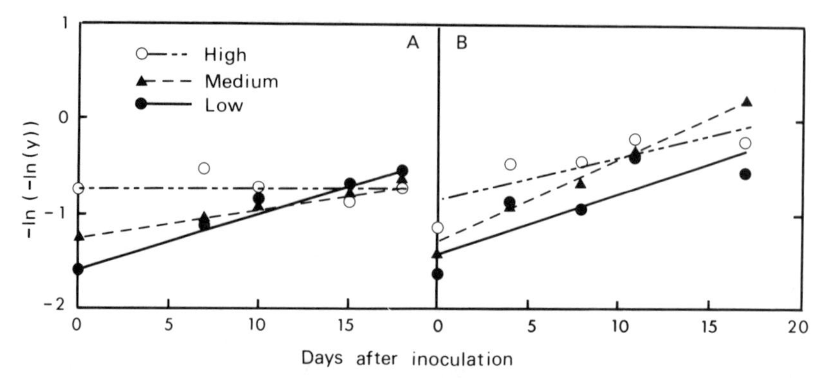

Figure 33. Progress of gray blight epidemics, caused by *Botrytis cinerea*, in *Begonia semperflorens* in the greenhouse. In two experiments (**A** and **B**) epidemics were initiated with disease severity at low, medium, and high levels. The values are the average of three replications, transformed by the Gompertz equation $Y = -\ln(-\ln y)$, and analyzed by least squares regression. Redrawn from Plaut and Berger (1981).

a high disease level. Evidently, many other limiting factors, such as available susceptible host tissue, spore interactions, and changing host resistance, contribute to rates of epidemic progress (Plaut and Berger 1981).

In order to reduce the initial inoculum, it is also important to recognize the site of its production and the conditions under which it is produced.

The movement of bacteria to the surface of a plant usually occurs at very high humidity when water films on the surface are continuous with water in the intercellular spaces of leaves via stomata or hydathodes or in waterlogged stem or fruit canker lesions. Bacteria exuded in mass from stomata are often mixed with a mucilaginous material, possibly polysaccharide, which may protect them against drying and aid in dispersal by water splash. They also exude from the cut surface of stems and leaves when cuttings are taken for propagation and are available for dispersal by water splash and on knives and fingers (Lelliott 1988a–c).

Most fungi pathogenic on aerial parts of plants sporulate profusely in moderate to high humidity. Pycnidial fungi generally sporulate better than the Hyphomycetes with humidity near saturation. *Didymella bryoniae* (Auersw.) Rehm (van Steekelenburg 1983, 1985b) and *D. lycopersici* Kleb. (Verhoeff 1963) produce tendrils of mucilaginous, hydrophilic pycnospores in very high humidity. Many fungi have mucilaginous and hydrophilic spores produced most abundantly under very humid conditions, including the hyperparasitic coelomycetes *Ampelomyces quisqualis* Ces. and *Coniothyrium minitans* Campbell, sporodochial fungi (microconidia of *Fusarium* spp., for example), and fungi forming acervuli (*Colletotrichum* spp., for example).

Aerial conidia of the Hyphomycetes are mostly produced in subsaturated atmospheres, at a vapor pressure deficit between 1.2 and 0.6 kPa. *Botrytis cinerea* produces long, indeterminate conidiophores with few conidia in saturated atmospheres and short conidiophores with profuse sporulation at moderate humidity (Hawker 1950; Paul 1929).

Conidia of the downy mildew fungi are generally produced on wet plant surfaces (Rotem et al 1978). *Pseudoperonospora cubensis* (Berk. & M. A. Curtis) Rostovzev sporulates on cucumber between 5 and 30° C (the optimum is about 15° C) when the leaves are wet for at least 6 h (Y. Cohen 1981). Duvdevani et al (1946) concluded that dew was a major factor in the sporulation of *P. cubensis*, which could be prevented by screening cucumber plants from dew at night. A temperature of not less than 18° C was required for sporulation. *Bremia lactucae* Regel on lettuce has a rather lower optimum temperature of 10° C (and a range of 0–21° C). It sporulates mostly at night when the relative humidity exceeds 95% and the leaves are probably wet (Crute and Dixon 1981).

By contrast, conidia of the Erysiphales are generally formed in dry atmospheres, at moderate temperatures, in moderate light, and on luxuriant plant growth (Yarwood 1957). Conidia are formed in chains, with a diurnal periodicity in spore maturation in most species (Butt 1978), so that those formed at night are ready for abstriction and spore release the next day. The optimum temperature usually lies between about 20 and 25° C, with spore production relatively sparse above 25° C. Sporulation of *Sphaerotheca fuliginea* (Schlechtend.:Fr.) Pollaci is most prolific at a vapor pressure deficit of about 1.4 kPa (Reuveni and Rotem 1974) and at 28°C (Abiko and Kishi 1979); it produces no conidia at very low vapor pressure deficits (Nagy 1976). Hammarlund (1925) reported the optimum value for the sporulation of *S. pannosa* (Wallr.:Fr.) Lév. also to be about 1.4 kPa, but Ragazzi (1980) could find no marked effect of either temperature or humidity on the sporulation of *S. pannosa* var. *rosae* Woronichin on roses. There are differences in interpretation of the evidence regarding the direct effects of water on powdery mildew colonies (Butt 1978), but it seems generally agreed that sporulation is inhibited when the colonies are wet, which is perhaps the basis of controlling them with a pressurized water spray (Abiko and Kishi 1979; Yarwood 1939a).

Limiting Inoculum Dispersal

Microorganisms are dispersed through greenhouse crops in one or more of three main ways: by water moving through the soil or splashed through the air; by air currents carrying them as discrete dry propagules; and on tools, machinery, fingers, and clothing.

At high humidity, most bacteria exude from stomata and pruning scars on infected plant parts and in guttation drops on hydathodes, whence they are readily dispersed by fingers, tools, and clothing (Lelliott 1988a–c). They are readily picked up in water droplets splashed from drips from the roof, overhead irrigation (Bradbury 1967; Hirano and Upper 1983; Kamerman

SUMMARY

Limiting Inoculum Production

Once a disease has appeared with a primary focus in a greenhouse, it is necessary to recognize the site where inoculum is produced and the conditions under which it is formed and then to limit its production. The progress of an epidemic is not necessarily related to the amount of inoculum but is affected by other factors as well—available susceptible host tissue, changing host resistance, and patterns of inoculum dispersal, among others.

The production of inoculum on infected tissue depends on the pathogen and the environment. Bacteria are mostly exuded on wet plants, and sporangia of the downy mildew fungi are also produced on wet plants. Conidia of the Hyphomycetes are formed in subsaturated atmospheres, whereas those of the Erysiphales are formed in dry atmospheres, at moderate temperatures, in moderate light, and on luxuriant plant growth. Depending on the pathogen, therefore, the greenhouse environment must be manipulated to reduce inoculum production.

COLOR PLATES

1. A poorly sited greenhouse with impeded natural ventilation. This house, of old design, had a long history of diseases in tomatoes. Courtesy of J. C. Fisher.

2. A tall Venlo style of greenhouse permits more photosynthetic area per unit of floor space and better air circulation around tall, trained vegetable plants.

3. The effect of cold walls on the growth of cucumbers. This problem was corrected by insulating the walls with straw bales.

4. The primary purpose of a fogging system is evaporative cooling of the air. Fogging may also lower the water vapor pressure deficit. Reprinted, by permission, from Jarvis (1989).

5. Fan-jet ventilation uses perforated plastic ducting above the crop canopy.

6. Perforated plastic ducting near the ground improves ventilation around the stems of tall crops. Courtesy of C. von Zabeltitz.

7. Plastic mulch reflects heat, which may damage seedlings.

8. Cucumbers growing hydroponically in bags of sawdust in British Columbia. Courtesy of Andrea Buonassisi.

9. The Guernsey arch, a traditional way of growing cucumbers on straw bales. The decomposing straw warms the roots and isolates them from groundbed soil pathogens.

10. Open-mesh benches permit through-the-bench ventilation, inducing thigmomorphogenesis and a drier microclimate in the canopy. Here, poinsettias are set on spots marked for adequate spacing.

11. Ebb-and-flow benches do not permit through-the-bench ventilation.

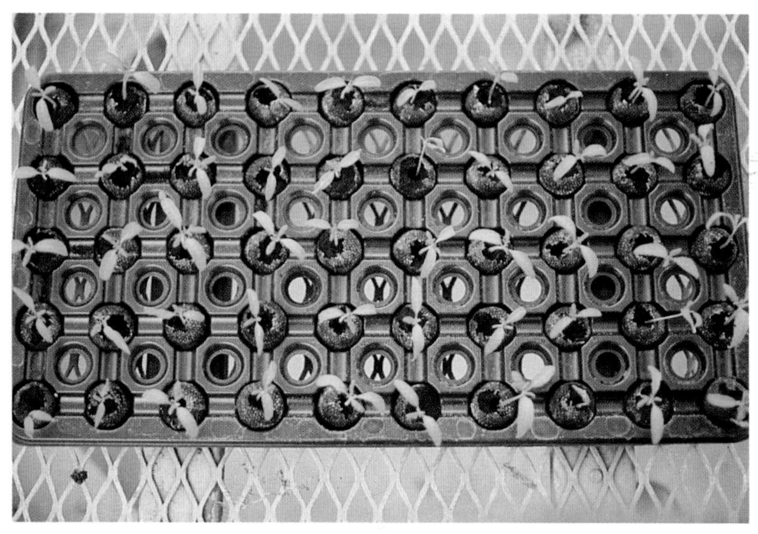

12. A transplant tray with ventilation holes alternating with plants in peat blocks.

13. A tomato stem infected with *Botrytis cinerea* in the humid, cool micro-climate of a straw mulch.

14. Heating pipes on the ground are used as railway lines for wheeled work platforms in tall crops.

15. A space heater. This one is very inefficient for warming seed trays on the ground.

16. Root-zone heating. Courtesy of Vary Industries Ltd.

17. A poorly sited aspirated psychrometer (arrow), monitoring the air some 1.5 m above the plants.

18. The Hoddesdon grid, by which steam is injected into the soil through hollow tines. The grid has to be moved after each small plot of soil has been sterilized.

19. Balloon-sheet steaming. The steaming has to be done in strips, so that when the cover is being moved, newly sterilized strips are readily recontaminated with adjacent nonsterile soil.

20. Watering by hose not only damages the plants but also disperses hydrophilic spores and bacteria.

21. Cucumbers planted on ridges to escape black root rot, caused by *Phomopsis sclerotioides*. The disease is more severe in cool soils. The ridges warm up quickly, promoting better root growth and lower fungal activity. Reprinted, by permission, from Jarvis (1989).

22. The dense canopy and convoluted habit of western hemlock seedlings make them very susceptible to gray mold, especially when closely packed in poorly ventilated containers. Reprinted, by permission, from Jarvis (1989). Photo by J. R. Sutherland.

23. Cucumber seedling infected with *Botrytis cinerea* after the cotyledon was damaged during transplanting (right) and a healthy seedling (left).

24. Deleafing the lower stems of tall vegetable crops improves ventilation and reduces the risk of infection by water-dependent pathogens.

25. Cucumber fruit infected with *Didymella bryoniae* after being pulled off the plant and damaged at the shoulder rather than cut cleanly.

26. Putting seedling and transplant flats on bare soil and beneath hanging baskets invites water-splashed soilborne pathogens.

27. Petal flecking caused by *Botrytis cinerea* is a quiescent state of gray mold.

1975; Lelliott 1988a–c; Yarwood 1956), or ineffectual pesticide sprays (Strider 1969a). In the field, windblown rain is an effective dispersal agent for bacteria (Faulwetter 1917; J. C. Walker and Patel 1964), a condition simulated in the greenhouse by watering with a moderate- or high-pressure hose (Plate 20). This method of irrigation has nothing to recommend it.

The hydrophilic conidia of the Coelomycetes similarly are easily wetted and picked up and dispersed by splash drops (Bilgrami 1963; Gregory et al 1959). Gregory et al investigated the dispersal of macroconidia of *Fusarium solani* (Mart.) Sacc. by water splash. A drop 5 mm in diameter falling onto a culture of *F. solani* broke up into about 5,200 splash droplets of diameters between 5 and 2,400 μm. The median for all droplets was 70 μm. About 3,000 of them, with a median diameter of 140 μm, carried one or more conidia. Splash droplets traveled distances of up to 100 cm. Conidia on acervuli, such as those of *Colletotrichum coccodes* (Wallr.) S. J. Hughes, are also dispersed by water splash (Davet 1971). The same fungus is readily spread from one tomato plant to another in the nutrient film technique (Schneider et al 1978).

The hydrophilic sporangia of the downy mildew fungi are formed on wet plant surfaces and are similarly dispersed by water splash (Y. Cohen 1981; Crute and Dixon 1981; Duvdevani et al 1946; Schnathorst 1962).

Notwithstanding that they are hydrophobic, conidia of *Botrytis tulipae* (Lib.) Lind are dispersed among tulips by splashing water drops from the greenhouse roof (Beaumont et al 1936). When water drips onto a mass of conidia of *B. cinerea*, composite projectiles are formed from splash droplets coated with a layer of dry conidia. These projectiles are remarkably stable and carry for distances of several centimeters (Jarvis 1962c).

The mechanism of dispersal of sporangia of *Peronospora hyoscyami* de Bary f. sp. *tabacina* (D. B. Adam) Skalický as discrete airborne units was described by Pinckard (1942). It is a hygroscopic mechanism depending on a drying environment, which may explain the observed peak of dispersal in early morning (Waggoner and Taylor 1958). It is not known whether *P. hyoscyami* is typical of other downy mildew fungi in its dispersal. Pinckard, observing that in a drying environment the sporangiophore becomes flat and twisted about its long axis, said that in response to changes in humidity, it twists violently enough to spin off sporangia centrifugally. Ingold (1971) was not convinced by this explanation but thought that the hygroscopic mechanism described by Jarvis (1960) for the dispersal of conidia of *B. cinerea* obtained. Jarvis also noted that the mature conidiophore of *B. cinerea* is a flat, twisted, ribbonlike structure, which twists rather gently but jerkily in response to changing humidity. The effect is to dislodge conidia from their fine attachment to the ampulla, a process Jarvis termed spore release, to be dispersed by air currents (Jarvis 1962a,b) or by splash droplets (Jarvis 1962c). In *B. cinerea* either rising or falling humidity, at a vapor pressure deficit between the approximate limits of 1.0 and 0.5 kPa, operates the hygroscopic mechanism and explains the observed two diurnal peaks in a raspberry plantation outdoors (Jarvis 1962a).

In order, therefore, to minimize spore dispersal by fungi of the *Pero-*

nospora and *Botrytis* type (the downy mildews and the Hyphomycetes), it is essential to maintain atmospheres of fairly even humidity and to refrain from any form of overhead irrigation.

Disturbances of infected crops by workers or irrigation often disperse clouds of spores of *B. cinerea* and other members of the Hyphomycetes, perhaps also by a hygroscopic mechanism (Gregory and Hirst in Hirst 1959; Hausbeck and Pennypacker 1987, 1991; Jarvis 1962b; Peterson et al 1988). Mechanical shock may also be a contributing factor, shaking spores free.

Jarvis (1962b) noted that the concentration of conidia of *B. cinerea* in the air increased during the activity of pickers in a strawberry plantation, and Hausbeck and Pennypacker (1987, 1991) also noted its conidia in a greenhouse of geraniums (*Pelargonium* × *hortorum* L. H. Bailey) when watering was being done, cuttings were being harvested, and pesticides were being applied. There was also a diurnal fluctuation resembling that occurring outside in a raspberry plantation (Jarvis 1962a). Peterson et al (1988) observed that the dispersal of conidia of *B. cinerea* in a greenhouse crop of *Pseudotsuga menziesii* (Mirb.) Franco coincided with irrigation, and they considered dispersal to be effected by mechanical means rather than by environmental changes.

Fungal spores occur in large numbers in undisturbed air in greenhouses (Gregory and Hirst in Hirst 1959), but the numbers caught in a spore trap vary according to disturbances in the crop. Gregory and Hirst found that tapping the stems of tomato plants in a poorly ventilated greenhouse increased the number of conidia of *Fulvia fulva* (Cooke) Cif. in the air but decreased that of *B. cinerea* (Table 22). Mechanical vibration of whole plants during the pollination of tomato flowers therefore might disperse conidia, but this has never been investigated. Sprinkling the foliage with water greatly increased the number of airborne conidia of *F. fulva* but reduced those of *B. cinerea* to zero.

It has proved impossible to consider the effects of humidity on sporulation and spore dispersal without considering the effects of wind, too. Jarvis (1962a) could not find an obvious effect of wind speed on the dispersal of conidia of *B. cinerea*, but Stepanov (1935) found them to be dispersed in a wind tunnel at wind speeds of 0.36–0.50 $m \cdot s^{-1}$. He also noted that

Table 22. Airborne conidia trapped in a poorly ventilated greenhouse containing a tomato crop subject to vibration and overhead irrigation (conidia per cubic meter)[a]

Fungus	Plants undisturbed	After tapping stems for 5 min	After sprinkling foliage and soil	1 h after disturbance	Outside the greenhouse
Botrytis cinerea	15,400	4,900	0	0	750
Cladosporium spp.	27,800	39,600	10,100	15,000	14,300
Fulvia fulva	11,300	132,000	1,275,000	1,100	1,500
Penicillium type	23,600	98,600	1,062,000	0	0
Others	4,950	6,400	25,000	9,750	1,900

[a] Adapted from Hirst (1959).

they could be dispersed in convection currents. C. M. Leach (1985) found that in still, saturated air *Drechslera turcica* (Pass.) Subramanian & P. C. Jain and *Peronospora destructor* (Berk.) Casp. in Berk. sporulated profusely on leaves of *Zea mais* L. and *Allium cepa* L., respectively, whereas in saturated air moving at 0.5 and 1.5 $m \cdot s^{-1}$, sporulation of *P. destructor* was inhibited. Leach thought it unlikely that electrostatic mechanisms operated here; on sunny, dry days, leaves become charged during daylight hours, but they are not normally charged at night when the humidity is high. He was unable to suggest an alternative reason for the inhibition of sporulation.

In wind tunnel experiments, Thomas et al (1988), determined that aerial mycelium of *B. cinerea* developed fastest from infected grapes at 21°C and 94% RH (vapor pressure deficit about 0.12 kPa) in still air; it did not develop in moving air at 69% RH (vapor pressure deficit of 0.70 kPa). On the other hand, the most conidia were produced at 21°C and 94% RH in air moving at 0.6 $m \cdot s^{-1}$ (Figure 34). The evaporative potential (EP)—the weight of water lost per hour from wet grape-sized plastic foam stoppers—was related to relative humidity and wind speed by the following equation:

$$EP = 0.2565 - 0.0028\ RH + 5.667\ WS - 0.055\ WS \times RH$$

where RH is percent relative humidity and WS is wind speed ($m \cdot s^{-1}$). Aerial conidiogenic mycelium decreased linearly as EP increased to 0.25 $g \cdot h^{-1}$ at 21 and 26°C; it did not occur at EPs greater than 0.25, 0.20, and 0.15 $g \cdot h^{-1}$ at 26, 21, and 16°C, respectively. The greatest number of conidia developed at EPs between 0.05 and 0.16 $g \cdot h^{-1}$, and the number increased as the temperature decreased.

Wind speed is thus an important factor in determining the growth and sporulation of *B. cinerea*, and a diurnal variation in EP in a vineyard closely paralleled the variation in wind speed. In a greenhouse, therefore, it would appear that a wind speed greater than 0.4 $m \cdot s^{-1}$ and a vapor pressure deficit greater than about 0.5 kPa would reduce fungal activity considerably, as well as evaporate inoculum drops. Wind speed is readily controlled by fans in the greenhouse, and the exposure of potential infection sites to wind can be greatly improved by manipulating crop habit and plant density.

Wind is also an important factor in the dispersal of conidia of the Erysiphales. Bainbridge and Legg (1976) found that air speeds of 0.6 and 0.5 $m \cdot s^{-1}$ were required for the release of conidia of *Erysiphe graminis* DC. f. sp. *hordei* Ém. Marchal from flapping and stationary barley leaves, respectively. Hammett and Manners (1973) gave rather higher speeds as necessary for *E. graminis* f. sp. *tritici* Ém. Marchal—a minimum of 1.14 $m \cdot s^{-1}$, with most conidia dispersed at 1.46 $m \cdot s^{-1}$. In still air, the chains of conidia in the pustules cracked and tangled, perhaps as the result of some hygroscopic changes, and no dispersal occurred.

Frinkling and Scholte (1983) observed that wind speeds in a greenhouse did not generally exceed 0.25 $m \cdot s^{-1}$ until the doors were opened, when they went up to 0.5 $m \cdot s^{-1}$. They found that most spores of *E. graminis* f. sp. *hordei* were liberated from colonies on potted barley leaves by air moving in convection currents and also by vibration of the leaves caused by people handling the crop. The air within the greenhouse could be purged of powdery mildew spores when there were 30–40 changes of air per hour

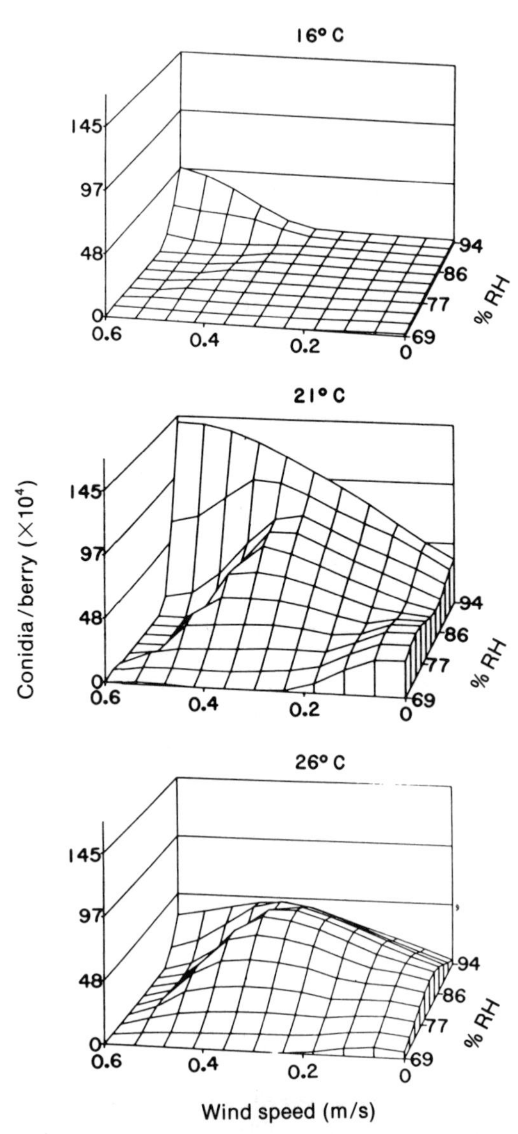

Figure 34. Effects of wind speed, temperature, and relative humidity on the development of conidia of *Botrytis cinerea* in 6 days on grape berries. Reprinted, by permission, from Thomas et al (1988).

with the ventilators fully open. They did not report whether the ventilation in their greenhouse was assisted by fans. In a cucumber greenhouse, Abiko and Kishi (1979) found peak concentrations of airborne conidia of *Sphaerotheca fuliginea* between 1200 and 1500 h.

It is interesting to speculate about the dissemination of hyperparasites and antagonists of the powdery mildew fungi. Speer (1978) discovered mycelium of the hyperparasite *Ampelomyces quisqualis* within conidia of *Microsphaera alphitoides* Griffon & Maubl., and infected conidia are probably dispersed with healthy conidia. It is also possiblc that the sticky pycnospores of *A. quisqualis* adhere to airborne conidia of *S. fuliginea.* Probably they are also dispersed by water sprays (Jarvis and Slingsby 1977). Conidia of *Stephanoascus* spp. (Traquair et al 1988) and a *Tilletiopsis* sp. (Hijwegen 1986, 1988; Hoch and Provvidenti 1979), antagonists of *S. fuliginea*, may also be dispersed on conidia of the host powdery mildew, but they are dry-spored and are probably independent constituents of the air microflora. W. R. Jarvis (unpublished results) trapped airborne conidia of *Sporothrix* spp. in an Anderson air sampler.

SUMMARY

Limiting Inoculum Dispersal

Because of the intensive nature of routine work in most greenhouses, the dispersal of pathogens on workers' fingers, clothing, and tools is a major factor in epidemiology. Hydrophilic propagules, such as bacteria, conidia of the Coelomycetes, and sporangia of the downy mildew fungi, are dispersed by water, and so irrigation has to be done with care. The hydrophobic conidia of the Hyphomycetes have a hygroscopic mechanism of spore release and are dispersed by air currents. In the case of *Botrytis cinerea*, disturbance of the leaf canopy over sites of sporulation operates the release and dispersal mechanisms, and in geranium houses, for example, high concentrations of airborne conidia are associated with routine operations, such as sticking, watering, applying pesticides, harvesting cuttings, and transplanting. Wind speed is an important factor in the dispersal of conidia of the Hyphomycetes and the Erysiphales, and leaf flutter may also play a part. This factor can be regulated by moderating the fan speed in the ventilation system. While it is necessary to reduce pathogen dispersal in the greenhouse by appropriate modification of work practices and environmental control systems, it may also be necessary to enhance the dispersal of biological control agents, which may be both water- and air-dispersed.

Controlling Pathogen Vectors

Pathogens of greenhouse crops are carried in from the outside and within greenhouses by several groups of insects and by nematodes and fungi (Harris and Maramorosch 1980), by workers, perhaps by pet cats and dogs (which are not uncommon in greenhouses), and, in the case of tomato mosaic virus, by birds (Broadbent 1965b). A number of examples are listed in Table 23.

Diseases initiated by insects from outside the greenhouse are often characterized by the development of symptoms on plants near doorways

Table 23. Vectors of some pathogens of greenhouse crops

Vector	Pathogen	Reference
Acalymma vittata	*Erwinia tracheiphila*	J. G. Leach (1964)
Aphis spp.	Several viruses	Kennedy et al (1962)
	Cucumber mosaic virus	de Brouwer and van Dorst (1979) Tomlinson et al (1970)
	Lettuce mosaic virus	Gonzalez and Rawlins (1969)
Bemisia tabaci	Tomato yellow top virus	H. Cohen et al (1961)
	Lettuce infectious yellows virus	J. K. Brown and Stanghellini (1988)
	Xanthomonas pelargonii	Bugbee and Anderson (1963)
Birds	Tomato mosaic virus	Broadbent (1965b)
Chytridiales	Various viruses	Teakle (1980)
Coleoptera	Various viruses	J. P. Fulton et al (1987)
Diabrotica undecimpunctata	*Erwinia tracheiphila*	Rand and Enlows (1916)
Frankliniella occidentalis	Tomato spotted wilt virus	Allen and Matteoni (1988) Matteoni et al (1988)
Heteroptera	Beet leaf curl virus	Proeseler (1980)
Lyriomyza trifolii	*Pseudomonas cichorii*	Matteoni and Broadbent (1988)
Micrutalis malleifera	Pseudo-curly top virus	Simons and Cox (1958)
Nematodes	Various viruses	Taylor (1980)
Olpidium brassicae	Lettuce big vein virus	Tomlinson and Garrett (1964)
	Tobacco necrosis virus	Teakle (1962)
O. radicale	Melon necrotic spot virus	Tomlinson and Thomas (1986)
Polymyxa betae	Beet necrotic yellow vein virus	Tamada (1975)
Spongospora subterranea	Potato mop top virus	Harrison (1974)
Thysanoptera	*Erwinia chrysanthemi* pv. *dianthicola*	Pussard-Radulesco (1930)
Trialeurodes vaporariorum	Beet pseudo-yellows virus	van Dorst et al (1983) Duffus (1965)
Trichodorus pachydermis, *T. similis*	Tobacco rattle virus	Cremer and Schenk (1967)
Xiphinema americanum	Tobacco ringspot virus	Fulton (1967)

SUMMARY

Controlling Pathogen Vectors

Many pathogens of crops are carried into the greenhouse by a wide variety of vectors. Viruses and some fungus spores are carried by insects, which are difficult to exclude completely. Tomato mosaic virus is even carried by birds. Windblown and waterborne fungus spores and bacteria are commonly the primary inoculum in the greenhouse, and many are carried in on clothing and machinery. It is essential for the prevention of diseases to recognize vectors and take steps to eradicate or exclude them. In addition the sources of pathogens in the vicinity of the greenhouse should be eradicated. Many pathogens, particularly viruses and mycoplasmas, have wild hosts. Greenhouses are best surrounded by a 10-m zone of well-mown, weed-free lawn.

and side ventilators. In the case of the viroid-incited pale fruit disease of cucumber, van Dorst and Peters (1974) deduced the likelihood of an insect vector from that pattern of symptom appearance, although the vector was not identified. This disease also followed main walkways, which suggests transmission on clothing or handcarts.

Many viruses and the cucumber pale fruit viroid (van Dorst and Peters 1974) are transmitted experimentally by dodder (*Cuscuta* spp.). Although occasionally found in poorly managed greenhouses, dodder is unlikely to be important in the spread of diseases through crops.

The implications of vector transmission of pathogens are that insects and nematodes should be controlled inside the greenhouse, in its immediate surroundings (within a 10-m buffer zone), and in adjacent field crops. Weeds, which are frequently reservoirs of pathogens, especially viruses, also should be rigorously controlled.

Limiting Pathogen Survival

After the death of the host, the response of many pathogens is dormancy, often by means of special structures—chlamydospores, sclerotia, resting mycelium, oospores, and so on—which are resistant to adverse environments, particularly to desiccation (D. Park 1965; Sussman 1965). Under suitable conditions, this microbial mass is reactivated and constitutes an important primary inoculum. The desired strategy is to eradicate dormant structures or to activate them in the absence of the host, so that the organisms become susceptible to biological, chemical, or physical controls.

It is important in this strategy to know where the resistant structures are and the period for which dormant microorganisms and viruses remain viable. Tomato mosaic virus can survive for many years in dry tomato roots in soil and on clothing (Broadbent 1960; Broadbent et al 1965; Fletcher 1969). Virus particles are difficult to inactivate by heat when they are present in thick roots, and so removal of the roots prior to heating the soil is important.

Pathogenic bacteria do not form spores, but they can survive for long periods in a dry matrix (Leben 1981). *Pseudomonas syringae* van Hall pv. *lachrymans* (Smith & Bryan) Young et al, for example, may survive in dry cucumber leaves for over 2 yr (Bradbury 1967). *P. caryophylli* (Burkholder) Starr & Burkholder may survive for several months in soil, probably in trash and rotted roots (Lelliott 1988b). *Clavibacter michiganensis* subsp. *michiganensis* (Smith) Davis et al survives in soil-filled crevices in wooden stakes and on twine and wire (Strider 1969b). It can also overwinter in plant debris in field soil, surviving from one growing season to another (B. N. Dhanvantari, personal communication).

In greenhouse soil, nematodes can survive at surprising depths, well below the reach of steam sterilization and fumigants. *Meloidogyne incognita* (Kofoid & White) Chitwood is able to follow tomato and cucumber roots (Bird 1969) to depths of 1.5 m and more than 1 m, respectively (Ward 1964, 1967). In sandy soils this nematode survives under concrete paths and reinfests crops very quickly, so that it is impossible to grow two crops a year without treating the soil in the intervening period (P. W. Johnson and McKeen 1973).

The survival period for soilborne fungi depends on many factors—the species, the soil type, temperature, moisture, cropping, and weed populations, among other factors. They are particularly persistent in dry soils (Ebben and Spencer 1976; D. Park 1965; Vaartaja 1964) and in dried-out soilless substrates (Orlikowski 1980). Soilborne pathogenic fungi have been classified into soil inhabitants, which are mostly unspecialized pathogens (such as *Pythium* spp. and *Rhizoctonia solani* Kühn), and soil invaders, which are associated fairly specifically with particular crops and their residues (such as the Fusarium wilt fungi) (Garrett 1970). The survival of soil invaders may be enhanced by their association with symptomless, resistant hosts and with nonhost crops and weeds. Wilhelm (1956) found the following pathogenic fungi on the roots of the weed hairy nightshade (*Solanum sarachoides* Sendt.): *Verticillium albo-atrum* Reinke & Berthier, *Colletotrichum coccodes, Pyrenochaeta terrestris* (E. M. Hans.) Gorenz, J. C. Walker, & R. H. Larson (*Phoma terrestris* E. M. Hans.), *Pythium ultimum* Trow, *Macrophomina phaseolina* (Tassi) Goidanich, *Fusarium roseum* Link:Fr., a *Phoma* sp., and a sterile fungus pathogenic to strawberry. Some resting structures do not germinate readily until stimulated by exudates from the host; this is the case with oospores of *Bremia lactucae*, which are stimulated to germinate by lettuce seedlings (Morgan 1983b; Norwood and Crute 1983).

Fungi pathogenic to foliage and fruit are also able to survive for long periods as sclerotia, chlamydospores, resting mycelium, and even conidia. Galea and Price (1988) found *Microdochium panattonianum* (Berl.) Sutton, Galea, & Price in Galea, Price, & Sutton (which causes lettuce anthracnose) to survive for 10–20 wk in lettuce debris buried in soil and for 58 wk in dry debris in air. Conidia of *Botrytis cinerea* can survive dry on the petals of greenhouse plants for very long periods (Salinas et al 1989; Verhoeff et al 1988), but not long in soil (Coley-Smith 1980). Bashi and Aylor (1983) found detached sporangia of *Peronospora destructor* and *P. tabacina* D. B. Adam (*P. hyoscyami* de Bary) to survive quite well. Those of *P. destructor* survived best at 10°C and poorest at 35°C. They survived poorly at 33% RH. Sunlight was the most important factor for both species. On cloudy days, the germination of detached sporangia of *P. destructor* was reduced from 83 to 68% in 6 h, whereas on clear, sunny days, it was reduced from 46 to 0% in 6 h.

Survival structures constitute a major part of the inoculum potential, which, according to R. Baker (1965), Garrett (1970), and Lockwood (1988), is determined by inoculum density, the endogenous and exogenous energy of the propagules, their genetic virulence, and the environment limiting their invasive force (Garrett 1970; Lockwood 1988). Two of these components, inoculum density and the environment, can be manipulated to effect disease escape. Manipulating the environment can also reduce exogenous energy sources for pathogenesis.

Inoculum density can be reduced by using pathogen-free planting material,

SUMMARY

Limiting Pathogen Survival

After the death of the host or following the trashing of crop residues, most pathogens are able to survive until at least the next crop by means of a variety of desiccation-resistant structures. Bacteria can survive in dry crop debris and soil for several months, as well as on stakes and twine. Root-knot nematodes can survive at depths in soil well below the level of penetration of steam and fumigants. Soilborne pathogenic fungi can be divided into two classes: soil inhabitants, which are relatively unspecialized, opportunist pathogens (such as *Pythium* spp. and *Rhizoctonia solani*), and soil invaders, which mostly enter the soil in the debris of their specific hosts (such as the Fusarium wilt fungi). Many fungi (such as *Verticillium* spp.) have symptomless, resistant hosts, and many (such as *Botrytis* and *Sclerotinia* spp.) have weed hosts, which may or may not have symptoms. Weeds are also reservoirs of viruses.

All of these intercrop survival structures constitute an important inoculum potential, compounded of inoculum density, endogenous and exogenous energy, the genetic virulence of the propagules, and the environment limiting their invasive force. It is important to be able to recognize this potential and eradicate the surviving inoculum.

by general hygiene, and by direct eradication from soil and debris by sterilization or pasteurization. It can also be reduced by providing an environment conducive to biological control or inimical to propagule germination and infection, as is discussed elsewhere.

Both the endogenous and the exogenous energy of survival propagules in soil can be reduced by keeping the soil moist and warm, encouraging the germination or formation of thin-walled and energy-poor structures, such as conidia or thin-walled mycelium. These are then more susceptible to antibiosis and competition from the surrounding active microflora as well as more susceptible to heat and pesticides.

Greenhouse Hygiene

Experienced extension personnel say that they can usually foretell the health of a crop even before they walk into the greenhouse simply from the state of hygiene of the greenhouse environs and the headerhouse. Weeds, algae, dirty machinery, trash piles of prunings and unsaleable produce, and sometimes field crops of the same plants as those growing inside all suggest scant regard for elementary greenhouse hygiene. Vanderplank (1963) calculated that destroying 90% of the inoculum reduces the incidence of disease from 62 to 9.2%, and elementary greenhouse hygiene centers on removing extraneous sources of inoculum.

Trash piles of prunings and cull piles of unsaleable produce have long been recognized as potent sources of inoculum for many diseases of many crops. For example, potato cull piles are sources of sporangia of *Phytophthora infestans*, the cause of late blight of the Solanaceae (Bonde and Schultz 1943). Jarvis et al (1983) readily isolated *Fusarium oxysporum* Schlechtend.:Fr. f. sp. *radicis-lycopersici* W. R. Jarvis & Shoemaker from trash piles of tomato prunings and uprooted dead plants outside greenhouses, as well as from dirt on headerhouse floors. Rowe et al (1977) regarded microconidia from those sources as the origin of inoculum for reinfestation of newly sterilized soil.

Ebben and Spencer (1976) found 10,000 propagules of *F. oxysporum* f. sp. *dianthi* (Prill. & Delacr.) W. C. Snyder & H. N. Hans. per gram of soil taken from walkways in a carnation greenhouse, and it is likely that infested dirt on walkways and headerhouse floors is responsible for much infestation of soil- and peat-based mixes put there for mixing. Favrin et al (1988) isolated four *Pythium* spp. from peat mixes in cucumber houses. Soil mixes should not be prepared on dirty floors or in the vicinity of a crop. Floors should be of washable concrete, and Tomlinson et al (1980) recommended 1% iodophor as a disinfestant for concrete troughs.

Plant debris left in the greenhouse as prunings on the floor or in the soil or fragments left hanging on training wires is frequently implicated as the source of inoculum of several pathogens, such as *Clavibacter michiganensis* subsp. *michiganensis* (McKeen 1973; Strider 1969a,b), *Pseudomonas caryophylli* (Lelliott 1988b), *Botrytis cinerea* (Jarvis 1977a),

Didymella bryoniae (van Steekelenburg 1983), *D. lycopersici* (Knight and Keyworth 1960; Verhoeff 1963), *Microdochium panattonianum* (Galea and Price 1988), tomato mosaic virus (Broadbent 1976), and pepper mild mottle and cucumber mosaic viruses (Pares and Gunn 1989).

Plants with highly contagious diseases, such as tomatoes with bacterial canker (caused by *C. michiganensis* subsp. *michiganensis*) or tomato mosaic virus, can be rogued out if not too numerous. They should be covered entirely by a plastic bag, cut out, and carried out without touching neighboring plants. Suspect areas should be worked last in cultural operations, to reduce the risk of plant-to-plant spread by contaminated fingers and tools.

It is not uncommon to see exotic plants such as fig (*Ficus carica* L.), grapevine (*Vitis vinifera* L.), and oleander (*Nerium oleander* L.) planted in corners of groundbeds in vegetable greenhouses, as well as hanging baskets of various ornamentals overhead. It is impossible to sterilize the soil around the perennials, and they and the hanging baskets are sources of inoculum of various soilborne pathogens, such as nematodes, and reservoirs of viruses, insects, and mites.

By the same token, seedlings should be raised absolutely separately from crops in production and preferably handled by different personnel and with separate tools and machinery, in order to maintain a high standard of hygiene.

Weeds and crops outside the greenhouse are also sources of viruses and insects. Tomlinson et al (1970), for example, listed 20 weed species as potential sources of cucumber mosaic virus and 60 species of *Aphis* that can transmit it. A number of virus diseases, including celery mosaic (Wellman 1937), tomato mosaic and other diseases caused by tomato mosaic virus (Broadbent 1976), and cucumber mosaic (Doolittle and Walker 1926), can be controlled by eradicating their nearby wild hosts.

It is therefore important to maintain a weed-free zone at least 10 m wide around the greenhouse, preferably in lawn kept disease-free and mown with a reel mower to keep down insect populations.

Field crops adjacent to a greenhouse are also sources of inoculum, particularly if they are botanically related to the crop in the greenhouse. *C. michiganensis* subsp. *michiganensis* on field tomatoes is believed to be carried into nearby greenhouses in airborne dust or perhaps as aerosols, and growers of greenhouse tomatoes are advised not to grow tomatoes in neighboring fields (McKeen 1973).

Some pathogens are transmitted in irrigation water; they include *Pythium* spp. (Hendrix and Campbell 1973), *Phytophthora* spp. (Erwin et al 1983), *Erwinia* spp. (Pérombelon and Kelman 1980), a Chenopodium necrosis strain of tobacco necrosis virus (Tomlinson et al 1983), cucumber green mottle mosaic virus (Lecoq 1988), melon necrotic spot virus (Bos et al 1984), tomato mosaic virus (Broadbent 1976; Broadbent and Fletcher 1963), and chrysanthemum stunt viroid (Monsion and Dunez 1971), as well as the virus vectors *Olpidium brassicae* (Woronin) P. A. Dang. and *O. radicale* Schwartz & Cook (Teakle 1962, 1980; Tomlinson and Garrett 1964; Tomlinson et al 1980). Most soil groundbeds have a water table continuous with that outside, and drainage water is liable to enter the greenhouse in times of high water unless the drainage is controlled to direct it away. Greenhouse siting is therefore important. Reservoirs and wells can similarly become contaminated by runoff, and these pathogens are readily transferred to greenhouse crops in irrigation water unless the water is disinfested.

Personal Hygiene

All of the pathogens transmitted by tools and machinery are equally well transmitted on the fingers, and hands should be washed frequently with hot, soapy water, particularly when crops with very contagious diseases are being handled.

Spores of fungi, bacteria, and viruses are transmitted on clothing and passed from house to house on wet and dirty footwear. Tomato mosaic virus is readily transmitted on clothing, and it survives for up to 3 yr on dry clothing stored in the dark (Broadbent and Fletcher 1963, 1966). In 1963, Broadbent and Fletcher found that overalls were seldom cleaned from year to year; in 1990 it is still possible to find sap-stained overalls in headerhouse closets. Clothing should be laundered daily in hot water with detergent to inactivate pathogens and dried in a hot-air dryer or in sunlight, which inactivates tomato mosaic virus within about 1 mo.

Tomato mosaic virus has been recorded in cigarette tobacco (Broadbent 1962), but Broadbent and Fletcher (1966) considered it to be a minor source of crop contamination. Nevertheless, the risk is there, and the handling of crops susceptible to the virus is best left to nonsmokers and nonchewers. Broadbent (1963) recommended rinsing the hands in 1 or 3% trisodium

SUMMARY

Greenhouse Hygiene

The culmination of all practices aimed at reducing the amount of inoculum and its dispersal is the formulation of a code of greenhouse hygiene. Destroying 90% of the inoculum has been calculated to reduce the incidence of disease from 62 to 9.2%. Sources of inoculum must be recognized and removed from the vicinity of the crop. They include trash piles, dirt on headerhouse floors and paths, weeds within at least 10 m of the greenhouse, perennials in vegetable and ornamental greenhouses, hanging baskets of flowers in vegetable houses, and open-water reservoirs receiving runoff water from fields. Since clothing contaminated with sap may constitute a source of viruses and bacteria, it should be laundered frequently, and tools and hands should be cleansed regularly during routine operations. Unessential personnel should be kept out of crops at risk from highly infectious pathogens.

phosphate before washing in hot, soapy water to cleanse them of tomato mosaic virus.

Broadbent and Fletcher (1966) also thought that casual visitors could introduce tomato mosaic virus into susceptible crops, and W. R. Jarvis and H. J. Thorpe (unpublished observations) thought that casual visitors in a densely crowded greenhouse district (Jarvis et al 1983) helped to spread *Fusarium oxysporum* f. sp. *radicis-lycopersici* among tomato crops. Fusarium crown and root rot, the disease caused by this pathogen, was new in the area surveyed by Jarvis et al, and many of the growers were related to each other and overly interested in their neighbors' crops.

Extension personnel, sales representatives, and other casual visitors should be encouraged to wear disposable overboots and overalls, particularly when visiting crops suspected of being or known to be diseased. A mat soaked in a disinfestant at each doorway helps to minimize the transfer of pathogens on footwear.

Sterilizing Hardware

A number of pathogens, particularly viruses and bacteria, are readily transmitted by pruning knives and other tools and can survive on pots, seed boxes, stakes, and machinery. Solutions of sodium hypochlorite (0.5%) or potassium hydroxyquinoline sulfate (0.04%) may be used to wash hardware. Neither of these disinfestants eradicates viruses from crop debris, and all debris should be scrupulously removed before disinfesting the interior of the greenhouse. Machinery should be washed down and disinfested like other hardware.

As a routine precaution, especially in situations involving contagious diseases such as bacterial canker in tomatoes, pruning knives and shears (secateurs) and electromechanical pollinators should be sterilized with ethanol or a solution of sodium hypochlorite at the end of every row. Secateurs are available that deliver a small volume of pesticide or a biological control agent at each stroke; they reduce the risk of transmitting fungi to pruning wounds (Grosclaude et al 1973). Trisodium phosphate at a concentration of 3% is generally recommended for inactivating tomato mosaic virus (Broadbent 1963) and was the only effective one of a number of agents tried by Pategas et al (1989) in a nutrient film system. Boxes, pots, and stakes can be treated with a solution of trisodium phosphate to inactivate tomato mosaic virus or with a detergent to control cucumber green mottle mosaic virus (Lecoq 1988).

The plastic gullies used in nutrient film culture are disposable but can be cleansed by sodium hypochlorite solutions with free chlorine at 100 $\mu g \cdot L^{-1}$ (Jenkins 1981). The system has to be flushed with water after disinfection, because the residues are phytotoxic. Concrete gullies can be cleansed with 1% iodophor (Tomlinson et al 1980), but sodium hypochlorite is commonly used.

Disease Escape

CHAPTER 7

Manipulating the Environment to Prevent Infection

For the greenhouse manager, disease escape means avoiding all conditions that predispose plants to infection as well as eliminating inoculum. Various environmental factors, such as temperature, water potential, crop nutrition, and light, can be manipulated to minimize stress on the plants and to avoid microenvironments that permit infection, such as those in which dew forms.

Obviously, the elimination of inoculum by sterilizing or pasteurizing soil, using pathogen-free seed and planting material, controlling vectors, and practicing strict hygiene in cultural operations are all disease-escape mechanisms of prime importance. However, notwithstanding the presence of inoculum or even incipient infection, it is still possible to escape the consequences by manipulating the crop and its environment to interrupt life-cycle pathways of a pathogen (see Figure 27) or to delay the onset of the next stage.

The pathways are affected differently by environmental conditions, such as temperature (see Table 26). The presence of a large quantity of airborne inoculum of *Botrytis cinerea* Pers.:Fr. does not guarantee infection (Jarvis 1977a). In a crop of gerbera daisies (*Gerbera jamesonii* H. Bolus ex J. D. Hook.), airborne conidia of *B. cinerea* were numerous during and after periods when the daily mean relative humidity in the canopy was 85–95% (Salinas et al 1989). The incidence of petal flecking was highest under those conditions, but only when the daily mean temperature was below 20° C. When the daily mean relative humidity was between 60 and 85%, the number of airborne conidia and the incidence of petal flecking were both low, and little petal flecking occurred at temperatures above 20° C. Surprisingly, the incidence of petal flecking was higher in areas of the greenhouse where the air was turbulent; it might have been supposed that inoculum droplets would dry out faster in turbulent air. The quantity of airborne inoculum was also higher in those areas, and it was higher within the plant canopy than at the top of it.

In unheated plastic houses in Israel, Yunis et al (1990) related the occurrence of gray mold in vegetables to the persistence of dew. In the early stages of an epidemic in cucumbers, the incidence of fruit infection was correlated with temperature in the range from 11 to 25°C, with 97–99% RH, and with persistent leaf wetness. At a later stage, disease progress was better correlated with temperature in the range from 11 to 16°C and 85% RH or more, but not with leaf wetness. On stems, early disease was associated with temperatures of 11–16°C, and late disease with 80–99% RH.

The complexity of quantifying the environment for determining proneness to disease is well illustrated by the studies of Marois et al (1988) on gray mold of rose flowers, caused by *B. cinerea*. First, the inoculum concentration (conidia per milliliter) was directly related to the number of lesions per flower 48 h after inoculation, but the correlation was different in the fall months (October and November) and in the winter (December–February). For the fall,

$$y = 13.0 + 0.011x \qquad r = 0.98$$

where y is the number of lesions per flower, and x the number of conidia per milliliter. For the winter,

$$y = 15.1 + 0.028x \qquad r = 0.98$$

Second, the overall mean temperature for fall was 20.1°C, significantly different from the winter mean of 19.4°C. There was a significant linear relationship between the overall mean temperature during the 5 wk before flower harvest and the slope of the IC/DS relation, where IC is inoculum concentration, and DS is disease severity. There was also a correlation between the mean temperature of the daylight hours only (between 0800 and 1900 h) and the slope of IC/DS. Overall, the mean relative humidity ranged from 80.0 to 84.3% during the 5 wk before harvest, and it was not affected by the season. However, the mean relative humidity between 0800 and 1900 h in winter (76.0%) was significantly higher than that in fall (70.2%). The linear correlation between the overall mean relative humidity and the IC/DS slope was not significant, but that between the mean relative humidity of the daylight hours and the slope was highly significant. The daily mean vapor pressure deficit ranged from 0.46 to 0.59 kPa. The mean in winter (0.48 kPa) was significantly less than that in fall (0.52 kPa), whereas the mean values for the daylight hours were 1.01 kPa in winter and 0.76 kPa in fall, a significant difference in the opposite direction. Marois et al (1988) pointed out that vapor pressure deficit, since it is independent of temperature, is by far more meaningful than relative humidity as a parameter of gray mold in roses.

The practical message for growers from this work is that *B. cinerea* should be prevented from producing conidia and that the disease, following the processes of spore production, liberation, dispersal, and infection, is temperature-dependent. Also, low vapor pressure deficits, with night or dawn deposition of dew, should be avoided, especially in winter. There is a need for heating and ventilating at sunset when the vapor pressure deficit is low (W. M. Morgan 1983a).

While gray mold is representative of several other diseases whose causal organisms are dependent on water for germination or entry into the host, the powdery mildews are generally regarded as dry-climate diseases (Yarwood 1957). Some powdery mildew fungi, at least, can tolerate very high vapor pressure deficits (Yarwood 1936). There is confusion, however, as to whether water is needed for the germination of conidia or is inhibitory. De Long and Powell (1988) developed a deterministic model for the epidemiology of rose powdery mildew (caused by *Sphaerotheca pannosa* (Wallr.:Fr.) Lév. var. *rosae* Woronichin). Leaf wetness sensors in the canopy of roses in a greenhouse indicated consistently wet conditions in a 9- to 12-h period at night during powdery mildew epidemics. Conidial germination was low when leaf wetness was either high and low, compared with germination when wetness was intermediate.

Yarwood (1939a) noted that several powdery mildews could be controlled with water sprays applied under considerable pressure, about 470 kPa, and Jarvis and Slingsby (1977) also controlled the cucumber powdery mildew fungus, *Sphaerotheca fuliginea* (Schlechtend.:Fr.) Pollaci, on cucumber with water sprays. Yarwood's objective was to achieve a washing effect; that of Jarvis and Slingsby was to enhance the activity of the hyperparasite *Ampelomyces quisqualis* Ces. Perhaps both effects operated in both sets of experiments, together with the enhancement of other antagonist populations (Jarvis et al 1989). Perera and Wheeler (1975) found that *S. pannosa* var. *rosae* was inhibited by water droplets on rose leaves. Similar numbers of conidia germinated on wet and dry leaves, but fewer developed into colonies on misted leaves; Perera and Wheeler thought that initial penetration was inhibited. By contrast, Yarwood (1978) recorded a case in which infection of cucumber by *S. fuliginea* was stimulated by water sprays. Hijwegen (1988) noted an increase in powdery mildew of cucumber when the leaves were sprayed with water after inoculation with the hyperparasites *Calcarisporium arbuscula* Preuss, *Cladobotryum varium* Nees, and *Sepedonium chrysospermum* (Bull.) Link:Fr. He attributed this to new colonies of *S. fuliginea*, which were perhaps initiated from water-dispersed conidia.

Evidently more critical work needs to be done on the role of water in infection by the Erysiphaceae, observing the strictures of Schein (1964) about the pitfalls of working with inadequate temperature controls at low vapor pressure deficits. It also seems likely that the autecology of the commensal microorganisms of the powdery mildews is intimately involved.

Managing Temperature

The phytopathological literature is full of "optimum temperatures" for most of the common greenhouse crop diseases (Tables 24 and 25). At best, these can only represent an empirical summation of all the temperature optima of the processes leading up to and including infection, affecting the development of the pathogen and also the defense reactions of the host. Both have to be qualified by water and nutritional relations. Empirical recommendations for optimum temperatures for growing a particular crop

Table 24. Optimum temperatures for disease expression in some greenhouse crops in various areas of the world[a]

Crop	Disease	Optimum temperature (°C)	Reference
Carnation	Alternaria blight	24	Strider (1978)
	Fusarium wilt	22–26	Harling et al (1988)
Chrysanthemum	Didymella spot	22	Punithalingam (1980)
	Fusarium wilt	29–35	Gardiner et al (1987)
	Rust	15–21	Punithalingam (1968a)
	Septoria spot	23–26	Punithalingam (1967)
	White rust	13–22	Punithalingam (1968b)
Cucumber	Angular leaf spot	24–27	Bradbury (1967)
	Black root rot	13	Ebben (1968)
	Cucumber mosaic	15	van Dorst (1975)
	Downy mildew	15	Palti and Cohen (1980)
	Gummy stem blight	23	van Steekelenburg (1983)
	Pale fruit viroid	30	van Dorst (1975)
	Powdery mildew	25	Abiko and Kishi (1979)
	Scab	17	Walker (1950)
Exacum	Botrytis blight	20, 30–35	Ploetz and Engelhard (1979)
Gerbera	Gray mold	18–25	Salinas et al (1989)
Gladiolus	Gray mold	13–18	McClellan et al (1949a)
	Smut	20	Schenk (1961)
Lettuce	Powdery mildew	18	Schnathorst (1959a)
Poinsettia	Pythium root rot	17	Bateman and Dimock (1959)
	Rhizoctonia root rot	30	Bateman and Dimock (1959)
	Thielaviopsis root rot	17	Bateman and Dimock (1959)
Rose	Gray mold	19–20	Marois et al (1988)
	Powdery mildew	18–21	Longrée (1939)
Snapdragon	Rust	21–24	Dimock and Baker (1951)
Spinach	Damping-off		
	Pythium aphanidermatum	23	Bates and Stanghellini (1984)
	P. dissotocum	17–22	Bates and Stanghellini (1984)
Strawberry	Powdery mildew	70	Peries (1962)
Tomato	Bacterial canker	28	Kendrick and Walker (1948)
	Didymella canker	15–20	Verhoeff (1963)
	Fusarium crown and root rot	18	Jarvis and Shoemaker (1978)
	Fusarium wilt	28	Clayton (1923a)
	Leaf mold	22–24	Holliday and Mulder (1976)
Tulip	Fire	10–15	D. Price (1970)
Several crops	Damping-off		
	P. aphanidermatum	25	Bouhot and Smith (1988)
	P. ultimum	10–15	Bouhot and Smith (1988)

[a] Local optima may vary.

Table 25. Optimum temperatures for some diseases of greenhouse ornamentals[a]

Pathogen	Host	Optimum temperature (°C)
Alternaria panax	*Brassaia actinophylla*	15–24
Botrytis spp.	Various hosts	18–22
Cylindrocladium spathiphylli	*Spathiphyllum* spp.	21–27
Drechslera setariae	*Calathea argentea*	15–21
Erwinia chrysanthemi	*Philodendron selloum*	28–34
Fusarium oxysporum	*Chrysanthemum × morifolium*	27–32
F. roseum	*Dianthus caryophyllus*	>21
Myrothecium roridum	*Dieffenbachia maculata*	21–27
Powdery mildew fungi	Various hosts	21–29
Pseudomonas cichorii	*Hibiscus rosa-sinensis*	21–27
P. solanacearum	*Pelargonium* spp.	22–35
P. syringae pv. *syringae*	*Hibiscus rosa-sinensis*	15–18
Rhizoctonia solani AG4[b]	*Nephrolepis exaltata*	<35 (air) <32 (soil)
Xanthomonas campestris		
pv. *hederae*	*Hedera helix*	20–30
pv. *malvacearum*	*Hibiscus rosa-sinensis*	24–33
pv. *pelargonii*	*Pelargonium* spp.	25–30
pv. *syngonii*	*Syngonium podophyllum*	26–30
pv. *zinniae*	*Zinnia elegans*	21–29
Unnamed pathovar	*Strelitzia reginae*	21–27

[a] Adapted from Chase (1991).
[b] Anastomosis group 4.

vary with different cultivars as well as different geographical areas and latitudes.

One of the principal pathogens of many greenhouse crops, *Botrytis cinerea*, has different temperature optima for its different growth processes (Table 26), all of which contribute to pathogenicity in the greenhouse. A similar disparity in temperature optima during pathogenesis was noted by Dimock and Baker (1951) in the case of snapdragon rust, caused by *Puccinia antirrhini* Dietel & Holw. on *Antirrhinum majus* L. Overall, the optimum temperature for disease expression was 21–24°C; uredospores germinated best at 10–13°C; many of the germ tubes aborted at 27°C; few infections survived 32°C; and aecidiospores were killed at 35°C. They also noted that pathogenesis was complicated by the presence of *Fusarium roseum* Link:Fr. At 10–16°C *F. roseum* invaded the rust pustules and reduced the inoculum of *P. antirrhini*, but at 21–32°C it moved through the pustules and into the snapdragon tissues.

It is evident from Table 26 that no one temperature can be identified as the one to avoid to effect disease escape. Further, the concept of L. D. Leach (1947), comparing the growth rates of host and pathogen, is of little use in this case, not least because the optimum temperatures for growth and for production may not be the same. In tomato crops in Ontario, for example, different optimum temperatures are recommended for different stages of crop growth and for day and night (I. Smith 1988). Most growers worry only about tomato productivity and not at all about pathogen biology.

Table 26. Optimum temperatures for growth phases of *Botrytis cinerea*

Growth phase	Optimum temperature (°C)	Reference
Mycelium growth	22	Jarvis (1977a)
Sporulation	15	Jarvis (1977a)
Spore germination	20	Hennebert and Gilles (1958)
Germ tube growth	30	Hennebert and Gilles (1958)
Appressorium formation	27–28	Morotchovski and Vitas (1939)
	15–20	Shiraishi et al (1970)
Sclerotium formation	11–13	Morotchovski and Vitas (1939)
Sclerotium germination	22–24	Morotchovski and Vitas (1939)

Nevertheless, the aggregate of all the processes of pathogenesis and of host defense mechanisms provides an empirical optimum temperature for the development of a disease. Therefore, an avenue of disease escape for the greenhouse manager is to grow the crop at a temperature as far as possible from that empirical optimúm, but consistent with productivity. Optimum temperatures for some greenhouse crop diseases are summarized in Tables 24 and 25. Comparison of data from various parts of North America and from different parts of the world shows that these optima cannot be generalized easily. Optimum temperatures in Florida are generally appreciably higher than in more northerly latitudes (J. P. Jones, personal communication). Greenhouse managers therefore have to be guided by local advisors and their own experience.

A useful way of effecting disease escape from root rots and basal stem lesions of cucumber and tomato (or, more correctly, of allowing diseased plants to survive them) is to take advantage of the tendency to produce adventitious roots in response to infection of cucumber by *Phomopsis sclerotioides* van Kesteren and infection of tomato by *Fusarium oxysporum* Schlechtend.:Fr. f. sp. *radicis-lycopersici* W. R. Jarvis & Shoemaker, *Didymella lycopersici* Kleb., and *Pyrenochaeta lycopersici* R. Schneider & Gerlach. These diseases are all more severe in cool soils (12–18° C). Last and Ebben (1966) and Jarvis et al (1983) recommended drawing aside any mulch material insulating the groundbed and mounding the base of the stem with a peat–soil mix to a height of about 15–20 cm. Dark in color, the mix warms up more quickly in the sun than the groundbed soil, and adventitious roots soon assume the function of the roots rotted off. Together with this procedure it is even more beneficial to warm irrigation water (Jarvis et al 1983). In soils known to be infested with *P. sclerotioides*, a difficult pathogen to eradicate, cucumbers are sometimes grown on ridges for early warming by insolation (Plate 21) or on straw bales (Plate 9). The microbial activity as the straw decomposes also adds to root-zone warming.

Managing Vapor Pressure Deficit

It has long been known that air circulation in the greenhouse and the ventilation of humid air to the outside is a primary control measure for leaf mold of tomato (caused by *Fulvia fulva* (Cooke) Cif.), Septoria leaf spot of tomato (caused by *Septoria lycopersici* Speg.) (Stair et al 1928), and gray mold (caused by *Botrytis cinerea*) of tomato (Bravenboer and Strijbosch 1975; Newhall 1928), geranium (Melchers 1926), *Saintpaulia* (Beck and Vaughn 1949), and *Stephanotis* (Tompkins and Hansen 1950).

Restricting ventilation for carbon dioxide enrichment can lead to an increase in the incidence of water-dependent pathogens, such as *B. cinerea.* Ghost spot, caused by *B. cinerea*, developed on the fruit of a tomato crop in such an environment (Anonymous 1967); in that case, the foliage was denser and had a higher sugar content than that of a comparable unenriched crop. There are a few possible explanations:

- A condensate and guttation could have persisted longer and more frequently, increasing the chances of infection.
- The dense foliage would certainly have restricted air movement and impeded evaporation.
- The observed higher sugar levels are in accord with the hypotheses of Horsfall and Dimond (1957) and Grainger (1962, 1968) that sugar-rich tissues are more susceptible to *B. cinerea.*
- The cuticle, like that of strawberry (Darrow and Waldo 1932), *Brassica oleracea* L. (E. A. Baker 1974), and tissue-cultured plants (Pierik 1988), may have been structurally defective at low vapor pressure deficits.

Whatever the explanation, carbon dioxide enrichment in a sealed environment has counterproductive risks.

Similar risks occur in plastic-covered greenhouses (Ferare and Goldsberry 1984a,b) and in normally leaky glasshouses insulated with plastic film for energy conservation (Watkinson 1975). These conditions, too, lead to high levels of infection by *B. cinerea*, and so it is necessary to ensure that ventilation is adequate to prevent the temperature from ever falling to the dew point. Early attempts to do that centered on ventilating when the relative humidity rose to a preset level. Winspear et al (1970) ventilated tomato crops by humidistat control at 75 or 90% RH in two temperature regimes: a constant 20°C and daytime and nighttime temperatures of 20 and 13°C, respectively. Both gray mold (caused by *B. cinerea*) and leaf mold (caused by *F. fulva*) decreased in incidence in the constant 20°C regime. Leaf mold incidence was 25% in the 20 and 13°C regime without humidity control, 2.8% when the house was ventilated at 90% RH, and absent when the house was ventilated at 75% RH. The incidence of ghost spot (caused by *B. cinerea*) on the fruit was 2.6% without ventilation, 1.6% with ventilation at 90% RH, and 0.2% with ventilation at 75% RH.

In addition to expelling masses of humid air by ventilation, heating the incoming replacement air is usually necessary. W. M. Morgan (1984a) maintained the night temperature of a tomato crop at a minimum of 16°C by heating, leaving the ventilators open from dusk to dawn. The incidence of ghost spot (caused by *B. cinerea*) was only 25% of that of a crop untreated with pesticides and unventilated at night. At a minimum night temperature of 13°C, however, even with the ventilators open, the incidence of ghost spot was twice that at 16°C. The vapor pressure deficit at 16°C was about

0.22 kPa in the ventilated house and only 0.12 kPa in the unventilated house, clearly low enough to allow the dew point to be reached frequently. Morgan (1984b) also found a considerable benefit in ventilating lettuce at night. The control of downy mildew (caused by *Bremia lactucae* Regel) was as good as that obtained by incorporating metalaxyl in the planting block. A heat-purge treatment, ventilating a mass of heated, humid air at dusk, increased the incidence of downy mildew. Morgan (1984b) considered continuous night ventilation to be the only effective control when legislation prevented the use of fungicides near harvest.

In a cucumber crop, van Steekelenburg (1984) set ventilation to begin by thermostat. In a heated crop (21° C during the day and 19° C at night) he found 0.5 to 3 times as many stem lesions caused by the water-dependent *Didymella bryoniae* (Auersw.) Rehm when the crop was ventilated at 26° C or above as when it was ventilated at 23° C or above, and internal fruit rot was two to three times higher at the 26° C set point. There was little difference in the incidence of gray mold (caused by *B. cinerea*) at the two ventilation set points. In this case, then, setting ventilators to open at a certain temperature was of dubious and haphazard value. Neither was there any great advantage in ventilating the headspace over the crop at a set rate of $12\ m^3 \cdot m^{-2} \cdot h^{-1}$.

A refinement of humidistat control was suggested by Bravenboer and Strijbosch (1975), who proposed that the electrical detection of water on a synthetic fruit could be used to switch on ventilation equipment. With computer control of the environment, however, accurate monitoring of a wet-and-dry thermistor can do the switching much more effectively before the dew point is reached.

The distribution of moving air masses within the crop is important in not allowing water drops and films on susceptible parts of plants (such as the petals of florist crops and pruning wounds on tomatoes and cucumbers) to persist long enough to permit infection. Whereas ventilation of the roof space may be needed to remove excess hot and humid air masses (Duncan and Walker 1973; Hanan et al 1978; Maher and O'Flaherty 1973; Mastalerz 1977; J. N. Walker and Duncan 1973a, 1974a), ventilation of susceptible parts of plants is essential to prevent the persistence of water (Augsburger and Powell 1986). For this purpose, air can be ducted among the lower stem areas of tomato plants (Plate 6), for example, and upward through potted plants on open-mesh or slatted benches (Hausbeck and Pennypacker 1987). This cannot be done on solid benches with subirrigation capillary systems, and in this case a horizontal airflow over the top of the bench is needed to prevent low vapor pressure deficits within the plant canopy. Plants with a dense canopy, such as seedlings of western hemlock (*Tsuga heterophylla* (Raf.) Sarg.), are very susceptible to *Botrytis cinerea* (Plate 22). Indeed, the method of training of row crops, row and plant spacing, and container spacing on the floor or on the bench are all important in determining the efficiency of crop ventilation and the evaporation of inoculum drops.

In diseases initiated in a water film on the surface of the plant, there is often a direct relationship between the period of time during which the plant remains wet and the success of infection. Such a relationship occurs in, for example, Alternaria blight of carnation (Strider 1978), gummy stem blight of cucumber (van Steekelenburg 1983, 1985a,b), gray mold of gladiolus (McClellan et al 1949a), and geranium rust (Spencer 1976). *Didymella bryoniae* achieves the first stage of penetration of cucumber in as little as 1 h at 23°C (van Steekelenburg 1985b), and *Botrytis cinerea* achieves it in *Exacum affine* I. B. Balf. ex Regel in about 4 h at 23°C (Ploetz and Engelhard 1979). Most other fungi probably take 5–8 h at their optimum temperature.

Therefore, if it is necessary to wet foliage for any reason, it is essential to maintain environmental conditions under which the foliage can dry out within a very short period.

There is no doubt that avoiding the dew point and proscribing overhead irrigation are vital in escaping infection from water-dependent canopy pathogens.

Managing Soil Water Potential

In contrast to the relative precision of temperature and atmospheric humidity as parameters of disease escape, there is a remarkable lack of precise information on water potentials to avoid in greenhouse crops. Obviously, avoiding waterlogging, which interferes with the oxygen supply to roots, constitutes disease escape in a general way. Similarly, drought is also to be avoided, as it impairs root function. However, there are few precise blueprint values for oxygen tension that permit disease escape, only generalizations for the most part.

In general, crops are grown in soils with constant optimum water potential for productivity. Most crops are predisposed to diseases by suboptimum high and low levels. Some diseases, however, are favored by the level of soil moisture that is optimum for the plant. One of them is bacterial canker of tomato, caused by *Clavibacter michiganensis* subsp. *michiganensis* (Smith) Davis et al (Kendrick and Walker 1948). Species in the *Pythium ultimum* group are generally most pathogenic in wet soils at 10–15°C, but *P. ultimum* Trow can infect plants in quite dry soils (Bouhot and Smith 1988).

In a mixture of sand, loam, and peat (1:1:1, v/v) at pH 7.0, with a moisture-holding capacity (mhc) of 56% (w/w), Bateman (1961) found that *Thielaviopsis basicola* (Berk. & Broome) Ferraris caused appreciable root rot of poinsettia when the soil water content was 36% of mhc. The damage increased as soil moisture increased to 70% of mhc. Root and stem rot caused by *Rhizoctonia solani* Kühn was most severe when the soil moisture was 40% of mhc, the severity decreasing as soil moisture increased to 80% of mhc. By contrast, *P. ultimum* caused only slight damage when soil moisture was 30–40% of mhc, which increased gradually as soil moisture increased to 70% of mhc. Damage became severe at levels greater than 70%.

Choosing a suitable soil water potential in which to grow poinsettia or indeed any other crop, therefore, depends largely on which pathogen is present. Raising soil moisture to escape Rhizoctonia root rot would provide more suitable conditions for Pythium and Thielaviopsis root rots. In cases like this, ensuring freedom from inoculum would be more appropriate than trying to manipulate soil water.

As a practical compromise, therefore, the water potential that is optimum for productivity would appear to be satisfactory for disease escape, too. The factors that determine irrigation demand in greenhouse crops, such as osmotic potential, radiation, nitrogen fertilizer, atmospheric vapor pressure deficit, substrate texture, and root-zone temperature (Papendick and Campbell 1975), can all be closely regulated.

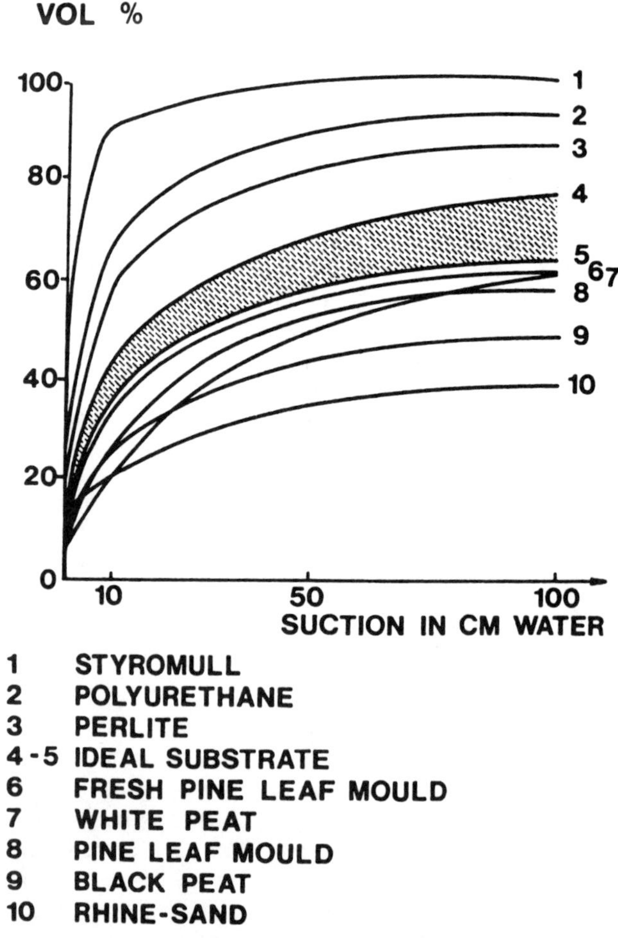

Figure 35. Water release curves for some potting media. Percentage of the volume of pore space occupied by water removable by suction pressure of 10–100 cm of water (approximately −1 to −10 kPa). Optimum water release to plants occurs in the shaded zone. Below 10 cm of suction pressure, an appreciable volume of air is considered to be present. Reprinted, by permission, from deBoodt and Verdonck (1972).

The water matric potential for a specific purpose can be selected from the known values of ψ_m of a number of potting mixes (Figure 35) (Bunt 1961, 1976; de Boodt and Verdonck 1972).

Automated tensiometers, with or without microprocessor integration of other environmental and cropping factors, can monitor and regulate irrigation demand continuously, eliminating the stressful wide fluctuations inherent in intermittent watering with a hose.

Managing Light

Near-UV light (320–380 nm) is routinely used to induce sporulation in cultures of *Botrytis cinerea* (Hite 1973) and other fungi (C. M. Leach 1962). Certain UV-absorbing plastic coverings for greenhouses absorb light at 340 nm and have been exploited to inhibit the sporulation of *Sclerotinia sclerotiorum* (Lib.) de Bary (Honda and Yunoki 1977) and of *Alternaria dauci* (Kühn) Groves & Skolko, *A. porri* (Ellis) Cif., *A. solani* Sorauer, and *B. squamosa* J. C. Walker (Sasaki et al 1985). In addition, sclerotium formation was inhibited in *B. squamosa* (Sasaki et al 1985). *Stemphyllium botryosum* Wallr. and *A. brassicae* (Berk.) Sacc. belong to another group of fungi controlled by filtering out UV radiation, to nullify sporulation induced by blue light (Sasaki et al 1985). Honda and Yunoki (1977) found 31% of flowers and fruits of cucumber and eggplant blighted by *S. sclerotiorum* and 13 stem lesions in a greenhouse covered by a UV-absorbing film, as against 85% of flowers and fruits blighted and 58 stem lesions in a house covered by conventional polyethylene film. Similarly, Sasaki et al (1985) obtained some control of Alternaria and Stemphyllium leaf spots of various vegetable crops and leaf blight of chinese chive caused by *B. squamosa* with a vinyl UV-absorbing film.

Peterson et al (1988) found that the mean intensity of light at wavelengths that inhibit sporulation of *B. cinerea* (430–490 nm) exceeded that at wavelengths inducing sporulation (300–420 nm) in greenhouses covered with fiber glass or polyethylene. Overall the polyethylene cover transmitted 16.5, 22.0, and 61.5% more light at 300–1,100, 430–490, and 300–420 nm, respectively. Conidia entered the greenhouses from outside sources, and thereafter airborne spore concentrations above benches of seedlings of Douglas fir (*Pseudotsuga menziesii* (Mirb.) Franco) were significantly higher in the fiber glass greenhouse. The difference was particularly marked on days when the seedlings were irrigated overhead, leading Peterson et al to conclude that the mechanical shock of impinging water drops liberate conidia. In multiple regression analysis, the environmental parameters that were significantly correlated with the concentration of airborne spores were within-canopy relative humidity and ambient above-canopy vapor pressure deficit. In addition, the combination of temperature and relative humidity 25 cm above the canopy of Douglas fir seedlings that they regarded as conducive to spore germination and infection (15–20°C and above 98% RH) persisted 14.5 times longer in the fiber glass house. In the fiber glass house, 40% of the seedlings were diseased; in the polyethylene house, 5% were diseased. The poor light transmission of the fiber glass promoted tall, succulent, and susceptible seedlings. The light intensity was even further reduced in such a canopy. This, in turn, promoted early senescence of needles, which provided an ideal substrate for *B. cinerea*. It seemed likely to Peterson et al that latent infections were established in this situation; the conditions conducive to infection occurred in July and August, but the symptoms did not appear until the autumn. A very complex set of conditions appeared to govern the incidence of gray mold: the covering material, light, humidity and temperature above and within the canopy, overhead irrigation, spore concentration outside the greenhouse, growth and senescence of the seedlings, and, probably, a quiescent phase in the disease cycle.

Jordan and Richmond (1972) investigated the effects of colored polyethylene films on strawberry production and on gray mold (caused by *B. cinerea*) and powdery mildew (caused by *Sphaerotheca macularis* (Wallr.: Fr.) Lind). Both diseases were more severe under colored than under clear plastic. Gray mold was the most severe under pink and blue plastic where vapor pressure deficits were the smallest, 0.411 and 0.638 kPa, respectively, compared with 1.139 kPa under clear plastic and 1.740 kPa under glass. There was therefore no advantage in colored plastics for the control of these two strawberry diseases.

Reuveni et al (1989) found that adding a UV-absorbing material (hydroxybenzophenone) to polyethylene film increased the ratio of blue

SUMMARY

Preventing Infection

The essence of disease escape is to manipulate the crop and its environment so that infection does not occur or, if it does occur and establishes a quiescent or latent phase, then to minimize its consequences. Various disease-predisposing environmental conditions can be avoided by the manipulation of such factors as temperature, light, water potential, and crop nutrition. Further, the conjunction of environmental factors that permit infection to occur, such as cold, wet, poorly aerated soil or the deposition of dew on leaves, can be avoided by proper regulation. In constructing a disease-escape strategy, it is useful to consider the autecology of a pathogen in all its phases in order to be able to interrupt as many of its vital pathways as possible. It is equally essential to understand the physiology of the crop so as to recognize suboptimum conditions that predispose it to disease.

There are in the literature reports of optimum temperatures for very many diseases. At best a single optimum value for a disease represents only an average temperature for all the processes of pathogenesis and host defense, and at worst it represents only the temperature at which symptoms develop fastest. In reality, temperature regulates a large num-

Table 27. Ratio of blue light to UV light transmitted and sporulation of *Botrytis cinerea* beneath polyethylene films treated with UV-absorbing hydroxybenzophenone and color filters[a]

Transmittance (%)			
Blue (480 nm)	UV (310 nm)	Blue–UV ratio	Relative sporulation
92.13	82.83	1.1:1	100.0
65.83	8.95	7.3:1	18.8
90.22	11.08	8.1:1	2.5
81.40	10.00	6.1:1	1.2
85.03	8.80	9.7:1	0.2

[a] Adapted from Reuveni et al (1989).

light to UV light transmitted and reduced the sporulation of *B. cinerea* in polystyrene petri dishes (Table 27). In a greenhouse, the treated polyethylene film reduced the number of infection sites on tomato and cucumber from 41 to 17 and from 29 to 15, respectively. This disease reduction, however, was not translated into yield increases, possibly because of the relatively low photosynthetically active radiation transmitted by this film. Evidently some compromise is necessary between light suppressing fungal sporulation and light optimum for yield.

ber of processes from sporulation to infection, host colonization, and host defense reactions, in the case of fungal diseases. In addition, the so-called optimum temperatures may differ between geographical areas.

Effecting disease escape by manipulating temperature requires that the component optimum temperatures of all the many processes involved in establishing a disease be recognized and counteracted at different stages of development of the pathogen and the crop. No generalized guidelines can be constructed for disease-escaping temperatures; each stage in each case has to be considered separately.

Grafting a scion of a susceptible but productive cultivar onto a resistant rootstock is a very effective form of disease escape commonly used against root pathogens, such as formae speciales of *Fusarium oxysporum, Verticillium* spp., and root-knot nematodes. Resistant rootstocks, however, are not resistant to all root pathogens, and so care must be exercised lest a second pathogen, to which the rootstock has no resistance, be invited. Pathogens such as tomato mosaic virus and various bacteria may also be spread on the knife and fingers in the process of grafting.

Manipulation of day length may serve as a disease-escape mechanism. Sironval (1951) found that growing *Begonia semperflorens* Link & Otto in long days enabled abundant adventitious roots to replace those lost to *Botrytis cinerea.*

Grafting

Grafting scions of a productive cultivar onto disease-resistant rootstocks has proved economical and effective as a disease-escape mechanism, particularly in vegetable production.

Bravenboer and Pet (1962) found a graft-compatible hybrid between *Lycopersicon esculentum* Mill. and *L. hirsutum* Humb. & Bonpl. that was resistant to corky root rot (caused by *Pyrenochaeta lycopersici*) and showed marked heterosis (hybrid vigor). When commercial cultivars were grafted onto a rootstock of this hybrid, they outyielded standard cultivars partly as a result of escape from disease and also because of heterosis.

Disease escape from Fusarium crown and root rot of tomato was achieved by Thorpe and Jarvis (1981) with the cultivar MR13 wedge-grafted on the Japanese breeding line IRB-301-31 (Yamakawa and Nagata 1975); by Thibodeau (1983) with the cultivars MR13 and Vendor on the resistant cultivar Larma; and by Kuniyasu and Yamakawa (1983) with the cultivars Walter and Fukuju 2 on the rootstock cultivars KNVF and KVF. These last rootstocks are routinely used in the Netherlands. N denotes resistance to the root-knot nematode *Meloidogyne incognita* (Kofoid & White) Chitwood; V, resistance to Verticillium wilt caused by *Verticillium dahliae* Kleb.; and F, resistance to Fusarium wilt caused by race 1 of *Fusarium oxysporum* f. sp. *lycopersici* (Sacc.) W. C. Snyder & H. N. Hans.

In cucumber production, rootstocks of *Cucurbita ficifolia* Bouché or *Benincasa hispida* (Thunb.) Cogn. (*B. cerifera* Savi) are used to escape Fusarium wilt (caused by *F. oxysporum* f. sp. *cucurbitacearum* Gerlagh & Blok). *B. hispida*, however, is susceptible to black root rot (caused by *Phomopsis sclerotioides*) and Verticillium wilt (caused by *V. albo-atrum* Reinke & Berthier).

In selecting a suitable rootstock for grafting, it is essential to know its susceptibility to other diseases. Tomato mosaic virus is a constant threat in tomato grafting, not least because of the risk of spreading the virus on the knife. Bravenboer and Pet (1962) found a line of *L. hirsutum* resistant to *Didymella lycopersici* but very susceptible to the root-knot nematode *M. incognita*. By crossing this line with lines of *L. esculentum* homozygous for resistance to the nematode, they obtained a rootstock resistant to the nematode and also to corky root rot. Similarly, resistance to *Verticillium* was added by crossing *Verticillium*-resistant lines of *L. esculentum* with *L. hirsutum.*

Plant Spacing and Habit

There has been very little study of plant spacing and pruning for disease escape in greenhouse crops and of ventilation for any purpose other than heat distribution. It is usually fortuitous that these arrangements may contribute to disease control. An open plant habit, which can confer disease escape in field crops such as beans, is disregarded in favor of the form of foliage plants or the productivity of vegetables.

There are some advocates of intercropping fruiting crops of tomato with young plants, to obtain a continuous yield (F. Richardson 1982). This presents enormous problems in the high density of roots and shoots and the perpetuation of inocula, should diseases break out. Richardson claimed that interplanting would reduce the likelihood of disease because tomato roots would not be in the soil long enough to become seriously affected by corky root rot (caused by *Pyrenochaeta lycopersici*); she thought that layering and training long-term crops by unskilled labor damaged the plants sufficiently to invite wound pathogens, such as *Botrytis cinerea*, and the layered stems lying on the ground would also form a microclimate favorable to *B. cinerea*. However, a doubled density of foliage in the first meter above the ground would appear to be a serious impediment to good air circulation and facilitate plant-to-plant disease spread. This would probably lead to more gray mold, not less, and would favor other stem and foliage pathogens, such as *Fulvia fulva* (causing leaf mold) and *Didymella lycopersici* (causing Didymella stem rot). There is, however, no experimental evidence one way or the other.

The horizontal and vertical spacing of plants is important in determining patterns of disease distribution (Burdon et al 1989). Patchy distribution occurs in the greenhouse, which Burdon et al ascribed to isolated sites of initial infection; patterns of pathogen dispersal, disease foci, and epidemic development; variations in the physical environment; genetically based variations in susceptibility; and ontogenetically based variations in susceptibility—the so-called windows of susceptibility—relating mostly to tissue age.

In order to prevent direct plant-to-plant spread of pathogens such as tomato mosaic virus or *Clavibacter michiganensis* subsp. *michiganensis* by contact, those such as *Didymella* spp. by water splash, or those such as *Pythium* spp. by movement through the soil, there would have to be an unrealistic spacing between plants (Burdon and Chilvers 1982; Thresh 1982). There is generally a direct relationship between close spacing and the incidence of disease, because of the availability of targets for pathogens and the reduced distance for their dispersal (Burdon and Chilvers 1982). At the same time, there are indirect effects of plant competition for light, water, and nutrients (P. C. Miller and Stoner 1979) and effects of the microenvironment and vector behavior. All these effects may combine and interact (Burdon and Chilvers 1982). In virus disease epidemiology in greenhouse crops, Thresh (1982) listed high temperature, rapid growth, close spacing, and frequent handling as important factors in plant-to-plant spread by contact. This last tends to be a more important means of spread than vector transmission.

Foliage density is also important in the epidemiology of the downy mildews. Splash dispersal occurs more easily if the target leaves are nearby, and the microclimate favors water-dependent sporulation, zoospore release, and infection (Palti and Cohen 1980).

There have long been recommendations to sow seed thinly to avoid damping-off (J. Johnson 1914). Burdon and Chilvers (1975a,b) rationalized these recommendations in studies on damping-off of seedlings of *Lepidium sativum* L., caused by *Pythium irregulare* Buisman. They established a negative correlation between the mean interplant distance and the rate of advance of a disease front and between that distance and the apparent infection rate *r* in a randomly inoculated plant stand (Figure 36) (Burdon and Chilvers 1975a). Varying the planting density and varying the density of applied inoculum had similar effects on the frequency of primary infection foci. In both cases, the spacing of the primary infection foci was proportional to the plant density, and a log-log plot of the data gave a linear regression with a slope of about 1 (Figure 37) (Burdon and Chilvers 1975b).

Huang and Hoes (1980) showed that sclerotia of *Sclerotinia sclerotiorum* in soil were the infective propagules in the development of Sclerotinia wilt of sunflowers (*Helianthus annuus* L.). The spatial distribution within the zone of lateral roots determined the distribution of hypocotyl infections. These constituted the primary infection loci; their incidence was highest when sclerotia were buried close to seeds, declining with increasing distance from the seed. Secondary disease spread occurred by root contact, so that plant spacing affected the efficiency and rapidity of spread.

Microclimate as affected by both spacing and habit determines the incidence of Botrytis blight in forest seedlings. The disease was severe in dense stands of Douglas fir seedlings (*Pseudotsuga menziesii*) when the relative humidity was consistently over 85% (Halber 1963), and the dense, convoluted growth of western hemlock seedlings (*Tsuga heterophylla*) makes

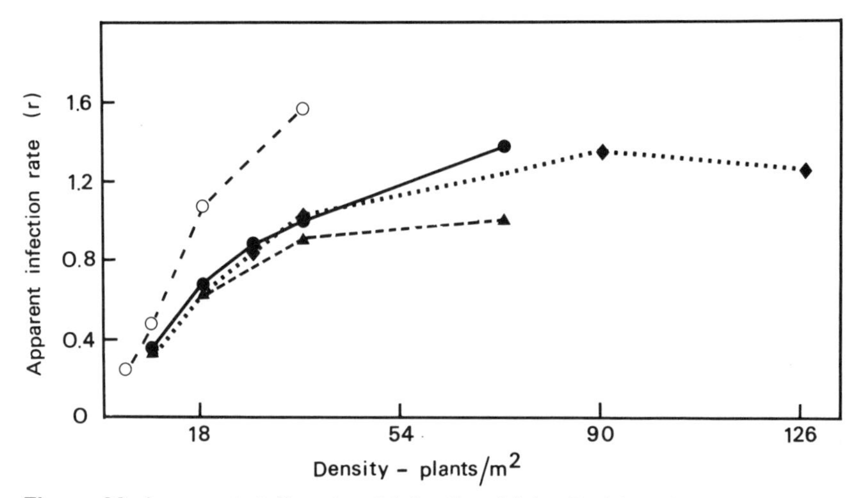

Figure 36. Apparent daily rate of infection (r) by *Pythium irregulare* in randomly inoculated plots of *Lepidium sativum* in relation to plant density in four experiments. Redrawn from Burdon and Chilvers (1975a).

them highly susceptible when grown on a bench without air movement through the bench (Plate 22) (Peterson et al 1988).

In Botrytis blight of *Exacum affine* (caused by *B. cinerea*), Trolinger and Strider (1984) found that infection of all floral parts (except the stamens) and leaves was enhanced when potted plants watered by overhead irrigation were crowded together on the bench until the leaves touched, compared with subirrigated plants with 5-cm gaps between them (Table 28).

English et al (1989) modified the microclimate around grape clusters by removing the leaves surrounding them and thereby significantly reduced the incidence and severity of gray mold (caused by *B. cinerea*) in the bunches. No single variable could be identified as the determining factor; squared canonical correlations for temperature, relative humidity, wind speed, and leaf wetness were less than 0.67 when each variable was considered singly, but they were generally greater than 0.58 when these variables were considered together. Wind speed was the parameter most affected by leaf removal; it was increased by a factor of three to four around exposed grape clusters, particularly in the afternoon and evening in a vineyard. It may be surmised that wind speed could be increased around the deleafed stems of cucumber and tomato growing in groundbeds, if the ventilation were correctly directed in that zone by means of perforated polyethylene ducting. English et al concluded that the impact of bunch rot in grapes may be related to interactions that distinguish canopies rather than to any single variable.

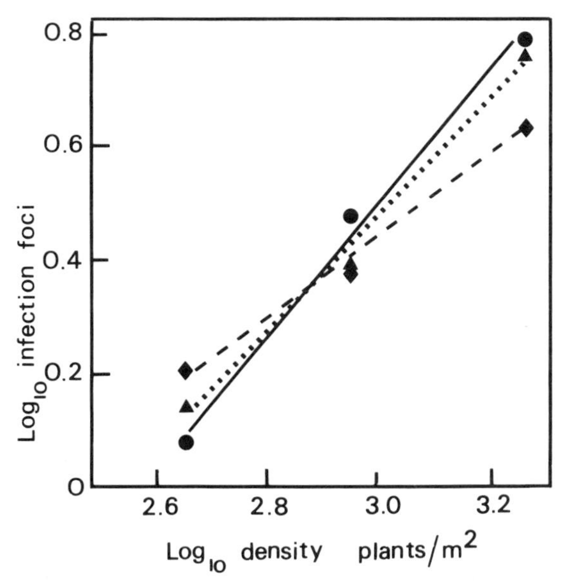

Figure 37. Regression of the number of successful infection foci of *Pythium irregulare* on the density of seedlings of *Lepidium sativum* (log-log transformation of data from three experiments). The slopes are 1.18, 1.02, and 0.76. Redrawn from Burdon and Chilvers (1975b).

Plant habit also has been shown to be important in allowing bean plants (*Phaseolus vulgaris* L.) to escape white mold (caused by *Sclerotinia sclerotiorum*) in the field (Blad et al 1978; Weiss et al 1980). The incidence and severity of white mold were significantly lower in the cultivars Aurora and Northern, which have an upright, open habit, than in the cultivar Tara, which has a luxuriant, dense habit. The disease was again worse in plots receiving overhead irrigation and where air, leaf, and soil temperatures were consistently lower. In lettuce, Abdel-Salem (1934) found gray mold (caused by *B. cinerea*) to be more prevalent in cabbage types, whose outer leaves spread to the ground, than in the upright cos types.

SUMMARY

Plant Spacing and Habit

There has been very little experimental work on the effects of plant spacing and habit on the incidence and severity of disease. Close spacing and dense habit would be expected to cause undue competition for nutrients, water, and light and so predispose the crop to infection. Patchy distribution of disease in the greenhouse may be attributed to isolated sites of initial infection; patterns of pathogen dispersal (often along rows), disease foci, and epidemic development; variations in the physical environment; genetic differences in susceptibility; and differences related to tissue age and crop development.

Some patterns of disease incidence and dispersal are related to the layout of greenhouses—concentration near doorways, where propagules spread through the air or fall from machinery; along pathways, where people and machinery move; along rows, where plants may be affected by routine daily operations; or along gutter rows, where dispersal by water splash from drips from the roof is characteristic of some bacterial diseases. In order to prevent plant-to-plant spread of pathogens by contact, vector, or water splash, there has to be an unrealistic, uneconomic spacing between plants. The microclimate of potential infection sites can be modified, however, by reasonable plant spacing and adequate ventilation with air at controlled temperatures and vapor pressure deficits. It can also be modified by avoiding overgenerous treatment with fertilizer, especially nitrogen, which promotes excessive vegetative growth. No breeding for disease-escaping open plant habit appears to have been done.

In general, crops grown with minimum stress from competition between plants and between fruit and foliage for light and nutrients are the least susceptible to diseases. Moreover, disease spread is generally slower under these conditions.

Table 28. Gray mold disease index in potted plants of *Exacum affine* crowded together or spaced apart[a]

	Spacing[b]	Disease index[c] Overhead irrigation	Subirrigation
Inoculated plants	Crowded	2.9 a	2.2 a
	Spaced	0.8 b	0.6 b
Uninoculated plants	Crowded	1.6 a	0.0 b
	Spaced	0.2 b	0.2 b

[a] Adapted from Trolinger and Strider (1984).
[b] Leaves of crowded plants were touching. Leaves of spaced plants were kept 5 cm apart.
[c] Disease index: 0 = no disease, 5 = plant dead. The plants were observed 14 days after inoculation. The data are means of 12 observations. Data in the same row followed by the same letter are not significantly different at $P = 0.05$.

In a similar way, mycelium of *S. sclerotiorum* is prolific beneath lettuce leaves lying on the ground, thus greatly facilitating infection and plant-to-plant spread. Unfortunately, few studies of this type appear to have been made in greenhouse crops.

Frequently, disease incidence is greatest near a doorway or unscreened ventilators, reflecting perhaps the entry of insect vectors or dustborne infective propagules. A higher incidence near pathways reflects dustborne propagules (Ebben and Spencer 1976) or perhaps plant damage by passing workers or sap transmission of viruses and bacteria. Dhanvantari and Dirks (1987) recorded a high incidence of bacterial stem rot of tomato in rows beneath the gutters connecting greenhouse bays, which they explained as the result of splash dispersal of *Erwinia carotovora* subsp. *carotovora* (Jones) Bergey et al. They also noted nonrandom, marked clumping of infected plants, which pointed to plant-to-plant spread during routine cultural operations. Dirks and Jarvis (1975) also presumed plant-to-plant spread of Fusarium crown and root rot of tomato to explain nonrandom chains of affected plants along rows, but in retrospect this distribution may have resulted from the planting pattern out of trays of latently infected transplants. This disease is also characterized by a higher incidence in doorways, probably resulting from foot- or machinery-borne or windblown inoculum (Rowe and Coplin 1976; Rowe et al 1977).

It is likely that, as in these cases, the probable source of inoculum and vectors and the presence of disease-predisposing microenvironments can be deduced from the patchy distribution of crops in greenhouses. Much of the patchiness is inherent in the close spacing of greenhouse crops and patterns of routine crop maintenance.

Cultural Operations

Moribund and dead plant tissue invites infection by necrotrophic pathogens, such as *Botrytis cinerea* and *Sclerotinia* spp. (Jarvis 1977a, 1980b;

Yarwood 1959a), and disease escape can often be effected simply by reducing damage to roots, stems, and foliage during cultural operations. A standard method of inoculating plants with *Verticillium* spp., for example, is to damage their roots before dipping them in a spore suspension. Therefore care must be taken not to damage roots during pricking-off, transplanting, and repotting operations. *B. cinerea* can often be found on broken cotyledons and pinch bruises on seedling stems (Plate 23) and on wounds made by careless pruning where the outer stem tissues have been stripped off by breaking or pulling. Dead flowers and leaves should be removed before they become a massive saprophytic base for inoculum (Beck and Vaughn 1949), and senescent lower leaves of tomato should also be removed, because they are an inoculum base for *B. cinerea*. The stem bases of both tomatoes and cucumbers are easier to ventilate with these lower leaves removed (Plate 24). Pruning should always be done with a sharp knife, to leave no snags. Similarly, cucumber fruits should be cut off, leaving a 2-mm stalk on the fruit, which is much less susceptible to *Didymella bryoniae* than a torn shoulder left after pulling (Plate 25) (van Steekelenburg 1986). Old leaves of cucumber become very susceptible to *D. bryoniae* when wounded and constitute an important inoculum base; they should be removed. Similarly, old flowers of cucurbits should be removed in areas where Choanephora blight (caused by *Choanephora cucurbitarum* (Berk. & Ravenel) Thaxt.) is endemic.

In order to promote the drying of tomato leaf scars after deleafing, leaves should be removed on bright, sunny days, and the plants should be irrigated early in the day rather than late. Tomato leaf scars heal rather slowly, and Ark (1944) showed them to be susceptible to *Clavibacter michiganensis* subsp. *michiganensis* for 72 h but resistant after 96 h. Verhoeff (1967) found no cambial activity in the petiole stump at the fifth node of 100-cm-tall tomato stems 3 days after deleafing. On the fifth day there was a cambium three to five cells thick on the pith and cortex parenchyma only, and the cambium was complete in 8 or 9 days. In 35-cm-tall plants, the cambium was complete in 7 days. A. R. Wilson (unpublished observation) found that the large vessels were never sealed and remained open to a type of infection by *B. cinerea* (Wilson 1963).

Verhoeff (1967) noticed that an abscission layer formed in unusually long petiole stubs inoculated at the distal end with *B. cinerea*, and they fell off to leave a resistant scar. This constitutes disease escape, but it is of dubious value to the grower.

Howard and Horsfall (1959) removed fruits and succulent cane tips of rose bushes to prevent the growth of *B. cinerea* through the pith and into the crown.

Overhead Irrigation

There are occasions when crops may be sprayed overhead for irrigation and the application of foliar nutrients and pesticides, but spraying to relieve plant stress by cooling and raising the humidity has little to recommend it except low capital outlay. Spray drops are relatively large (20–200 μm or more in diameter), and when impelled through nozzles by water pressures on the order of 500 kPa, they can damage delicate floral structures, leaf hairs, and so on. Sufficiently forceful deposition of water may break off trichomes to allow the entry of such pathogens as *Clavibacter michiganensis* subsp. *michiganensis* into tomato tissue (Kontaxis 1962) and *Botrytis cinerea* into antirrhinum stems (McWhorter 1939). Large spray drops take longer than mist drops to evaporate, and if they persist for more than a very few hours, most contained fungal spores readily germinate, often stimulated by the exosmosis of nutrients into the inoculum drop (W. Brown 1922).

Rotem and Palti (1969) showed there is often a direct relation between the method of irrigation and the severity of diseases, particularly the downy mildews. Palti and Cohen (1980) discussed the role of water in the biology of downy mildew of cucumber, caused by *Pseudoperonospora cubensis* (Berk. & M. A. Curtis) Rostovzev. Infection can occur in dew, and thereafter not only is water essential for sporulation, the liberation of zoospores, and infection, but dispersal depends very largely on irrigation. Rotem and Cohen (1966) also demonstrated a direct relationship between the method of irrigation of tomato crops and the severity of the foliage diseases caused by *Stemphyllium botryosum* and *Xanthomonas vesicatoria* (Doidge) Dows. Subirrigated crops yielded 30–50% more fruit than overhead-irrigated crops. A similar relationship was established in bacterial canker of tomato (W. P. C. Smith and Goss 1946).

Many other organisms are dispersed in overhead irrigation, including *Didymella lycopersici* (Verhoeff 1963) and *Clavibacter michiganensis* subsp.

SUMMARY

Cultural Operations

Many greenhouse crops receive intensive and daily attention in such operations as irrigating, applying pesticides, deleafing, trimming, tying, and harvesting, and most of these operations are potentially predisposing to disease unless carefully done. High-pressure watering with overly cold water stresses seedlings, and careless pricking off and transplanting damages both roots and tender stems, leaving the roots susceptible to pathogens such as *Pythium* and *Verticillium* spp. and the stems susceptible to necrotrophic pathogens, such as *Botrytis cinerea*. Pesticide sprays applied at high pressure break off trichomes and permit the entry of pathogens, and since pesticides are rarely completely effective or, as in the case of bacterial diseases, are mostly ineffective, the disease may be exacerbated rather than controlled. Any routine task that leaves necrotic tissue or stresses the plant is best left to skilled labor, to minimize the risks of infection by necrotrophic pathogens.

michiganensis (Strider 1969a), both in tomato. Conidia of *Botrytis tulipae* (Lib.) Lind are splash-dispersed among tulips by water drips from the greenhouse roof (Beaumont et al 1936).

Soil splash from overhead irrigation readily disperses such fungal pathogens as *Rhizoctonia solani, Pythium* spp., and *Erwinia* spp. (Augsburger and Powell 1986), and seedlings and transplants in flats on bare ground should never be watered by hose (Plate 26).

Pesticide sprays can be as bad as overhead irrigation in enhancing the pathogenicity of nontarget pesticide-tolerant organisms. Strider (1969a) showed that *Clavibacter michiganensis* subsp. *michiganensis* not only survived in maneb (Manzate), tribasic copper sulfate, and anilazine (Dyrene) but also could infect tomato plants when it was present in pesticide deposits. The insecticides acephate and oxamyl reduced the severity of Alternaria leaf spot (caused by *Alternaria panax* Whetzel in Whetzel & Rosenb.) of schefflera (*Brassaia actinophylla* Endl.), oxamyl about as well as some fungicides (Osborne and Chase 1984), and an insecticidal soap containing potassium salts of fatty acids also reduced the severity of that disease and Bipolaris leaf spot (caused by *Bipolaris setariae* (Sawada) Shoemaker) of areca palm (*Chrysalidocarpus lutescens* H. Wendl.). By contrast, the insecticidal soap significantly increased the severity of Fusarium leaf spot (caused by *Fusarium moniliforme* J. Sheld.) of *Dracaena marginata* Lam., as well as Myrothecium leaf spot (caused by *Myrothecium roridum* Tode:Fr.) of *Dieffenbachia maculata* (Lodd.) G. Don (Chase and Osborne 1983). Because both *F. moniliforme* and *M. roridum* were suppressed in culture by the soap, Chase and Osborne thought that its effect was on the respective hosts, although neither showed phytotoxicity. These and similar results point to the need for a better understanding of the ecology of pathogens on plants in the presence of pesticides.

There is a common belief that wetting plants in full sunlight causes damage and necrosis to delicate tissues; drops of water lying on a surface are thought to act as tiny lenses focusing radiant energy on the surface. Kramer (1939) examined this hypothesis, however, and could find no supporting evidence. The focal point of a 1-mm drop lay about 1.5 mm beyond the surface, and the leaf temperature immediately below a drop was 4–12°C lower than that of the surrounding surface.

If crops must be irrigated with water or sprayed with pesticides from overhead, then the droplet size must be made as small as possible, and the water must be warmed, to avoid a temperature shock to the plants and to avoid lowering the temperature of the substrate. The crops should be sprayed as gently as possible and only when surface water will evaporate within an hour.

Fertilizers, pH, and Salinity

Inorganic fertilizer amendments and pH adjustments in the rhizosphere affect not only the productivity of greenhouse crops but also their susceptibility to diseases of both roots and shoots, through stress imposed on both the pathogen and the host. There are a large number of practical recommendations for fertilizing crops for disease escape (Engelhard 1989).

In general, increasing the nitrogen supply to the plant results in increased susceptibility to *Botrytis cinerea* and incidence of gray mold, not least because the usual increased leafiness makes for a more humid microclimate, favorable for the germination of conidia and infection. A typical example was described by Hobbs and Waters (1964). The incidence of gray mold in chrysanthemum flowers cut from plots receiving nitrogen at 1.5, 3.8, or 6.0 $g \cdot m^{-2}$ increased in a quadratic manner with increasing nitrogen. Neither potash (K_2O) nor the potassium–nitrogen ratio had significant effects in this case. Hobbs and Waters were unsure about the effect of the increased nitrogen—whether it improved the nutrient status of the chrysanthemum tissue available to *B. cinerea*, whether it made the plants more susceptible because of increased succulence, or both.

Chase and Poole (1985a) provided a similar example in the leaf spot of *Dieffenbachia maculata* caused by *Myrothecium roridum*. The number of leaf spots rose linearly (in four of five tests) as the amount of slow-release nitrogen fertilizer was increased from 112 to 896 $g \cdot m^{-2}$.

The literature on the effects of different forms of nitrogen on Verticillium wilts is confused and often difficult to interpret (Pennypacker 1989), probably because of complicated interactions with other nutrition factors. Generally, reduced rates of nitrogen give increased resistance, depending on the host. Ammonium nitrogen increases resistance in antirrhinum and possibly tomato, however.

Nitrate nitrogen fertilizer together with liming, and particularly along with chemotherapy, has given outstanding control of Fusarium wilts of several crops (J. P. Jones et al 1989). For example, Woltz and Engelhard (1973) obtained complete control of the Fusarium wilts of chrysanthemum and aster with a regime of nitrate nitrogen, lime, and benomyl.

Verhoeff (1965, 1968) examined the interactions of nitrogen, phosphorus, potassium, magnesium, and lime as soil amendments and their effect on tomato gray mold. For a given level of nitrogen, higher rates of potassium decreased the rate of lesion extension in tomato internodes. This potassium–nitrogen ratio in tomato nutrition was also shown to be important in bacterial stem rot, caused mainly by *Erwinia carotovora* subsp. *carotovora* (B. N. Dhanvantari, personal communication). The incidence of the disease was very low at a potassium–nitrogen ratio of 4:1 and increased at ratios of 2:1 and 1:1.

Calcium generally enhances resistance, because of its role in the integrity of the cell wall. In general, high or adequate calcium levels, with the usual accompanying relatively high pH from liming, are associated with reductions in the severity of Fusarium wilts of carnation, chrysanthemum, cucumber, gladiolus, tomato, and several other crops (J. P. Jones et al 1989). Stall (1963) and Stall et al (1965) similarly found a low incidence of gray mold in tomatoes at high calcium levels when the calcium–potassium ratio was relatively high. Plants with a low ratio at either low or high rates of calcium and potassium were equally or more susceptible.

In practical horticultural terms, the salinity of the soil or a hydroponic

Table 29. Salinity levels reducing crop yields[a]

Crop	Electrical conductivity ($mS \cdot cm^{-1}$) at which yield is reduced by:		
	10%	20%	50%
Spinach	5.5	7.0	8.0
Tomato	4.0	6.5	8.0
Lettuce	2.0	3.0	5.0
Bell pepper	2.0	3.0	5.0
Green bean	1.5	2.0	3.5

[a] Adapted from Bernstein (1964).

solution is measured by electrical conductivity (EC), expressed in siemens per centimeter ($S \cdot cm^{-1}$). Optimum EC values for vegetable production in greenhouses are about 2.0–3.5 $mS \cdot cm^{-1}$. Roots are damaged at values of 3.5–5.0 $mS \cdot cm^{-1}$ or higher: salinity at levels associated with high EC kills root tips and root hairs and impedes the uptake of water and nutrients (Bernstein 1964; I. Smith 1988; van den Ende et al 1975). Salinity levels at which some crop yields are reduced are given in Table 29.

In an inert medium, such as rock wool, crops generally tolerate a somewhat higher EC, on the order of 3.0 $mS \cdot cm^{-1}$, than in soil. For tomatoes in nutrient film systems, an EC of 2.2 $mS \cdot cm^{-1}$ is recommended (I. Smith 1988). For cucumbers, values of 1.5–1.8 $mS \cdot cm^{-1}$ are recommended for soil and up to 3.0 $mS \cdot cm^{-1}$ for nutrient film systems.

Growers manipulate EC as a powerful tool for controlling growth. Growth is retarded and "hard" plants are produced at high EC, particularly when it is combined with restrained irrigation, lowered temperature, and a reduced nitrogen supply or an increased potassium–nitrogen ratio (I. Smith 1988). This tool, however, has to be used with caution, since damage to root tips and root hairs encourages infection by necrotrophic pathogens.

Beach (1949) noted that *Rhizoctonia solani* and *Fusarium oxysporum* f. sp. *lycopersici* (Sacc.) W. C. Snyder & H. N. Hans. could grow in Knop's solutions up to 16 times normal strength (with an EC of 30.4 $mS \cdot cm^{-1}$), which seriously impaired the germination of cucumber and tomato seeds and restricted seedling growth. *Pythium ultimum* did not grow well at 4.3 $mS \cdot cm^{-1}$, and infection of cucumber and tomato by *P. ultimum* declined with increasing EC. By contrast, infection of cucumber and tomato by *R. solani* increased with increasing EC, as did infection of tomato by *F. oxysporum* f. sp. *lycopersici*. High rates of fertilizer and salts also increase the severity of bacterial canker (caused by *Clavibacter michiganensis* subsp. *michiganensis*) of tomato (Kendrick and Walker 1948).

When tomato roots were stressed by high salt levels (50 $meq \cdot L^{-1}$) before inoculation with *Phytophthora parasitica* Dastur, severe root rot developed, but when they were stressed after inoculation, there was less root rot (van Bruggen and Bouchibi 1989). The higher salt concentration and higher sodium–calcium ratio caused zoospores of *P. parasitica* to burst or encyst.

High EC can also affect diseases of ornamental foliage plants. In *Schefflera arboricola* Hayata inoculated with the leaf-spotting bacterium *Pseudomonas cichorii* (Swingle) Stapp, the number and size of lesions decreased as the amount of complete fertilizer added to the potting mix was increased to levels up to six times the commercially recommended amount (Chase and Jones 1986), and the fertilizer rate did not affect plant height, color, number of leaves, or fresh weight of the tops. Lesion number and size also decreased as leaf age increased from the top to the bottom of the plant. Light levels did not affect the disease. In *Ficus lyrata* Warb. attacked by the same pathogen, the percentage of leaf area affected was correlated with the amount of fertilizer (19:2.6:10, NPK) between 4 and 16 g per 12.5-cm pot (Chase 1988), about the optimum level for plant growth.

Optimizing EC for foliage quality in *Pilea spruceana* Wedd. similarly brought an increase in the severity of leaf spot caused by *Xanthomonas campestris* (Pammel) Dowson (Chase 1989). Chase counseled growers to rely on pathogen exclusion and preventive applications of bactericides. In *Syngonium podophyllum* Schott attacked by *X. campestris* pv. *syngonii* Dickey & Zumoff, the percentage of blighted leaf area decreased linearly as the fertilizer rate was increased, and so growers can be advised to increase the rate of complete fertilizer or phosphorus or potassium in moderate amounts.

It is evident that no general practical recommendations can be made for controlling diseases by adjusting the fertilizer levels supplied to plants. Each host–pathogen combination reacts differently.

SUMMARY

Fertilizers, pH, and Salinity

Every crop species and cultivar requires a special fertilizer regime in order to obtain maximum productivity and to prevent stress on the plant. Fertilizer requirements change as the crop ages from seeding to harvest. In general, excessive nitrogen leads to excessive foliage that is intrinsically more succulent and susceptible to damage and necrotrophic pathogens. Nitrogen generally has to be balanced with potassium; for many diseases, susceptibility decreases as the potassium–nitrogen ratio increases. The form of nitrogen, nitrate or ammonium, is also important. Nitrate nitrogen together with soil liming, and especially with benomyl treatment, is particularly useful in controlling the Fusarium wilt diseases. Calcium generally enhances resistance, with its role in the integrity of the cell wall. The overall concentration of soluble salts in the root zone has marked effects on water and salt uptake. Too high a concentration, as measured by high electrical conductivity, damages the root tissue, impedes its physiological activity, and makes it more susceptible to pathogens. Salt concentrations in the root zone may also affect the incidence and severity of some shoot pathogens.

Quiescent Infections

A quiescent infection is one in which the pathogen has ceased, either temporarily or permanently, to be aggressive. Symptoms may be visible or not, and in the latter case the infection is often referred to as latent, meaning hidden.

Latency has a number of shades of meaning in the literature. Gäumann (1951) defined latency as a quiescent or dormant parasitic relationship, which after a time can change into an active one. Verhoeff (1974) accepted this definition in his review of latent infections by fungi, and he excluded cases in which a spore has alighted on a host but has not germinated, in which an appressorium with a penetration peg has formed, or in which infection has occurred and vigorous invasion of the host is occurring without symptoms. This symptomless invasion occurs during the growth of hyphae of *Botrytis aclada* Fresen. in onion leaves (Tichelaar 1967) and of *B. narcissicola* Kleb. in narcissus flower stalks (W. R. Jarvis, unpublished observation).

Vanderplank (1963), discussing epidemiology, defined the latent period as the time required for newly infected tissue to become infectious, by the production of infective propagules. However, tissue-to-tissue hyphal growth of a fungus such as *B. cinerea* (for example, between two fruits touching) makes this definition sometimes difficult to interpret.

For bacterial infections, Hayward (1974) said that Gäumann's definition of latency (Gäumann 1951) depended on the sensitivity of the physiological or histological techniques by which the state of the host is assessed. The distinction between internal and external bacteria and the part they play in any subsequent aggression is not clear.

A somewhat different shade of meaning is given to latency in describing infections that remain quiescent for long periods. Infections by *B. cinerea* may remain quiescent for 3–5 wk in strawberry flowers (Jarvis 1962d; Powelson 1960) and 10–12 wk in tomato stems (Wilson 1963). It is these infections that have important implications in disease control (Hayward 1974; Jarvis 1965, 1977a, 1980b, 1985; Verhoeff 1974, 1980). They are particularly important in bacterial soft rots (Hayward 1974; Samish and Dimant 1959; Samish and Etinger-Tulczynska 1963) and diseases caused by *Botrytis* and *Colletotrichum* spp., symptoms of which may not appear until after harvest (Verhoeff 1974).

Quiescence is a temporary dynamic equilibrium between the pathogen's infective processes and the host's defenses. The pathogen can also be held quiescent by antagonistic microorganisms on the host surface (Blakeman 1980), although Verhoeff (1974) did not recognize this as a latent infection. Quiescence can occur in almost any stage of pathogenesis (Table 30), and the quiescent pathogen remains demonstrably viable. Indeed, quiescent infections need not be latent in the sense of hidden. Chocolate spot of *Vicia faba* L., caused by *Botrytis fabae* Sardiña, and brown speck of banana fruit, caused by *Colletotrichum musae* (Berk. & M. A. Curtis) Arx, are examples of this type. Like symptomless infections, they can become aggressive under suitable conditions. Ghost spot of tomato fruit, caused by *B. cinerea*, is an example of a quiescent infection that rarely, if ever, becomes aggressive (Verhoeff 1970). Most petal-flecking diseases caused by *Botrytis* spp., however, can become aggressive (Plate 27) (Jarvis 1977a).

Quiescence has two important implications for disease control and escape strategy. The first is that prophylaxis has to be applied long before symptoms appear, sometimes weeks before, although most growers act only when symptoms develop. Recognizing the conditions under which infection is likely to occur is important in the correct timing of prophylaxis.

The second implication is that when infection has occurred, the quiescent period can be extended to a point at which the aggressive state is never triggered or is triggered too late to affect yield. The factors triggering this change are not well understood (Hayward 1974; Jarvis 1977a; Verhoeff 1974) but may include temperature, water potential of the host tissues, osmotic relations, the state of pectic materials in the host cell walls, enzyme activation, and chemical changes associated with ripening.

Wilson (1963) described the mechanism whereby *B. cinerea* establishes quiescent infections in tomato stems (Figure 25) and showed that the onset of aggression could be delayed by refraining from overwatering (Wilson 1964). In plots of quiescently infected tomatoes, heavy watering resulted in the early appearance of symptoms in most plants, whereas watering only enough to maintain growth delayed symptoms, and lesions developed in fewer of the plants that were not overwatered.

Table 30. Examples of quiescent infection by bacteria and fungi, in order of increasing symptom expression

Parasite–host relationship	Parasite	Host	Reference
Epiphytic bacteria	*Pseudomonas syringae* pv. *tomato*	Tomato	Ercolani and Casolari (1966)
Epiphytic fungal conidia	*Botrytis cinerea*	Rye, strawberry	Kerling (1964)
Internal bacteria	Several bacteria	Tomato fruit	Samish and Etinger-Tulczynska (1963)
Internal bacteria	Several bacteria	Cucumber fruit	Samish and Dimant (1959)
Conidia in petiole vessels	*B. cinerea*	Tomato petiole stubs	Wilson (1963)
Internal bacteria	*Xanthomonas campestris* pv. *vesicatoria*	Tomato fruit	Ercolani and Casolari (1966)
Internal pectolytic bacteria	*Pseudomonas, Bacillus, Xanthomonas*	Cucumber fruit	Meneley and Stanghellini (1972)
Internal bacteria	Unidentified	Tomato fruit	Stall and Hall (1969)
Petal flecking	*Botrytis narcissicola*	Narcissus	W. R. Jarvis (unpublished observation)
Petal flecking	*B. cinerea*	Cyclamen	Wenzl (1938a)
Petal flecking	*B. cinerea*	Rose	Wenzl (1938b)
Ghost spot	*B. cinerea*	Tomato fruit	Verhoeff (1970)
Petal spot	*Botrytis tulipae*	Tulip leaves	D. Price (1970)
Root rot	*Colletotrichum coccodes*	Tomato roots	Schneider et al (1978)

Verhoeff (1965) showed that in quiescent infection of tomato by *B. cinerea*, the physiological age of the stem tissue affects the rate of mycelial growth. The mycelium grew faster in lower, older internodes than in upper, younger ones. This rate of growth and also the time and frequency of appearance of symptoms in plants growing in the groundbed were affected by the level of nitrogen fertilizer (Verhoeff 1968). Infected plants had more lesions, and the lesions developed earlier, with nitrogen applied at the rate of 45 $mg \cdot kg^{-1}$ of soil than at 165 $mg \cdot kg^{-1}$. Verhoeff (1968) thought that the higher level of nitrogen might delay senescence in the tomato stems.

Colletotrichum coccodes (Wallr.) S. J. Hughes is another fungus latent in tomato. Schneider et al (1978) noted that symptoms in a hydroponic crop in beds of loose gravel appeared several weeks after infection. Since the conidia of *C. coccodes* are waterborne, they recommended multiple nutrient tanks, to confine the disease to only a part of the hydroponic system.

Since quiescent infections, established preharvest, may not be triggered to become aggressive until produce is in storage or being marketed, it is imperative to harvest before the produce becomes overmature and to ensure that storage and shipping conditions maintain the infection in a quiescent state. This mostly means that the temperature should be as low as possible without lowering produce quality. Fruit and flowers and leafy vegetables

SUMMARY

Quiescent Infections

Infections that enter an apparently quiescent state, either with or without macroscopic symptoms, pose special problems for disease control. Symptoms may be delayed for several days or weeks, as in the case of tomato stems infected by *Botrytis cinerea*. Fungicides traditionally applied when symptoms appear are far too late to be effective against the superficial spore and its exposed germ tube; this type of prophylaxis must be imposed when environmental conditions and the presence of the pathogen and susceptible host tissue make initial infection imminent. Quiescent or latent infections are common in diseases caused by *Botrytis* and *Colletotrichum* spp., and some bacteria are also quiescent in host tissue, becoming aggressive or multiplying only when some (usually poorly defined) physical or physiological change occurs in the host tissue. It is believed that overwatering, excessive nitrogen fertilization, or exposure to ethylene may hasten the onset of the aggressive state of *B. cinerea*, and so one method of control is to water and fertilize in moderation and prevent the exposure of plants and harvested produce to ethylene. The quiescent state is thus prolonged to a point at which yield is not affected, despite the tissue's being infected.

should be harvested early in the day, before their temperature rises, and the greenhouse heat should be removed promptly from cucumbers and tomatoes, to cool them to 12 or 13°C.

Flowers and fruit should never be stored with heavy producers of ethylene, such as apples. Ethylene can enhance the pathogenicity of *B. cinerea* in carnations (W. H. Smith et al 1964), and infected flowers add to ethylene production (Nichols 1966). In the presence of ethylene, *B. cinerea* also becomes more aggressive on rose flowers (Elad 1988a,b), and cucumber fruit becomes more susceptible to flower-end infection by *Didymella bryoniae*. Further, careless handling causes wounds and stimulates ethylene production; in tomatoes this enhances infection by *B. cinerea* and *Geotrichum candidum* Link (Barkai-Golan et al 1989). All stores and packing sheds should therefore handle only one type of produce at a time, and they should be cool and adequately ventilated.

Resistant Germ Plasm

CHAPTER 8

Genetic resistance of plants to pathogens is the prime means of raising disease-free crops, or, to put it in a more correct way, breeders look for nonsusceptible plants. Plants are nonhosts to the majority of pathogens; they are susceptible only to relatively few (Vanderplank 1984).

Flor (1942, 1971) put forward a gene-for-gene hypothesis to explain the specific pairing of host and parasite, first that of *Linum usitatissimum* L. and *Melampsora lini* (Ehrenb.) Desmaz.: for every gene that conditions a reaction in the host, there is a corresponding and specific gene that conditions pathogenicity in the parasite. This is well illustrated in the case of downy mildew of lettuce, caused by *Bremia lactucae* Regel, in which pathogenicity is controlled by independent single loci, with nonpathogenicity dominant over pathogenicity (Crute and Dixon 1981; Farrara et al 1987; Norwood and Crute 1983).

The case for adequate genetic analyses of both host and parasite has been stated by Crute (1984), who also pointed out the pitfalls of inadequate analyses where close linkage of genes is involved.

Vanderplank (1984, 1989) has argued that gene-for-gene specificity lies in susceptibility, not in resistance, which is wholly unspecific. For example, in poinsettia and hundreds of other hosts, the genes conferring resistance to *B. lactucae* are identified individually and specifically, but the resistance they impart is unspecific. They are recognized by the corresponding virulence allele, not an avirulence allele, in the pathogen. Vanderplank (1984) modified Flor's hypothesis (Flor 1942) to state that gene-for-gene recognition is, in fact, protein-for-protein recognition, so that specific susceptibility, or mutual recognition in a compatible host–parasite combination, implies a specific molecular association. It requires a specific inducer produced by the pathogen and a specific receptor in the host.

There are two bases for resistance—horizontal and vertical—which are defined in terms of interactions of host and parasite; as a corollary, there are two bases for pathogenicity—aggressiveness and virulence, respectively (Vanderplank 1984). Horizontal resistance is spread evenly against all races of a pathogen, and races that do not interact differentially with host cultivars differ in aggressiveness. Isolates of *Fusarium oxysporum* f. sp. *lycopersici* that differ only in the degree of reaction against the same tomato cultivars differ in aggressiveness. Differences between races are quantitative rather than qualitative, with aggressiveness varying continuously. On the other hand, if a cultivar is more resistant to some races than others, the resistance is vertical, and races that interact differentially and clearly with host cultivars differ in virulence. Resistance in any particular host can be a mixture of horizontal and vertical resistance in any proportion, and aggressiveness and virulence can be mixed in any proportion in the pathogen. The inheritance of virulence is most likely oligogenic in pathogens that have gene-for-gene relationships, whereas the inheritance of aggressiveness is generally polygenic.

Although resistance may be dominant or recessive (or susceptibility recessive or dominant), environmental variance can impose powerful effects on host–parasite genetic relationships. Cirulli and Ciccarese (1975) reported that the tomato gene *Tm-1* for resistance to tomato mosaic virus behaved as a dominant gene against 53% of isolates of the virus at 17°C but against none at 30°C. The gene *Tm-2* was dominant against 80% of isolates at 17°C but against 34% at 30°C. Temperature also determined the expression of resistance of cucumber cultivars to *Cladosporium cucumerinum* Ellis & Arth. (Walker 1950). Below 17°C all material was susceptible, but above 21°C lesion development was rapidly impeded by a cicatrizing tissue reaction.

Genes do not always segregate independently but are linked to various degrees. Gene action is additive in the resistance of *Gerbera jamesonii* H. Bolus ex J. D. Hook. to *Phytophthora cryptogea* Pethybr. & Lafferty (Sparnaaij et al 1975). The resistance of cucumber to cucumber mosaic virus (CMV) is determined by three major genes and possibly some modifiers as well (Sitterly 1972).

Resistance, in a loose sense, does not necessarily reside in the intrinsic resistance determined by direct protein-for-protein relationships, but it may reside instead in a plant habit that confers disease escape or in insect repellency that prevents virus infection. Disease escape from CMV can

Table 31. Some pathogens existing in cultivar-specific pathogenic races

Host	Pathogen	Reference
Carnation	*Fusarium oxysporum* f. sp. *dianthi*	Demmink et al (1989)
Cucumber	*Glomerella orbiculare*	S. F. Jenkins et al (1964)
	Erysiphe cichoracearum	Kooistra (1968)
	Sphaerotheca fuliginea	Kooistra (1968)
Cucurbits	Cucumber mosaic virus	Devergne and Cardin (1970)
Rose	*Diplocarpon rosae*	Bolton and Svejda (1979)
	Sphaerotheca pannosa var. *rosae*	Bender and Coyier (1984)
Snapdragon	*Puccinia antirrhini*	Aitken et al (1989)
Tomato	*Clavibacter michiganensis* subsp. *michiganensis*	de Jong and Honma (1974)
	Fusarium oxysporum f. sp. *lycopersici*	Scott and Jones (1989)

be attributed to resistance to the vector *Aphis gossypii* Glover (Lecoq et al 1979). A CMV-resistant line of *Cucumis melo* L. was a less effective source of the virus for the aphids *Myzus persicae* (Sulzer) and *A. gossypii* than the CMV-susceptible muskmelon Charentais. The resistant line of *C. melo* was also less susceptible than Charentais to CMV strain 14 when transmitted by *M. persicae* but virtually resistant to that strain when transmitted by *A. gossypii.*

Many pathogens exist in a number of pathogenic races determined by the spectrum of cultivars and breeding lines of hosts they infect (Table 31). It is incumbent on the grower to be aware of the particular races present in the greenhouse and to be alert to the unexpected appearance of disease in hitherto resistant plants, which may indicate the presence of a new or different race. Reputable seed and plant catalogues give details of resistance to major diseases and nematodes.

One of the most cosmopolitan pathogens, *Botrytis cinerea* Pers.:Fr., appears not to exist in host- or cultivar-specific pathogenic races (Jarvis 1977a, 1980a). There is very little known immunity to *B. cinerea*, which has wide genetic plasticity. Hausbeck and Pennypacker (1988), for example, found only differences in susceptibility among cultivars of *Pelargonium* × *hortorum* L. H. Bailey and *P. peltatum* (L.) L'Hér. ex Aiton. This is a common experience (Jarvis 1977a). Campbell (1984), however, reported that seedling resistance in *Hibiscus cannabinus* L. was conditioned by two

SUMMARY

Resistant Germ Plasm

Genetic resistance is the first line of control for any plant disease and is to be preferred over all other means. The genetic basis of resistance is determined either by the main effect of variation between cultivars (the so-called horizontal resistance), the main effect of variation between pathogen isolates (aggressiveness), or by an interaction of cultivars and isolates (vertical resistance and, correspondingly, virulence in the pathogen). Monogenic gene-for-gene relationships are all of the latter type. Environment has powerful effects on host–parasite relationships, which can be considerably modified by temperature, for example. Many pathogens exist in a number of pathogenic races, determined by the cultivars and breeding lines they infect, and it is always necessary to be alert to the occurrence of disease on hitherto resistant host material, which may signal the appearance of a new race. Genetic resistance is not common in ornamental crop production but is essential in the production of some vegetable crops, especially lettuce in the presence of *Bremia lactucae*. There is little or no genetic resistance to some pathogens, such as *Botrytis cinerea*.

Table 32. Genetic relationships in reactions of tomato to tomato mosaic virus[a,b]

Tomato mosaic virus strain	Resistance genes in tomato							
	Tm-1/+	*Tm-1*/*Tm-1*	*Tm-2*/+	*Tm-2*/*Tm-2*	*Tm-2*^{2}/+	*Tm-2*^{2}/*Tm-2*^{2}	*Tm-1*/+, *Tm-2*/+	*Tm-1*/+ *Tm-2*/*Tm-2*^{2}
0	T	T	R*	R	R*	R	R	R
0Y	T	T	R*	R	R*	R	R	R
1	S	S	R*	R	R*	R	R*	R
2	T	T	S	S	R*	R	R	R
1.2	S	S	S	S	R*	R	S	R*

[a] Adapted from T. J. Hall and Bowes (1980).
[b] R = normal resistance reaction, with no symptoms; R* = resistance reaction in which a deleterious systemic hypersensitive necrotic reaction may occur; S = normal susceptible reaction; T = tolerance reaction, with mild mosaic symptoms and little or no effect on growth.

genes, the genotype *R-ss* conferring resistance, and *R-S-*, *rrS-*, and *rrss* conferring susceptibility. Major gene resistance in chickpea (*Cicer arietinum* L.) was also suspected by Rewal and Grewal (1989), but with an epistatic interaction in which an *S* gene suppresses the expression of an *R* gene, even when present in a heterozygous condition. Good resistance to *B. cinerea* remains elusive for most crops.

The complexity of genetic relationships in other diseases is illustrated by downy mildew of lettuce (Crute and Dixon 1981; Farrara et al 1987), in which genotypes are used to differentiate races of *Bremia lactucae*. The genetic relationships in tomato mosaic (T. J. Hall and Bowes 1980), caused by tomato mosaic virus, are illustrated in Table 32.

Pathogen populations can be kept down, in a way analogous to control by pesticides, by exploiting the durability of host resistance (M. S. Wolfe 1981). This is most conveniently done by concurrent cropping of cultivars with different factors for resistance. Mixtures of cultivars commonly have three components, which differ in resistance characters of major and minor effect. Each component tends to select for its corresponding pathogen race, which is less well adapted to the other components. The selection of cultivar-specific races of pathogens is thereby decreased. There is, however, some risk that complex races adapted to more than one component of resistance will emerge (Wolfe 1981), and new components have to be added to maintain a different composition in the spectrum of host resistance. This system interacts with pesticide strategies (Wolfe 1981), but there seems to have been no work on this aspect in greenhouse crops.

Biological Control

CHAPTER 9

Biological control was defined by Cook and Baker (1983) as the reduction of the amount of inoculum or disease-producing activity of a pathogen accomplished by or through one or more organisms other than humans. A voluminous literature describes a wide variety of mechanisms whereby organisms impede the pathogenic activity of others, but in general they may be distilled into antibiosis mediated by specific or nonspecific metabolites of microbial origin, parasitism and predation, and competition (Fravel 1988).

It has long been recognized that biological control occurs naturally without the intervention of agriculture and horticulture (K. F. Baker 1987; K. F. Baker and Cook 1974; Cook and Baker 1983). Conscious biological control in agriculture and horticulture, recognizing the mechanisms of interaction between a pathogen and its commensal microorganisms, seeks to modify the equilibrium between populations (Anderson 1979; R. Baker 1986; J. C. Miller 1963). In the greenhouse, the environment can be manipulated to enhance natural biological control, or antagonistic organisms can be introduced and maintained by appropriate technology.

Linderman et al (1983) enumerated some strategies for recognizing and characterizing biological control when it occurs, the better to exploit it in comparable situations. The telltale signs, with examples, are the following:

- Absence of disease where the pathogen has been introduced (as in suppressive soils)
- Absence of disease where the pathogen is present (as in the suppression of crown gall)
- Pathogen decline with monoculture (as in take-all decline)
- Physicochemical treatment affecting antagonist or pathogen growth (as in the suppression of antagonists to damping-off pathogens by soil pasteurization)
- Recovery of mycoparasites by baiting (as in the recovery of antagonists by baiting soil with sclerotia of a pathogen)

One of the key strategies of Linderman et al (1983), which has been elaborated upon by R. Baker and Scher (1987), is to determine whether the activity and population of a biocontrol microorganism can be enhanced.

An obvious way to do this is to isolate the microorganism, increase its population in culture, and then reintroduce it to a crop under threat from a pathogen. This practice, however, frequently runs into problems with registration authorities, particularly if the organism's activity has been enhanced by mutagenesis, genetic engineering, or even reselection. In greenhouse crops, with excellent opportunities to modify cultural conditions and the environment, enhancement of indigenous populations of biocontrol microorganisms may be a better strategy.

For the purposes of discussion, it is convenient to expand Fravel's three modes of biocontrol (Fravel 1988)—antibiosis, parasitism, and competition—into eight variants of these, namely, allelopathy, antibiosis, competitive saprophytic ability, amendment of soils, cross protection, hyperparasitism, hypovirulence, and suppressive soils. It is immediately apparent that the distinctions between these topics are blurred so that they overlap. Indeed, many biocontrol organisms operate in more than one mode. *Trichoderma* spp. have a role in antibiosis, competition, hyperparasitism, and suppressive soils (Chet 1987). The integration of more than one mode of biological control and the further integration of biological control with all other methods of control (genetic resistance, disease escape, pesticides, clean planting material, etc.) is, as Bruehl (1989) pointed out, analogous to incremental, nonspecific, quantitative horizontal resistance, in contrast to vertical resistance, which is determined by a single factor. Thus, for example, Harman et al (1989) combined effective biocontrol strains of *T. harzianum* Rifai with a solid-matrix seed-priming treatment to improve seed emergence in the presence of *Pythium ultimum* Trow, and Paulitz

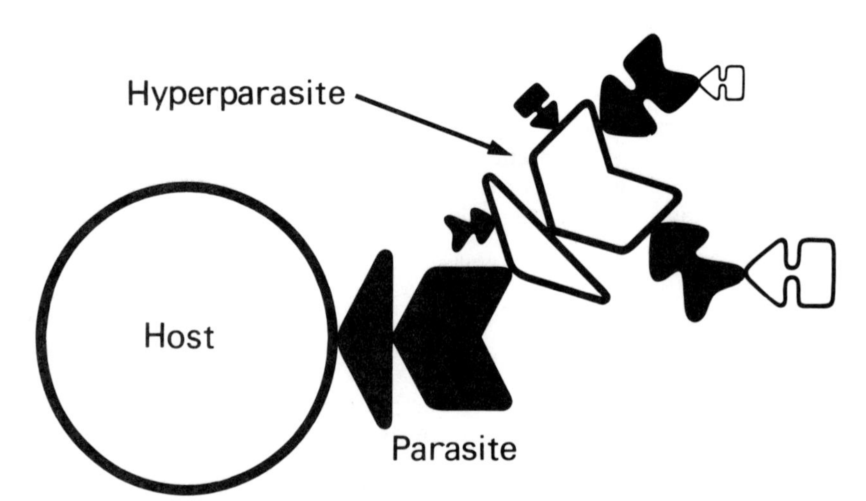

Figure 38. Hyperparasites and antagonists do not act unimpeded on a pathogen; there is a chain of other microorganisms affecting their activity, all of which are affected adversely or benignly by each other and by the physical and chemical environment. Reprinted, by permission, from Jarvis (1989).

et al (1990) combined *P. nunn* Lifshitz, Stanghellini, & Baker with a rhizosphere-competent mutant of *T. harzianum* to improve the control of Pythium damping-off of cucumber.

It should be further borne in mind that biocontrol microorganisms have their own environmental adversities and biological antagonists, which act contrary to plant disease control (Figure 38). Thus, it is incumbent on the pathologist to ensure, for example, that a biocontrol organism is phyllosphere- or rhizosphere-competent and that it tolerates pesticides deemed necessary to control other disease organisms, weeds, and insects (Papavizas 1987). The aim of conscious biological control in the greenhouse is to enhance only organisms that control plant pathogens.

Allelopathy

The term *allelopathy*, derived from Greek words meaning "mutual harm," has come to mean any harmful effect on a plant resulting directly or indirectly from chemical compounds produced by another plant and released into the environment (Putnam and Duke 1978; Rice 1974). The chemical compounds are secondary metabolites called allelochemicals, and in the broadest sense they may include insect and herbivore repellents and toxins, phytoncides, phytoalexins, and microbial antibiotics (Whittaker and Feeny 1971). In horticulture, the most familiar examples of allelopathy are chrysanthemum autoallelopathy (Kozel and Tukey 1968) and that exerted by *Juglans nigra* L. against many crop plants, especially apples (Massey 1925; Schneiderhan 1927) and tomatoes (E. F. Davis 1928; Massey 1925). The distinction between allelochemicals and phytoncides is blurred; juglone and hydrojuglone glycoside are implicated in the resistance of young leaves of *J. nigra* to *Gnomonia leptostyla* (Fr.:Fr.) Ces. & De Not. (S. Cline and Neely 1984).

Allelopathy is implicated in the replant problems of peach and apple (Traquair 1984) and in the predisposition of plants to root rots following root damage by allelochemicals from decomposing plant residues (Patrick and Toussoun 1965), but it can be turned to advantage in the biological control of some diseases (Patrick 1986). Apart from antibiosis (i.e., the microorganism-versus-microorganism interaction) as a form of allelopathy,

Table 33. Populations of *Pratylenchus* spp. in samples of roots of *Rosa canina* and *Tagetes patula* and associated soil[a]

Sample and host	June	July	Aug.	Sept.	Oct.	Nov.	Dec.
100 ml of soil							
R. canina	375	125	170	115	100	85	190
T. patula	318	240	50	0	0	0	25
10 g of roots							
R. canina	4,500	5,400	4,013	1,720	3,630	2,575	110
T. patula	200	60	90	0	5	0	0

[a] Adapted from Oostenbrink et al (1957).

Table 34. Effects of four green manure crop residues on Fusarium crown and root rot of tomatoes[y]

Preceding crop	Mean disease rating[z]	
	Trial 1	Trial 2
Lactuca sativa	0.00 a	0.35 a
Lepidium sativum	0.20 ab	1.81 bc
Brassica juncea	0.45 bc	1.42 b
Spinacia oleracea	0.68 c	2.19 c
Fallow	1.20 c	2.94 d

[y] Adapted from Jarvis and Thorpe (1981).

[z] 0 = No symptoms; 5 = severe symptoms and death. The data are means of 40 plants in trial 1 and 48 plants in trial 2. The plants were grown in potting soil infested with *Fusarium oxysporum* f. sp. *radicis-lycopersici* and rated when they were 90 days old. Ratings in the same column followed by the same letter are not significantly different at $P = 0.05$.

secondary metabolites from higher plants can also act against pathogenic microorganisms. Conversely, secondary metabolites from fungi, including ectomycorrhizal fungi, can act against higher plants (Chaumont and Simeray 1985).

Allelochemicals are often derived from the shikimic acid pathway (Whittaker and Feeny 1971) and include phenols, phenylpropanes, acetogenins, flavonoids, terpenoids, alkaloids, and polyacetylenes. Many of these substances, some perhaps originally phytoncides, are found in decomposing plant residues (Patrick and Toussoun 1965; Tukey 1969). They may stimulate resting structures such as sclerotia, chlamydospores, and oospores to germinate and then, in the absence of a host, cause germ tubes to lyse. Synthetic onion oils induce the germination of sclerotia of *Sclerotium cepivorum* Berk., which are subsequently decayed by other microorganisms (Merriman et al 1981). Other secondary metabolites formed or released during plant decomposition may kill or inhibit the growth of soilborne microorganisms. Butyric acid and other phenolic acids formed upon the decomposition of residues of ryegrass (*Lolium perenne* L.) and timothy (*Phleum pratense* L.) inactivated plant-parasitic nematodes but not saprophytic ones (Sayre et al 1965).

Tagetes patula L. and *T. erecta* L. also secrete nematicides, in the form of polythienyls, principally α-thienyl and 5-(3-buten-1-ynyl)-2,2-bithienyl (Oostenbrink et al 1957; Uhlenbroek and Bijloo 1959; Winoto Suatmadji 1969). The *Tagetes* spp. suppress populations of *Pratylenchus, Tylenchorhyncus*, and *Rotylenchus* in roots and soils, including those of companion plants (Table 33).

A practical application of allelopathy is in the control of Fusarium crown and root rot of tomatoes, caused by *Fusarium oxysporum* Schlechtend.:Fr. f. sp. *radicis-lycopersici* W. R. Jarvis & Shoemaker. Since the disease had been shown to be amenable to biological control (Jarvis 1977b), Jarvis and Thorpe (1981) attempted to accelerate the colonization of newly sterilized groundbed and potting soils with antagonistic microorganisms by growing green manure crops and returning the residues to the soil. Crops were selected that would be cheap to grow between spring and fall crops of tomatoes and might even realize a profit. They were spinach (*Spinacia oleracea* L.), leaf mustard (*Brassica juncea* (L.) Czernj. & Coss.), cress (*Lepidium sativum* L.), and lettuce (*Lactuca sativa* L.). All of the residues reduced the incidence and severity of the disease in pot tests and in the groundbed and increased the populations of microorganisms over those in uncropped sterilized soils, but lettuce was significantly better (Table 34). Residues of dandelion (*Taraxacum officinale* Wigg.) were even more effective (W. R. Jarvis and H. J. Thorpe, unpublished observations). Lettuce and dandelion are both members of the tribe Lactuceae of the Compositae, noted for a wide range of biologically active secondary metabolites (Gonzalez 1977; Towers and Wat 1978). Kasenberg and Traquair (1987, 1988, 1989) found high levels of *o*-diphenols in root and leaf tissues of lettuce and dandelion, particularly in those challenged by *F. oxysporum* f. sp. *radicis-lycopersici*, and postulated that these materials act as catecholic siderophores, sequestering iron necessary for conidial germination, hyphal growth, and pathogenesis. With slight modifications in cropping sequences, a catch crop of lettuce is an effective means of controlling Fusarium crown and root of tomatoes, and in British Columbia companion planting of lettuce or dandelions with tomatoes has also been effective in sawdust bag culture (Plate 8) (Jarvis 1988).

Allelopathy has been implicated in the control of cabbage yellows, caused by *F. oxysporum* f. sp. *conglutinans* (Wollenweb.) W. C. Snyder & H. N. Hans. Like the Lactuceae, some members of the Cruciferae upon decomposition release biologically active volatile metabolites (Lewis and Papavizas 1970, 1974) that are toxic to *Aphanomyces euteiches* Drechs. and *Rhizoctonia solani* Kühn. These metabolites include sulfur-containing substances

SUMMARY

Allelopathy

Allelopathy is the direct or indirect harmful effect of one plant on another through the production of chemical compounds, allelochemicals, which escape into the environment. This concept has been expanded to include the action of allelochemicals on pathogens as a means of biological control. Allelopathy in disease control has not been reported very often, but it does offer a cheap and easy control. Examples include the action of residues or companion plantings of the Lactuceae on Fusarium crown and root rot of tomato and the action of residues of some Cruciferae against *Fusarium oxysporum* f. sp. *conglutinans, Aphanomyces euteiches*, and *Rhizoctonia solani*. In this case activity can be enhanced by solarization of soil containing the residues. The action of *Tagetes* spp. against nematodes is also attributed to allelochemicals.

and ammonia. Ramirez-Villapudua and Munnecke (1987) found that when volatiles were partially confined by plastic sheets during solarization of field soil containing residues of cabbage (*Brassica oleracea* L.), populations of *F. oxysporum* f. sp. *conglutinans* were significantly lower than in soil containing no cabbage residues or soil containing residues but not covered. The incidence of yellows was also significantly reduced. Ramirez-Villapudua and Munnecke thought that volatile metabolites might also be responsible for the fungistasis of *Fusarium* spp. observed by Zakaria and Lockwood (1980) in soils amended with soybean meal. Reduction in populations of *Fusarium* spp. occurred only in soil in the laboratory, not in the field; this result was not explained by Zakaria and Lockwood.

Antibiosis

In the literature on biological control and fungistasis in the phyllosphere and rhizosphere, the term *antibiosis* has come to mean a general antagonism between microorganisms, suppressing the activity of some of them. A more restricted definition was published by R. M. Jackson (1965) and accepted by Fravel (1988): strictly, antibiosis is antagonism mediated by specific or nonspecific metabolites of microbial origin, including lytic agents, enzymes, volatile compounds, and other toxic substances. Many cases of fungistasis in soil and of competition for nutrients in the phyllosphere (Blakeman and Fokkema 1982) are not instances of antibiosis by this definition because no toxic metabolite has been identified. It is therefore prudent to treat antagonism in two parts: antibiosis sensu stricto (Fravel 1988; Jackson 1965) and antibiosis sensu lato, in which antagonism occurs with no clearly defined mode of action save competition for nutrients at the infection site or saprophytism. The role of siderophores, which are not themselves toxic but deprive microorganisms of iron, falls somewhere between these two forms of antibiosis.

Often, mixed microbial populations are screened for biological control microorganisms simply by growing randomly selected fungi or bacteria, one at a time, in dual culture with a pathogen, to select those producing a zone of growth inhibition. Those showing antagonism in this way are then typically used in mixed inocula to assess their biocontrol potential. This approach was used, for example, by Boland and Inglis (1988) to find fungal antagonists of *Sclerotinia sclerotiorum* (Lib.) de Bary, by Ebben and Spencer (1978) for *Phomopsis sclerotioides* van Kesteren, and by Newhook (1951a,b) for *Botrytis cinerea* Pers.:Fr. The candidate microorganisms are usually isolated from the phylloplane or soil in the rhizosphere and therefore tend to be the faster-growing ones, and the dual cultures are usually made on rich laboratory media, where antibiotics may or may not be produced. It is likely, then, that many promising antagonists sensu stricto, and even sensu lato, may be missed and that those selected by this rather artificial means may not be well adapted to the ecological niche at the infection site (M. E. Brown and Beringer 1983). For example, *Bacillus subtilis* (Ehrenberg) Cohn, which has been used in the control of Pythium and Rhizoctonia damping-off in ornamental plants in pasteurized soil (Broadbent et al 1971), produces effective amounts of the antibiotic subtilin only on an appropriate substrate (Weinhold and Bowman 1968), in this case, on soybean material.

Rather than use a hit-and-miss method of screening for antibiotic producers, it is better to make isolations from diseased plants, particularly where the disease is not progressing as well as might be expected (K. F. Baker and Cook 1974). This strategy was used to isolate *Stephanoascus* spp. from powdery mildews (Traquair et al 1988; Jarvis et al 1989) and a strain of *B. subtilis* from rust-infected leaves of geranium (*Pelargonium* × *hortorum* L. H. Bailey) (Rytter et al 1989).

Antibiosis sensu stricto. A number of cases of biological control of pathogens appear to involve the production of antibiotics, although the antibiotic is not always identified. Among them are cases of lysis of mycelium and nondormant spores of fungi. *B. subtilis* causes mycelium of *Rhizoctonia solani* to lyse in soil (Olsen and Baker 1968), and the process is affected by the carbon–nitrogen ratio of the soil (Papavizas 1963). A high carbon–nitrogen ratio in straw-amended soil enhanced lysis, and the addition of nitrogen decreased it.

A number of organisms lyse conidia and mycelium of *Sphaerotheca fuliginea* (Schlechtend.:Fr.) Pollaci on cucumber without penetrating them (Hijwegen and Buchenauer 1984). A *Tilletiopsis* sp. killed cells of *S. fuliginea* without penetration, in what Hoch and Provvidenti (1979) called mycoparasitism. Jarvis et al (1989) thought that a toxin was responsible for the rapid collapse of *S. fuliginea* in the presence of *Stephanoascus* spp., since they could find no evidence of penetration.

Some of the earliest known antibiotics are produced by *Streptomyces* spp. Tahvonen (1982a,b) and Tahvonen and Uoti (1983) isolated a number of these species from light-colored peat cut from the surface layers of a bog. Dark-colored peat from deeper layers yielded fewer *Streptomyces* spp. The light-colored peat in mixes strongly suppressed damping-off and root rot of cucumbers, an effect duplicated by the isolated *Streptomyces* spp. The suppressive peat and *Streptomyces* spp. also suppressed Fusarium wilt of carnation, caused by *Fusarium oxysporum* f. sp. *dianthi* (Prill. & Delacr.) W. C. Snyder & H. N. Hans., and gray mold of lettuce, caused by *Botrytis cinerea.* Unsterilized peat also suppressed Fusarium crown and root rot of seedling and transplant tomatoes, caused by *F. oxysporum* f. sp. *radicis-lycopersici* (Jarvis 1977b). On the other hand, streptomycetes did not consistently protect tomato seedlings from this pathogen in rock wool and recirculating hydroponic systems (van Steekelenburg and van der Sar 1987).

Trichoderma spp. are also frequently isolated from peat, and Chérif et al (1989) found two species that were antagonistic to *F. oxysporum* f. sp. *radicis-lycopersici* through their production of chitinases and β-(1→3)-glucanases.

Although antibiosis is exploited in biological control, very few antibiotics have been used commercially as pesticides in greenhouse production. An exception was griseofulvin, produced by *Penicillium griseofulvum* Dierkx, which was active against *B. cinerea* (Brian 1949), but it no longer seems to be used in plant protection.

Bacteria produce antibiotics (bacteriocins), some of which have been characterized. That produced by strain K84 of *Agrobacterium radiobacter* (Beijerinck & van Delden) Conn is a water-soluble 6-*N*-phosphoramidate of an adenine nucleotide and was named agrocin 84 (Kerr 1980). Although strain K84 of *A. radiobacter* applied in suspension as a root dip for woody nursery cuttings is very successful in controlling crown gall, caused by *A. radiobacter* var. *tumefaciens* (Smith & Townsend) Keane, Kerr, & New (Kerr 1980), agrocin 84 has not been developed as a formulated biological pesticide. Another very effective antifungal bacterium is *Pseudomonas cepacia* (Burkholder) Palleroni & Holmes. Its antibiotic is a chlorinated phenylpyrrole (Roitman et al 1990).

Antibiosis sensu lato. The infectivity of pathogens can be reduced by antimicrobial activity of other microorganisms, and though this is often reported as antibiosis, mechanisms other than the intervention of an antibiotic material may be involved. Blakeman and Fokkema (1982) noted that these mechanisms include competition for nutrients at the infection site, hyperparasitism, and contagious hypovirulence, as in the case of hypovirulent strains of the chestnut blight fungus, *Cryphonectria parasitica* (Murrill) Barr (Van Alfen 1982).

In the soil, antibiosis is probably also affected by microbial antibiotics as well as by soil toxins (Papavizas and Lumsden 1980). In addition, fungistasis may be enhanced by the addition of energy sources and other organic

SUMMARY

Antibiosis

Antibiosis strictly is antagonism mediated by specific or nonspecific metabolites of microbial origin, including lytic agents, enzymes, volatile compounds, and other toxic substances. Antibiosis, however, has also come to mean a general antagonism between microorganisms, and the term has been applied, for example, to competition for nutrients in the rhizosphere or on the phylloplane. The selection of potential antagonists by the behavior of dual cultures on artificial media does not guarantee useful biological control of a plant disease; potential antagonists must also possess rhizosphere or phylloplane competence, and they must also be able to produce the active metabolites under natural conditions. It is more appropriate to look for antagonists under natural conditions in which a disease is not progressing as fast as might be expected. Despite many years of research, the exploitation of antibiosis has rarely reached commercial levels. An exception is the use of strain K84 of *Agrobacterium radiobacter* to control crown gall of nursery stock. Evidently many problems remain to be solved.

materials to the soil. Sometimes it is mediated through volatile metabolites released from organic residues. Ammonia released from chitin-amended soil was thought to cause fungistasis of *Fusarium oxysporum* f. sp. *cucurbitacearum* Gerlagh & Blok (Schippers and Bouman 1973), and sulfur-containing volatiles released from cruciferous crop residues inhibit *F. oxysporum* f. sp. *conglutinans* (Ramirez-Villapudua and Munnecke (1987) and *Aphanomyces euteiches* (Lewis and Papavizas 1971).

In general, antibiosis has not been very successfully exploited for the control of soilborne pathogenic fungi (Papavizas and Lumsden 1980), because of the failure to translate known mechanisms of fungistasis into cultural practices. Often the mode of action of so-called antagonists remains obscure, and the practicalities of introducing microorganisms into cropping practices after they have shown promise in the laboratory (M. E. Brown and Beringer 1983) and maintaining effective populations of them (Ebben and Spencer 1978) remain major problems to be solved.

Competitive Saprophytic Ability

Inoculum potential is a function of not only the concentration and vigor of inoculum in contact with the host but also the collective effect of environmental conditions that determine the realized energy of growth (Garrett 1970), whether in the rhizosphere or on the phylloplane. This last factor acting at or near the infection site is largely determined by the competition of commensal microorganisms. Several root diseases of greenhouse crops are caused by microorganisms that are capable of returning quickly to newly sterilized substrates and building up large populations in the biological near-vacuum. *Pythium* spp. are capable of this, as is the tomato crown and root rot pathogen, *Fusarium oxysporum* f. sp. *radicis-lycopersici.*

Marois and Mitchell (1981a,b) and Marois et al (1981) quantified the saprophytic and parasitic abilities of *F. oxysporum* f. sp. *radicis-lycopersici* in fumigated soil amended with a natural microflora or with conidia of fungi selected for antagonism against the pathogen. Following fumigation, the population (colony-forming units) of the pathogen and the incidence of infection of the susceptible tomato cultivar Bonny Best were both inversely related to the logarithm of the number of saprophytic fungal propagules (Marois and Mitchell 1981b). *Penicillium funiculosum* Thom, *Trichoderma harzianum*, and *Aspergillus ochraceus* K. Wilh. were selected as particularly antagonistic to this pathogen (Marois and Mitchell 1981a). When chlamydospores of *F. oxysporum* f. sp. *radicis-lycopersici* were introduced into fumigated soil, 6,500 chlamydospores per gram were required for infection of 50% of Bonny Best tomato plants, as against 900 per gram in nonfumigated soil and only 300 per gram in newly fumigated soil. In fumigated soil that was initially infested with 1,000 chlamydospores per gram and then allowed to be recolonized naturally over a 46-day period, there was a parallel decrease in the incidence of tomato root infection and in the population of the pathogen recoverable from soil sampled every 7 days (Figure 39) (Marois and Mitchell 1981b). *F. oxysporum* f. sp. *radicis-lycopersici* is a rapid colonizer of sterile soils (Rowe et al 1977) but has poor competitive saprophytic ability (Jarvis 1988).

In natural cotton soils, the population of *Pythium ultimum* in crop residues declined exponentially after an initial increase (Hancock 1981). This rapid decline lasted 2–3 mo, with a population half-life of about 30 days, and was followed by a period of stable or increased population as oospores became exogenously dormant. With inoculum in the form of sporangia, the decline in inoculum density over the first few weeks was more prominent in previously sterilized soils to which spores of *P. ultimum* had been subsequently added. The decline appeared to be mediated, in part, by hyperparasitic microorganisms.

The inoculum potential of *P. aphanidermatum* (Edson) Fitzp. was found by Elad and Chet (1987) to be determined, in part, by bacteria that compete for nutrients. These bacteria included *Pseudomonas putida* (Trevisan) Migula, *Pseudomonas cepacia*, and an *Alcaligenes* sp. Established on the roots of cucumber by means of seed treatment or by direct application, they reduced the incidence of infection by *P. aphanidermatum*.

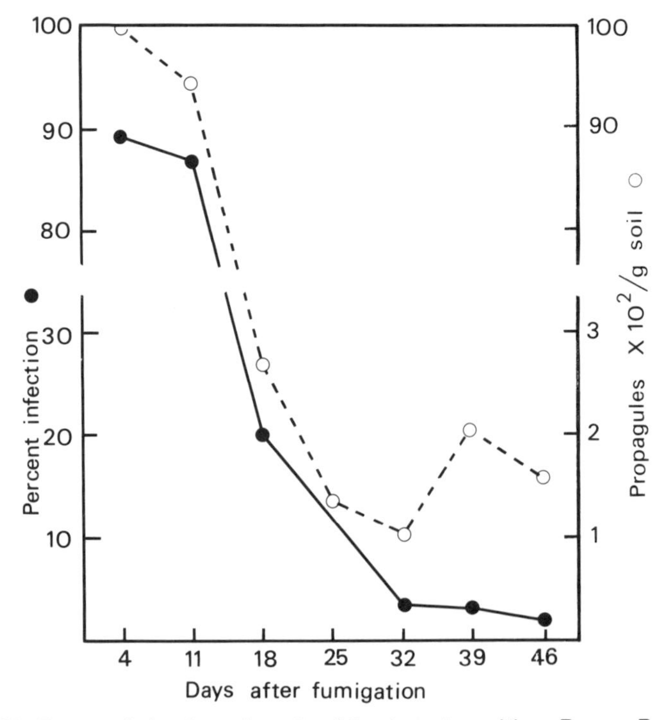

Figure 39. Percent infection of roots of the tomato cultivar Bonny Best (solid line) and inoculum density of *Fusarium oxysporum* f. sp. *radicis-lycopersici* (dashed line) in soils that were fumigated and then allowed to be recolonized naturally. Inoculum was added at the rate of 1,000 propagules per gram to 1,500 g of soil seeded with tomato and sampled at 7-day intervals. Redrawn from Marois and Mitchell (1981b).

Funck-Jensen and Hockenhull (1983a,b) considered that although *Pythium* spp. have the potential to cause considerable damage in hydroponic systems, they rarely do so because the dilution of plant exudates and the low concentration of microbial nutrients deprive them of the exogenous energy required for infection. The addition of 1% sucrose to the hydroponic solution provided that energy and resulted in destructive infection. Funck-Jensen and Hockenhull (1983a) thought that removing microbial nutrients by biological filtration or establishing microbial populations that would utilize them would diminish the risk of root infection by *Pythium* spp.

Blakeman (1978, 1985) and Blakeman and Fokkema (1982) also thought that the competition for exogenous nutrients by scavenging saprophytes can exert biological control on the phylloplane, whereas aphid honeydew (Fokkema et al 1983), pollen, and other organic materials (Blakeman 1980, 1985) facilitate the germination of pathogen spores by providing exogenous nutrients.

Bacteria are normal inhabitants of the phylloplane, fast-multiplying and effective scavengers of exogenous nutrients (Leben 1965a). Leben (1964)

SUMMARY

Competitive Saprophytic Ability

In order to attain an effective inoculum potential, a pathogen must be able to compete successfully as a saprophyte or survive as a dormant propagule when it is not colonizing a plant. An important strategy for biological control is to prevent the pathogen population from building up to dangerous levels or otherwise to reduce its inoculum potential by manipulating its microbial and physical environment. Some pathogens, such as *Pythium* spp. and *Fusarium oxysporum* f. sp. *radicis-lycopersici*, are rapid colonizers of sterile soil but are sensitive to microbial antagonism sensu lato. Often the antagonists can be returned to the soil and their populations enhanced by residues of nonhost crops. A suggested strategy for reducing the pathogenic activities of *Pythium* spp. in hydroponic systems is to use other microorganisms to compete for nutrients, especially sugars, but this strategy seems not to have been tested. On the phylloplane, saprophytic bacteria compete with fungi for nutrients, but their effect is overridden by deposits of insect honeydew or pollen. This mechanism resembles fungistasis in soil. The prior establishment of saprophytes at infection sites has proven promising in some cases of flower infection. The biological control value of antagonistic microflora can be seriously impaired by the wrong choice of pesticide; iatrogenic diseases are partly the result of differential killing of antagonist and pathogen, and they are particularly severe if the pathogen has developed resistance to the pesticide.

and Leben and Daft (1965) found a bacterium, designated A180, on the phylloplane of cucumber that exerted some antagonistic control over *Colletotrichum lagenarium* (Pass.) Ellis & Halst., the anthracnose pathogen. A180 had no effect, however, against *Sphaerotheca fuliginea*, the cucumber powdery mildew fungus.

Bacteria exert a powerful effect on the saprophytic existence of conidia of *Botrytis cinerea* on the phylloplane; paradoxically, the population of epiphytic bacteria can itself be increased by the presence of conidia (Blakeman 1972, 1978; Blakeman and Fraser 1971). When the population of a *Pseudomonas* sp. was selectively increased on senescing beetroot leaves, the bacterium inhibited the germination of conidia of *B. cinerea* (Blakeman 1972; Blakeman and Brodie 1977; Brodie and Blakeman 1975), by a mechanism preventing or delaying germination and infection (Blakeman 1980). This form of fungistasis on the phylloplane can be attributed to competition for nutrition. Many bacteria can also act antagonistically to *B. cinerea.* Newhook (1951a) found that *Bacillus, Pseudomonas*, and *Chromobacterium* spp. isolated from lettuce leaves inhibit germination and lyse germ tubes in vitro and furthermore make the medium alkaline and inimical to the growth of *B. cinerea.*

Newhook (1951b) also found that actinomycetes and fungi from senescent lettuce leaves could inhibit the saprophytic phase of *B. cinerea* sufficiently to prevent infection of healthy lettuce leaves. Protection was achieved only if the competitive saprophytic organisms were encouraged by periods of free water or high humidity and temperatures between 15 and 25°C prior to a challenge by *B. cinerea.* Protection against *B. cinerea* and *Rhizoctonia solani* was less successful in the field, where the optimum conditions were provided only haphazardly (Wood 1951a,b). The establishment of a saprophytic population of *Cladosporium herbarum* (Pers.:Fr.) Link, *Aureobasidium pullulans* (de Bary) G. Arnaud, and a *Penicillium* sp. on young strawberry fruits delayed or prevented infection by *B. cinerea* (Bhatt 1962).

The relative competitive saprophytic ability of pathogens and the restraining effects of competitors and antagonists may be completely altered by fungicides that selectively kill components of the phylloplane microflora that exert biological control. As a consequence, unrestrained iatrogenic diseases are common (Griffiths 1981). For example, when *Penicillium* spp. were killed by benomyl, benomyl-resistant strains of *B. cinerea* were able to parasitize cyclamen (Bollen and Scholten 1971).

Although there are many examples of means by which the saprophytic existence of pathogens on the phylloplane can be impeded by competitors and antagonists, none has been developed as an effective control measure on a commercial scale (Blakeman and Fokkema 1982).

Composts and Other Soil Amendments

Farmyard and other manures have long been used to augment the natural fertilizer in soil, and many of these amendments have marked effects on the incidence and severity of plant diseases. Not all of the effects are beneficial, however. Some crop residues added to the soil when they are relatively fresh enhance root rots through the action of toxins on roots and increases in pathogen populations (Patrick and Toussoun 1965; Schippers et al 1987). Some amendments are beneficial in some cases and not in others (Linderman 1989). Thus Kim et al (1975) noted that of 52 samples of commercial horticultural peat, all contained *Fusarium* spp., 15 contained various pathogenic *Pythium* spp., but none contained *Rhizoctonia solani* or *Verticillium* spp. On the other hand, Tahvonen (1982a,b) and Wolffhechel (1989) found peat to be disease-suppressive through the antagonism of streptomycetes and *Trichoderma harzianum* and other fungi.

Straw, another common groundbed amendment, has complex effects. It is not easily decomposed in soil or even in mushroom composts (Lynch 1979). It suppresses damping-off of radish caused by *R. solani* (Blair 1943) but enhances Fusarium crown and root rot of tomato, caused by *F. oxysporum* f. sp. *radicis-lycopersici* (Jarvis et al 1983). Prior colonization of straw by *Trichoderma viride* Pers.:Fr., *T. album* G. Preuss, *Penicillium* spp., and *F. culmorum* (Wm. G. Sm.) Sacc. prevents the establishment of would-be colonizers (Bruehl and Lai 1966). Prior colonization is achieved by composting. Ebben (1987) thought that 50 years ago the liberal incorporation of farmyard manure and straw rich in actinomycetes and saprophytic fungi and bacteria was a major factor in suppressing black root rot of cucumbers (caused by *Phomopsis sclerotioides*) in raised beds, a major problem in soils receiving little organic amendment in contemporary flat-ground production. One of those suppressive fungi is *Paeciliomyces lilacinus* (Thom) R. A. Samson (*Penicillium lilacinum* Thom in older literature), a common soil fungus, which is antagonistic to some soilborne pathogens. It helped to reduce the quantity of inoculum of *P. sclerotioides* in pot experiments but was of little value once the pathogen had entered the roots (Ebben and Spencer 1978).

The mechanisms of control of soilborne pathogens by composted agricultural, forestry, and municipal wastes has been comprehensively reviewed by Hoitink and Fahy (1986). In addition to the destruction of pathogens by heat and toxic metabolites, there is a residual microflora that exerts biological control on pathogens in soils amended with composts. As well, there are alterations of significant chemical and physical factors, such as particle size, nitrogen content, cellulose and lignin content, soluble salt content, pH, and toxins. In greenhouse media, hardwood bark composts, with a complex array of chemical and biological effects, are particularly valuable (Hoitink 1980; Hoitink and Fahy 1986).

Composted conifer bark, like composted hardwood bark, suppresses a number of nematode species, including *Meloidogyne hapla* Chitwood and *M. incognita* (Kofoid & White) Chitwood, which are stimulated by peat amendments (McGrady and Cotter 1989). Fresh conifer bark (up to 3 yr old) also suppressed galling of tomato roots by *M. incognita* in groundbeds amended with 10–20% bark, by volume. Aged bark, stored over 5 yr, was less effective in suppressing galling, perhaps because of an increase in pH, which diminished nematicidal activity in continuous cropping.

Composted organic household waste is also a useful adjuvant in potting media. Schuler et al (1989) added 8, 10, or 30% compost (by volume) to sand infested with *Rhizoctonia solani* or *Pythium ultimum* and obtained good control of damping-off of beans (*Phaseolus vulgaris* L.) and beetroot (*Beta vulgaris* L.). Peas (*Pisum sativum* L.) were less successfully protected, possibly because germinating pea seeds produce copious exudates that stimulate *R. solani*. Composted household waste retained its suppressive properties over 46 wk of storage.

Chitin substrates have been added to soil in attempts to enhance populations of chitinolytic microorganisms that might act against pathogens. Van Eck (1978) found that chitin induced autolysis in chlamydospores of *Fusarium solani* (Mart.) Sacc. f. sp. *cucurbitae* W. C. Snyder & H. N. Hans. Schippers and de Weyer (1972), however, found that although a chitin amendment delayed the lysis of its macroconidia and inhibited the formation of chlamydospores, there was an overall rise in the number of viable propagules for the first 6 wk after amendment. Thereafter most of the partly lysed macroconidia died, resulting in a reduction in the overall population of propagules; however, enough viable propagules survived to constitute an effective inoculum. Chitinolytic bacteria have been used to control *F. oxysporum* f. sp. *dianthi* (Sneh 1981) and f. sp. *callistephi* (Beach) W. C. Snyder & H. N. Hans. (Pfister and Peterson 1990).

Spent mushroom compost, rich in chitin, contains a microflora generally antagonistic to *Phomopsis sclerotioides* (Ebben 1987). Ebben used it to make raised beds on groundbed soil amended with *Paeciliomyces lilacinus* in a cucumber house. The combination of mushroom compost and *P. lilacinus* reduced the incidence of root lesions in 120-day-old plants to 0.4 on a scale of 0 to 5; the score for mushroom compost alone was 1.5. These figures compared with 2.9 for raised beds of peat with *P. lilacinus* and 1.6 for peat alone.

Spent mushroom compost as raised bed material was also effective in combination with a carbendazim drench (250 mg a.i. per kilogram in 5 $L \cdot m^{-2}$) in the groundbed: 7% of cucumber roots had black root rot lesions, as against 15% growing in mushroom compost alone, 50% in a peat raised bed without carbendazim, and 44% in peat with carbendazim (Ebben 1987). In addition to a rich microflora, spent mushroom compost also contains large amounts of fungal chitin (Stoller 1954), which may well encourage the activity of chitinolytic antagonists.

Diseases caused by formae speciales of *F. oxysporum* are generally amenable to control by soil amendment. Sun and Huang (1985) devised a commercial mixture of materials that suppressed several of those diseases. The mixture comprised bagasse (sugarcane waste), rice husks, ground oyster shells (which are rich in chitin and calcium), urea, potassium nitrate, calcium superphosphate, and mineral ash from steel mills. This mixture inhibited spore germination and enhanced germ tube lysis. Its effects were enhanced by the alkaline pH of amended soils. Populations of fungi and actinomycetes were increased by factors of 25 and 2, respectively, but bacterial populations were not affected. However, when bacterial inhibitors were added to amended soil, the germination of pathogen propagules increased.

Another commercial compost, Biobalans, containing a high population of *Trichoderma* spp., controlled gray mold (caused by *Botrytis cinerea*) in strawberry (Svedelius 1989).

Other organic amendments have complex and unpredictable effects, because of their great variability in chemical and physical properties. Wicks et al (1978) found that fowl manure suppressed cucumber wilt, caused by *F. oxysporum* f. sp. *cucurbitacearum*, if it was added to soil after fumigation, but not if it was added before fumigation. Pathogen propagules occurred in the rhizospheres of both healthy and diseased plants, mostly around diseased plants, but they were more numerous around healthy roots in soils amended with fowl manure after fumigation. Homma et al (1979a) found that propagules of *F. oxysporum* f. sp. *lycopersici* (Sacc.) W. C. Snyder & H. N. Hans. decreased in soil amended with powdered crab shell (another source of chitin) or rapeseed oil cake, but not fowl manure. In soil amended with fowl manure, the germination of microconidia and chlamydospores was inhibited by antifungal substances from increased numbers of other microorganisms (Homma et al 1979b).

Greenhouse vegetable crops are seldom rotated, and crop residues are usually returned to groundbeds and worked in with a rotary tiller. It might

SUMMARY

Composts and Other Soil Amendments

The disease-controlling effects of composts and other organic materials added to soils vary widely and are often unpredictable. Some crop residues enhance root rots, because of metabolites that are toxic to the roots. Peat can sometimes be beneficial, because it contains streptomycetes and other antagonists, and can sometimes be harmful, because it contains pathogens, particularly *Pythium* and *Fusarium* spp. Similarly, the addition of straw may suppress the activity of *Rhizoctonia solani*, but it enhances the activity of *F. oxysporum* f. sp. *radicis-lycopersici*. Prior colonization of straw by antagonists, however, may annul such harmful effects. Composting of crop residues, properly done, generates lethal temperatures for most pathogens except some viruses. In general, composted hardwood bark has proved very beneficial in the control of damping-off diseases. Composted conifer bark generally suppresses nematodes, but it has not been so successful in controlling fungi.

Composts rich in chitin, such as spent mushroom compost or crustacean residues, have also proved promising, because they encourage chitinolytic antagonists, which lyse delicate fungal structures in soil. This type of compost may be improved by the addition of calcium and other nutrients to enhance saprophyte populations.

be supposed, therefore, that disease levels would increase over the years (Ebben and Last 1975; Last et al 1969). Ebben (1987) retarded the buildup of damaging levels of *Pyrenochaeta lycopersici* R. Schneider & Gerlach, the fungus causing of corky root rot of tomato, over a 5-yr repeated cropping sequence of tomatoes by adding leaf prunings and end-of-season debris to the soil and incorporating it. Out of concern for the integrated management of other diseases, she flash-steamed some of the debris to inactivate *Didymella lycopersici* Kleb. and used trickle irrigation to restrict the saprophytic activity of *Botrytis cinerea* in prunings left in situ all season. In other treatments, unsteamed debris was returned, or no debris. The prunings remained dry and disintegrated quite quickly. A bonus from returning crop debris to the soil was a marked reduction in the amount of phosphate and potassium fertilizers required; a total of 1,356 $kg \cdot ha^{-1}$ of triple superphosphate fertilizer was required in the no-debris treatment, as against 170 $kg \cdot ha^{-1}$ with unsteamed debris and 338 $kg \cdot ha^{-1}$ with steamed debris. Ebben (1987) did not report on the effect of these fertilizers and organic amendments on the microflora, though the populations of soil fungi and bacteria were higher in the debris-amended soils. The fungi commonly occurring were species of *Penicillium, Acremonium, Trichoderma, Fusarium*, and *Cladosporium herbarum*, generally in similar proportions in all three soils.

Cross Protection, Induced Resistance, and Passive Exclusion

Cook and Baker (1983) found it difficult to distinguish clearly between cross protection and induced resistance. Cross protection occurs when an organism first to arrive at an infection site acts directly and antagonistically against a second, more virulent pathogen arriving later. The first arrival may, alternatively or in addition, induce a reaction in the host that imparts resistance against the more virulent pathogen. Passive exclusion is a case of cross protection in which the first organism to arrive simply occupies the infection site, without antagonism except probably competition for physical space and nutrients, and without inducing a host reaction. Sometimes cross protection is effected by a less virulent or an avirulent strain of a pathogen, such as a symptomless strain of tomato mosaic virus widely used to protect tomatoes from virulent strains, and this biological control can also be regarded as exploiting hypovirulence.

Cross Protection

In greenhouse crops, cross protection has been successfully used in two main areas: the control of tomato mosaic virus (ToMV) in tomatoes and the control of *Fusarium*-incited diseases.

Control of ToMV. Rast (1972) obtained a mutant form of ToMV by treating the virus with nitrous acid. This form was designated MII-16, but actually it is a mixture of a symptomless strain and a strain resembling the parent strain (Rast 1975). It causes some stunting and a mild mosaic, but it generally protects ToMV-susceptible tomato cultivars from later infection by yellowing and distorting strains of ToMV as well as the parent strain. MII-16 causes no loss of yield or quality and has been successfully used on a commercial scale (Anonymous 1973; Channon et al 1978; Fletcher and Butler 1975; Fletcher and Rowe 1975; Mossop and Procter 1975). The attenuated virus is applied to young plants by airbrush in a suspension incorporating a little Carborundum powder, but not within a day or two of pricking off. Resistant cultivars are not inoculated (Anonymous 1973; Fletcher and Rowe 1975). Rast (1979) inoculated seed with MII-16. In the United Kingdom, the use of MII-16 induced changes in the populations of strains of ToMV (Fletcher and Butler 1975). Whereas ToMV strain 1 accounted for 12% of the population in 1971, prior to the introduction of MII-16, it accounted for 64% in ToMV-susceptible cultivars by 1974. In greenhouses where MII-16 had been used, 94% of the isolates of ToMV were strain 1, as against 39% from other greenhouses. No new strains were found; other strains were either strain 0 or intermediate between strain 0 and strain 1.

Similar control of ToMV was achieved with a heat-attenuated strain of the virus (produced at temperatures over 34°C), but it did not protect tomatoes inoculated with a mixture of severe ToMV and potato virus X (Paludan 1973).

Using other attenuated and challenger strains of ToMV, Marrou and Migliori (1971) concluded that cross protection was successful only when the protective strain is host-adapted and the challenger is of the same strain group. Their strain SP controlled tomato strain SPI well in the first few weeks, but later 16–30% of protected plants showed infection.

Summarizing the experiences with attenuated ToMV, R. W. Fulton (1986) emphasized that protection is incomplete; the symptoms are slower to develop, but there is no immunity. Protection can be overcome with a very high level of inoculum. There is also the risk that the attenuated form could spread to other hosts, on which it might be more dangerous. The treatment is relatively expensive.

Control of *Fusarium* diseases. D. Davis (1967) showed that seedlings can be protected from their specific wilt-causing forma specialis of *Fusarium oxysporum* by prior inoculation with a nonpathogenic forma specialis; for example, tomato seedlings were protected from *F. oxysporum* f. sp. *lycopersici* by inoculation with f. sp. *dianthi*, and carnation was protected from f. sp. *dianthi* by f. sp. *lycopersici*. The degree of protection was independent of the interval between the two inoculations, but roots injured at the second, challenge inoculation were less resistant. Formae speciales of *F. oxysporum* were more effective protectants than some other fungi, such as *Penicillium notatum* Westling, *Neurospora crassa* Shear & B. O. Dodge, *Verticillium albo-atrum* Reinke & Berthier, and *Rhizoctonia solani*. In a second paper (D. Davis 1968), however, he was less sanguine about the possibilities of cross protection, concluding that neither increases in the concentration of inoculum of the protectant nor treatment with formae speciales other than the ones he tried (*F. oxysporum* f. sp. *tracheiphilum* (E. F. Sm.) W. C. Snyder & H. N. Hans., f. sp. *vasinfectum* (Atk.) W. C. Snyder & H. N. Hans., and f. sp. *batatas* (Wollenweb.) W. C. Snyder & H. N. Hans.) were likely to induce more than a delay in the appearance of symptoms of tomato wilt caused by *F. oxysporum* f. sp. *lycopersici*. A very large amount of inoculum of the protectant was needed, far greater than could be expected to provide cross protection in unamended soils in the field, and Davis (1968) thought this form of wilt control to be impracticable. He thought that mutant formae speciales might be better protectants, but this suggestion seems never to have been followed up.

Wymore and Baker (1982) examined cross protection against Fusarium wilt of tomato using *F. oxysporum* f. sp. *dianthi* as the protectant. It required a large quantity of inoculum, comparable to that of the pathogen challenge, and it was only effective if the protectant was applied a few days before the challenge. The protection lasted for only 3 or 4 wk, and symptoms appeared by day 34. Wymore and Baker found it difficult to explain this in terms of antibiosis or competition; they suggested rather that phytoalexin production might be the basis of protection. This, then, appeared to be a case of induced resistance.

Rattink (1988) explored the possibility of cross-protecting carnation cuttings against *F. oxysporum* f. sp. *dianthi* by avirulent isolates of the same fungus. This effort was not successful when the protectants were added to the rooting medium but was successful if roots had formed on the cuttings prior to treatment. This might again suggest an induced resistance in root tissue, absent from stem tissue.

Copeman et al (1988), on the other hand, obtained considerable success in cross-protecting tomato grown in sawdust against the crown and root rot pathogen, *F. oxysporum* f. sp. *radicis-lycopersici*. Nine-day-old seedlings were treated with a mixture of three nonpathogenic isolates of *F. oxysporum* and were inoculated with the pathogen on the 26th day. Over a 20-wk harvest, a 55% increase in yield was obtained, with a significant increase in the number of fruit and in the percentage of fruit in the premium size range. A second protectant treatment at 58 days gave no better yield increase.

Gladiolus corms being put into storage were protected from the corm-rotting *F. oxysporum* f. sp. *gladioli* (L. Massey) W. C. Snyder & H. N. Hans. by prior treatment with *F. moniliforme* J. Sheld. f. sp. *subglutinans* Wollenweb. & Reinking and *F. solani* (Magie 1980). Isolate M-685 of *F. moniliforme* f. sp. *subglutinans* in fact protected freshly harvested corms as well as or better than benomyl. Magie explained this on the basis of the prior occupation of infection sites, and so it apparently amounts to a case of passive exclusion. He thought that this treatment would be particularly valuable where natural competitors had been eliminated by tissue culture propagation or by hot-water treatment.

Avirulent mutants of *Pseudomonas solanacearum* (Smith) Smith protected tomato from bacterial wilt when roots were inoculated with them simultaneously or sequentially with the original virulent strain (Trigalet and Trigalet-Demery 1990). Provided that it did not constitute more than

10% of the inoculum, the virulent strain did not multiply in the plant under conditions favorable for the disease.

Induced Resistance

Nobécourt (1928) put forward the idea that plants could be immunized against infection by *Botrytis cinerea* by pretreatment with culture filtrates, an idea taken up without much progress by many investigators in the next 10 years. Later, Heale and Stringer-Calvert (1974) reported that symptoms of gray mold on conidium-inoculated carrot callus tissue were delayed 1–7 days if the tissue was first treated with fluid in which conidia had germinated. The onset of symptoms was associated with increasing levels of insoluble invertase activity in the callus tissue.

It has since been shown that resistance to anthracnose caused by *Colletotrichum lagenarium* can be induced in cucumber (*Cucumis sativus* L.) by prior challenge with *C. lagenarium* (Kuć 1987; Kuć et al 1975). The protection, evident about 72 hr after the first challenge, is systemic and lasts through fruiting if a booster treatment is given (Kuć and Richmond 1977). Resistance induced by inoculation of the first true leaf is transmitted to a scion of the same cultivar grafted on the infected plant above the first leaf (Dean and Kuć 1986; Jenns and Kuć 1979). A signal generated in the inoculated leaf rapidly activates the cucumber defense mechanism.

Similarly, resistance to *Pseudomonas syringae* van Hall pv. *lachrymans* (Smith & Brian) Young et al can be induced in cucumber by prior inoculation with *C. lagenarium* or *P. syringae* pv. *lachrymans* (Caruso and Kuć 1979), and resistance to cucumber mosaic virus can be induced by prior inoculation with *C. lagenarium, P. syringae* pv. *lachrymans*, or tobacco necrosis virus (Bergstrom et al 1982; Jenns and Kuć 1980). Not only can resistance to leaf pathogens be induced in cucumber, but also resistance to Verticillium wilt (caused by *Verticillium albo-atrum*) can be induced by prior treatment of the roots with a culture filtrate of *V. albo-atrum* (Tjamos 1979), and resistance to Fusarium wilt (caused by *Fusarium oxysporum* f. sp. *cucurbitacearum*) can be induced by culture filtrates of nonpathogenic formae speciales of *F. oxysporum* (Gessler and Kuć 1982).

The nature of the elicitor or elicitors of this induced resistance is unknown, but Dean and Kuć (1986) considered the possibility of an electrical signal and the induction of lignin, perhaps with changes in protein and carbohydrate. Salt et al (1988) noted that soluble carbohydrates increased in stems and leaves of tobacco systemically protected against *Peronospora tabacina* D. B. Adam (*P. hyoscyami* de Bary) by *P. tabacina*.

Resistance to virus infection can also be systemically induced in plants inoculated with some viruses, fungi, and bacteria, perhaps involving the transcription of cellular DNA to RNA.

Although there is considerable academic interest in the host defense mechanisms elicited by inoculation with various pathogens, none has yet been exploited as a horticulturally important disease control measure. Kuć (1987), reviewing progress in this field, considered that the signal inducing systemic and persistent resistance could eventually be characterized, and he suggested that it could be used in a spray or seed treatment or that synthetic chemicals might be able to mimic it.

Passive Exclusion

A number of pathogens fail to achieve infection in the presence of large numbers of saprophytes. *Botrytis cinerea* is a component of the epiphytic flora of the leaves of many plants (Blakeman 1980), but it is usually prevented from infecting them, because of competition from its commensal microorganisms for exogenous nutrients needed for the germination of its conidia. W. Brown and Montgomery (1948) noted that the incidence of gray mold in lettuce was lower in hollows between soil ridges than on the tops of the ridges; they thought this was because the fallen-in soil provided a large source of saprophytes that excluded *B. cinerea* from the susceptible, moribund leaf bases at soil level. Newhook (1957) was able to protect tomato flowers from infection by *B. cinerea* by inoculating the senescent flowers with species of *Cladosporium, Penicillium*, and *Alternaria*. Fruit rot initiated from a saprophytically based inoculum in the senescent flowers was considerably reduced.

Exclusion by mycorrhizae. Ectomycorrhizal fungi form a thick mantle around the roots of many shrub and tree seedlings and present few niches for pathogens. Zak (1964) and Marx (1972) reviewed the role of ectomy-

SUMMARY

Cross Protection, Induced Resistance, and Passive Exclusion

It is almost impossible to distinguish clearly between cross protection, in which an organism first to arrive at an infection site acts directly or indirectly against a pathogen arriving later, and induced resistance, in which the first arrival induces a defense reaction in the host. Passive exclusion is simply prior occupation of an infection site by an innocuous organism, sometimes an avirulent strain of a pathogen.

Cross protection has been successfully used to control tomato mosaic virus, by inoculation of the host with an innocuous mutant form of the virus prior to exposure to the virulent form. This cross protection is only successful when the protective strain is host-adapted and the challenge virus is in the same strain group. Even so, the protection is incomplete; symptoms are slower to develop, but there is no immunity. Cross protection has also been used to control diseases caused by formae speciales of *Fusarium oxysporum*; plants are preinoculated with alien formae speciales that protect them from a later challenge by the host-specific forma specialis. Again, however, protection is incomplete, symptoms being only delayed. Further, an impractically large quantity of inoculum of the protectant forma specialis is needed, and

corrhizae in excluding pathogens and concluded that *Rhizoctonia solani* and species of *Cylindrocarpon, Fusarium, Phytophthora*, and *Pythium* are excluded from the feeder roots. Ectomycorrhizal fungi also produce antibiotics, modify the exudations of roots into the rhizosphere, and form a mechanical barrier against pathogen penetration. Trappe (1977) considered the exclusion of pathogens and the protection of tree seedlings from allelochemicals to be valid factors in the selection and supply of ectomycorrhizal fungi in forest nurseries.

Szaniszlo et al (1981) identified hydroxamate siderophores in 14 strains of ectomycorrhizal fungi representing five genera in three families, which they thought were liberated into the rhizosphere and so provided iron to the host. By the same token, the siderophores undoubtedly affect the supply of iron to pathogens.

Vesicular-arbuscular mycorrhizae have similar rhizosphere-modifying properties, though they pose less of a mechanical barrier than ectomycorrhizae. Caron et al (1985a,b, 1986) found that root necrosis caused by *F. oxysporum* f. sp. *radicis-lycopersici* was reduced in tomato roots previously inoculated with the mycorrhizal fungus *Glomus intraradices* Schenck & Smith, and populations of the pathogen were suppressed in these roots at all soil phosphorus levels tested. There was no interaction between mycorrhizal colonization and the potting medium, and the same amount of root

the treatment is effective only when applied a few days before the challenge.

Induced resistance to a disease results from prior challenge of the host with the pathogen causing the disease, with another pathogen, or even with a culture filtrate. In cucumber, this induced resistance is systemic in the plant. It has been hypothesized that the induction of resistance involves an electrical signal, the formation of a lignin defense mechanism, and perhaps changes in protein and carbohydrate.

Passive exclusion, the prior occupation of an infection site by an innocuous microorganism, has been successfully, but not commercially, used to prevent infection of flowers by *Botrytis cinerea*. Ectomycorrhizal fungi appear to exclude pathogens from access to roots and also produce allelochemicals. There may be a siderophore mechanism that sequesters iron, which is required by a number of pathogens. Similarly, although not presenting a physical barrier, the endomycorrhizal fungus *Glomus intraradices* excludes *Fusarium oxysporum* f. sp. *radicis-lycopersici* from tomato roots, perhaps because of a competition for nutrients or an induced defense reaction in the host.

necrosis occurred in five different media. Caron et al (1986) concluded that there was probably a direct interaction between extramatrical hyphae of *G. intraradices* and *F. oxysporum* f. sp. *radicis-lycopersici*, perhaps in competition for nutrients, or that *G. intraradices* induced a defense reaction in the roots.

Hyperparasitism

In the wild, hyperparasitism is very common indeed. For example, *Ampelomyces quisqualis* Ces. can be found on established colonies of most powdery mildew fungi throughout the year, but it is rarely found controlling them to any appreciable extent in cultivated crops. Exploiting hyperparasitism for biological control implies altering environmental conditions in favor of the hyperparasite or applying it in a bulk inoculum as a biological pesticide. The latter case requires some form of registration to ensure the safety of the crop, the operator, bystanders, consumers, and the environment. It also requires good knowledge of the autecology of the hyperparasite and, as Jarvis (1989) pointed out, a good knowledge of naturally occurring antagonists, sensu lato, of the hyperparasite itself.

Ebben (1987) thought that different hyperparasites might well have to be used in different ways, depending on the etiology and epidemiology of different parasite–hyperparasite interactions. She used *Coniothyrium minitans* Campbell and *Gliocladium roseum* Bainier to control white mold of lettuce, caused by *Sclerotinia sclerotiorum*, out of concern for increasing difficulties with fungicide residues on lettuce as well as on celery and Chinese cabbage. Wheat-grain inocula of the two hyperparasites were raked into groundbed soil, and sclerotia of *S. sclerotiorum* were scattered over the soil before lettuce was planted in successive crops in each of two years. The incidence of white mold was divided into primary infections initiated from ascospores and secondary infections caused by mycelial growth from adjacent plants. In the third crop of the second year, secondary spread accounted for a sharp increase in disease incidence 45–58 days after planting (Figure 40), but secondary infections were reduced in the plots treated with *G. roseum*. In this trial, as well as in pot experiments, *G. roseum* was the only hyperparasite to give a lasting reduction in disease, improving with each successive crop. *C. minitans* gave control equivalent to that from repeated fungicide sprays and appeared to delay the appearance of symptoms. It did not persist well between crops and was reapplied for each new crop. It appeared to reduce the number of sclerotia on infected plants (Ebben 1987; Trutman et al 1982). Ebben concluded that *G. roseum* acted by destroying mycelium of *S. sclerotiorum* and that *C. minitans* acted by destroying sclerotia. That being so, it would seem beneficial to apply *C. minitans* to lettuce residue immediately before sclerotia are returned to the soil just after the crop is harvested, so that the sclerotia enter the soil complete with inoculum, rather than to rely on a more haphazard contact of hyperparasite and sclerotia in an uncontrollable soil environment.

W. R. Jarvis (unpublished observations), in an attempt to circumvent the registration problems of a one-on-one relationship in the control of cucumber powdery mildew (caused by *Sphaerotheca fuliginea*) with *Ampelomyces quisqualis* (Jarvis and Slingsby 1977), sought to enhance the indigenous population of the hyperparasite with microbial nutrients. These comprised mixtures of sugars and peptone, together with gelatin as a sticker and glycerol as a hygroscopic material to keep the leaf surface moist. These treatments did not enhance the population of *A. quisqualis*, but excellent control of powdery mildew was obtained through the stimulation of a yeastlike fungus (Figure 41) (W. R. Jarvis and P. T. Atkey, unpublished results), which was later characterized as a probable *Sporothrix* sp. (Jarvis et al 1989; Traquair et al 1988). Jarvis and Berry (1986) subsequently noted that glycerol alone provided some control over powdery mildew. It would appear likely, then, that the particular microbial nutrients used in this treatment enhanced not the target hyperparasite but another, unexpected one. In addition, some control may have been provided by the physical effects of spraying, as Yarwood (1939a) believed to be the case with water sprays.

In most cases in which hyperparasites have been used for biological control, results have been variable, probably because the autecology of hyperparasites is poorly understood, and inappropriate methods have been used. *A. quisqualis* is a pycnidial fungus, and it is generally assumed that its pycnospores are water-dispersed or perhaps dispersed as surface contaminants on the relatively larger, airborne conidia of its hosts (Philipp et al 1984). Its activity is enhanced under humid conditions (Philipp and Crüger 1979; Philipp et al 1984; Sundheim 1982), and Jarvis and Slingsby (1977) obtained improved control of powdery mildew and increased yields of

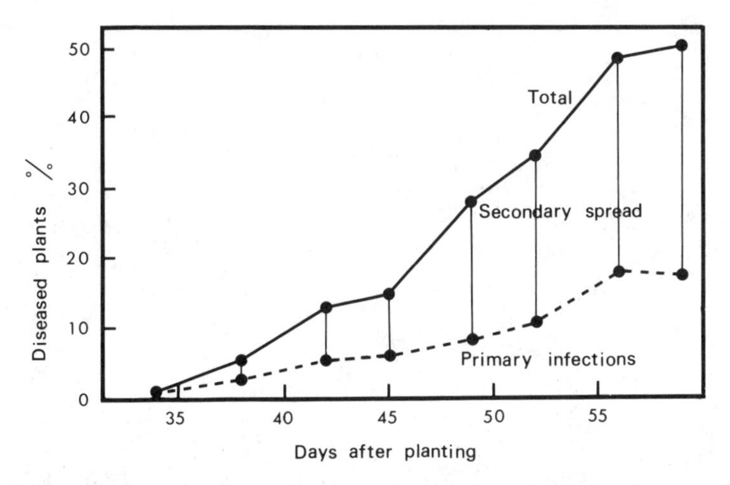

Figure 40. Primary infection of lettuce by *Sclerotinia sclerotiorum* was well controlled by application of the hyperparasites *Gliocladium roseum* and *Coniothyrium minitans*, but secondary spread of lettuce drop increased sharply in the third crop of the second year after treatment. Redrawn from Ebben (1987).

cucumbers when sprays of spore suspension were interspersed with water sprays after a few days' interval. In retrospect, it seems likely that these treatments also encouraged an increase in the population of *Sporothrix* spp. (Jarvis et al 1989) and perhaps some other antagonistic organisms (Diop-Bruckler and Molot 1987; Hijwegen 1986, 1988; Hijwegen and Buchenauer 1984; Hoch and Provvidenti 1979). *S. flocculosus* Traquair, Shaw, & Jarvis and *S. rugulosus* Traquair, Shaw, & Jarvis require a vapor pressure deficit of less than about 0.6 kPa and have an optimum temperature of 26° C for pathogenicity (Jarvis et al 1989). Unlike *A. quisqualis*, whose mycelium invades the mycelium and conidia of powdery mildew fungi (Speer 1978), *Sporothrix* spp. do not appear to invade the host but kill it very quickly, possibly disrupting membrane activity.

Other fungi reported as hyperparasites of powdery mildew fungi, though again without evidence of hyphal penetration, include *Acremonium alternatum* Link:S. F. Gray (Malathrakis 1985), *Aphanocladium album* (G. Preuss) W. Gams, *Paeciliomyces farinosus* (Holmsk.) A. H. S. Brown & G. Sm. (also an insect parasite), *Tilletiopsis minor* Nyland (Hijwegen 1986; Hoch and Provvidenti 1979) and other *Tilletiopsis* spp., *Verticillium lecanii* (A. Zimmerm.) Viégas, *V. fungicola* (G. Preuss) Hassebr., and several other hyphomycetes (Diop-Bruckler and Molot 1987; Hijwegen 1986, 1988; Hijwegen and Buchenauer 1984). All seem to require high humidities, but little else is known of their autecology. *A. alternatum* took 3 days to parasitize 70% of the thallus of *Sphaerotheca fuliginea* at temperatures of 24° C and

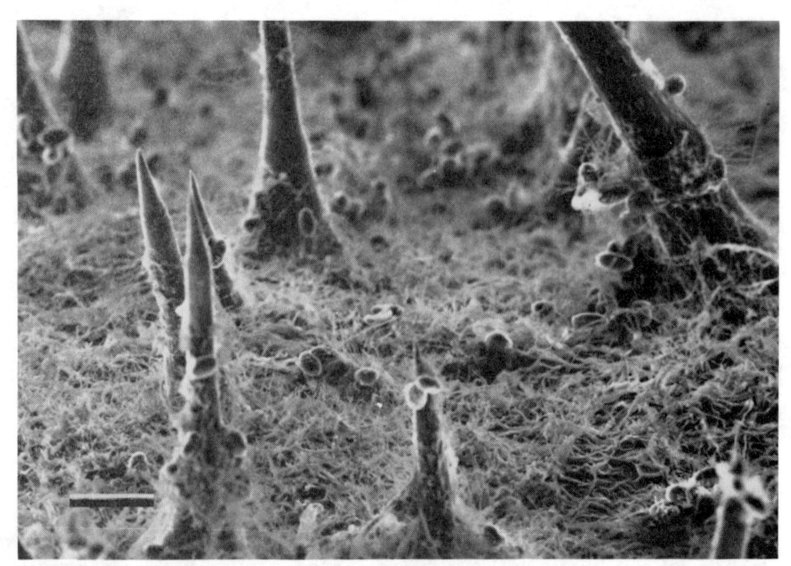

Figure 41. Powdery mildew (*Sphaerotheca fuliginea*) on a cucumber leaf controlled by a yeastlike fungus, the growth of which was stimulated by application of a microbial nutrient spray. Bar = 100 μm. Courtesy of P. T. Atkey.

above (Malathrakis 1987), whereas *Sporothrix* spp. destroyed it in as little as 8 h at 26° C (Jarvis et al 1989).

Philipp et al (1982, 1984), Sundheim (1982), and Sundheim and Amundsen (1982) found that integrated control of *Sphaerotheca fuliginea* by *Ampelomyces quisqualis* was possible because the hyperparasite tolerated the fungicide triforine and a number of insecticides. Only one third as much of the fungicide was needed to achieve control when it was alternated with sprays of a spore suspension of *A. quisqualis*. Sundheim and Amundsen recorded a 50% increase in yield over that of a no-pesticide check, a result better than that obtained with triforine alone.

Besides the powdery mildew fungi, other pathogens of leaves and shoots have hyperparasites with the potential for biological control, but none has been exploited. *Hansfordia pulvinata* (Berk. & M. A. Curtis) S. J. Hughes is a common phylloplane mycoparasite whose potential for biological control has been little investigated. Tirilly et al (1987) found up to 80% of tomato leaf mold spots, caused by *Fulvia fulva* (Cooke) Cif., were parasitized by *H. pulvinata*, but it failed to make close contact with its host and was not readily disseminated. Combination with fosetyl-Al improved the control of leaf mold.

A strain of the insect pathogen *Verticillium lecanii* was noted as a hyperparasite of the carnation rust fungus, *Uromyces dianthi* (Pers.:Pers.) Niessl (Spencer 1980; Spencer and Atkey 1981), but has not been exploited commercially. Conidia of *V. lecanii* had to be mixed with uredospores of *U. dianthi* to reduce the incidence of sori to 10–16% of the check level, hardly a practicable form of biological control. *Puccinia violae* (Schumach.) DC., the cause of rust of violet (*Viola odorata* L.), was parasitized by *Cladosporium uredinicola* Speg. (Traquair et al 1983) but again has not been exploited in biological control. There are very many other reports of phylloplane hyperparasitism without pointers to biological control (Blakeman and Fokkema 1982).

V. lecanii has also been recorded as a parasite of cysts of the nematode *Heterodera schachtii* Schmidt (Hannsler and Hermanns 1981).

Hyperparasitism is common in soil, particularly by *Gliocladium* and *Trichoderma* spp., which also exert considerable antibiosis, and fungi in both genera may kill the host without penetration (Papavizas 1985).

G. virens J. H. Miller, J. E. Giddens, & A. A. Foster inhibited the formation of sclerotia of *Sclerotinia sclerotiorum* (Tu 1980) and parasitized those already formed. It also parasitized mycelium of *Rhizoctonia solani* but not *Pythium ultimum* (Tu and Vaartaja 1981), which it inhibited by antibiosis (Howell 1982). *G. virens* has considerable possibilities for biological control in soilless potting mixes. Lumsden and Locke (1989) incorporated strain G20 of *G. virens* into a sodium alginate formulation, in which it maintained high populations for 2 mo at 4 and 20° C, long enough to control damping-off caused by *Rhizoctonia* and *Pythium*.

Moody and Gindrat (1977) recovered the mycoparasite *G. roseum* from a black organic soil baited with mycelial mats of *Phomopsis sclerotioides*, the cucumber black root rot pathogen. They had noted that this soil, artificially infested with *P. sclerotioides*, had a lower disease potential than a clay loam soil with a comparable inoculum. When *G. roseum*, cultured in a peat–soil medium, was introduced to a mineral soil infested with *P. sclerotioides*, it significantly decreased the severity of the disease. An inoculum prepared in conifer bark pellets lost its efficacy with time. The best control of black root rot by *G. roseum* occurred in wet or dry soil and following long periods of incubation at 30° C (Gindrat et al 1977).

G. catenulatum Gilman & E. Abbot parasitized mycelium of *Botrytis cinerea* in culture (Simay 1988) but has not been used commercially in biological control.

Species of *Trichoderma* are also important hyperparasites in the soil, as well as having other antibiotic properties. *T. harzianum* degrades fungal cell walls by the lytic action of β-(1→3)-glucanases and chitinase (Sivan and Chet 1989a). *T. hamatum* (Bonord.) Bainier produces cellulase (Chet and Baker 1981) and parasitizes *Pythium ultimum* and *Rhizoctonia solani* (Chet et al 1981). *T. viride* degrades hyphae of *Sclerotinia sclerotiorum* (D. Jones and Watson 1969). Certain *Trichoderma* spp. and strains, but not all, decay sclerotia of *Botrytis cinerea* (Köhl and Schlösser 1988, 1989).

A valuable attribute of *Trichoderma* spp. is that they can be genetically manipulated and selected for fungicide resistance and enhanced hyperparasitic and antagonistic abilities and can be adapted to particular rhi-

SUMMARY

Hyperparasitism

Hyperparasitism is a means of controlling pathogens in both their saprophytic and their parasitic states. Under experimental conditions, it has appeared to be very promising in disease control, but examples of commercial application are rare. The main reason seems to be that the autecology of hyperparasites is poorly understood for the most part, so that they are expected to be effective under conditions that are less than optimum for them. *Ampelomyces quisqualis*, a parasite of the Erysiphales, has some promise, and it has the advantage that it is compatible with some pesticides used on greenhouse crops. Its activity is enhanced when applications of pycnospores are interspersed with water sprays, which perhaps improve the conditions for hyperparasitic infection as well as disperse its hydrophilic pycnospores. *Trichoderma* spp. are active against both phylloplane and rhizosphere fungi. They are hyperparasitic, invading the mycelium and spores of pathogenic fungi; they also secrete chitinases and other lytic enzymes, such as glucanase and cellulase. Biotechnology has developed pesticide-tolerant strains of *Trichoderma* spp. and strains better adapted to specific rhizospheres. *Trichoderma* spp. are prolific colonizers of sterilized soils and enable many crops to survive the early, most pathogen-vulnerable stages.

zospheres (Papavizas 1985; Papavizas and Lewis 1981; Papavizas et al 1982). Along with *Gliocladium* spp. they are aggressive and prolific colonizers of sterilized soils, competing successfully for nutrients needed by pathogens (Sivan and Chet 1989b). They are thus particularly valuable in integrated control programs. For example, Cole and Zvenyika (1988) infested newly fumigated tobacco seedbeds with a triadimenol-resistant isolate of *T. harzianum*. It prevented the buildup of populations of *Rhizoctonia solani* and, to a lesser extent, *Fusarium solani*. In addition there was better growth and leaf yield, which could not be entirely connected with better disease control. Triadimenol enhanced the disease control given by *T. harzianum*.

Although the effect of *T. harzianum* persisted from the seedbed to the harvest of the tobacco leaf (Cole and Zvenyika 1988), Sundheim (1977) noted that the control of *Phomopsis sclerotioides* (the cucumber black root rot pathogen) by a bark pellet inoculum of *T. viride* and *T. polysporum* (Link) Rifai in a cucumber groundbed lasted for about 2 mo and was substantially lost by the third month. The *Trichoderma* spp. may have succumbed to competition from latecomers in the soil, or *P. sclerotioides* may not have been susceptible to them once it was in the cucumber roots. The fate of *Trichoderma* over long periods in different soils is not well understood.

Hypovirulence

Castanho and Butler (1978a,b) noted a disorder in *Rhizoctonia solani* that reduced its pathogenicity. They called the disorder Rhizoctonia decline and suggested that it had a potential use in the control of *Rhizoctonia*-induced diseases. The hypovirulent *R. solani*, however, had poor fitness for survival in the soil, and its potential in biological control could not be realized. The Rhizoctonia decline syndrome was associated with double-stranded RNA (Castanho et al 1978).

A hypovirulent strain of *Fusarium oxysporum* f. sp. *cucumerinum* J. H. Owen (*F. oxysporum* f. sp. *cucurbitacearum*) was found by Ishiba et al (1981). When the cut ends of cucumber leaf petioles were dipped for 2 h in a suspension of microconidia (2×10^4 per milliliter) of this strain, systemic resistance to a virulent strain of the pathogen was induced in the first true leaves, and the number and size of lesions caused by *Colletotrichum lagenarium* were reduced in a challenge inoculation 2–4 wk later. The inducer fungus survived in the hypocotyl but was not detected at the challenge sites in the leaves. The inducer appeared to be associated with an insoluble cell wall fraction of *F. oxysporum* f. sp. *cucumerinum*.

In neither case can hypovirulence be said to be a practicable possibility for biological control on a commercial scale.

Several early reports of the exploitation of hypovirulence in the control of virus diseases referred not so much to the cross-protectant effect of an attenuated virus as to the resistance of plants to strains of the same virus (R. W. Fulton 1986). Cross protection of a host by inoculation with a mild strain of a virus induces other risks: the protection may be incomplete; the mild inoculant strain may spread to other hosts in which its effects

SUMMARY
Hypovirulence

Except in the case of attenuated strains of tomato mosaic virus used for protection against virulent strains, hypovirulence is not used in practical biological control of greenhouse crop diseases, although it is recognized in *Rhizoctonia solani*. A hypovirulent strain of *R. solani* with double-stranded RNA had reduced pathogenicity but poor fitness for survival in soil. Other reported cases of hypovirulence may be interpreted as induced resistance conferred by a less pathogenic strain of a pathogen. Hypovirulence is not an inherited condition that leads to disease control, as in the well-known example of chestnut blight. The protection given by attenuated strains of a virus is not the result of true hypovirulence either, and the use of such strains has attendant risks. The protection may be incomplete; the mild strain may spread to more susceptible hosts; the host could be predisposed to attack from synergistic action of a severe, unrelated virus; and the mild strain could mutate to a severe strain.

could be more severe; the host could be predisposed to infection by a severe, unrelated virus through synergistic action; and the mild strain could mutate to a severe strain (Fulton 1986).

Nevertheless, some mild strains of viruses have been successfully used to protect crops from severe strains. In greenhouse crops, the most notable examples are tomato mosaic virus strain MII-16, produced by Rast (1972), and the heat-attenuated strain L_{11}. After passage through resistant hosts, the latter was reselected as strain L_{11A}, which protected many tomato cultivars lacking the *Tm-1* gene. Eventually, further passages resulted in strain L_{11A237} (Oshima et al 1978), which multiplied successfully in the host and protected cultivars containing *Tm-1*.

Suppressive Soils

It has long been recognized that some field soils bear crops that have less disease than neighboring crops growing in a different soil. K. F. Baker and Cook (1974) defined the soils with less disease as suppressive; even though a pathogen and a susceptible host are present, disease development is suppressed. The nature of the suppression is often poorly defined and may include fungistasis, poor competitive saprophytic ability of the pathogen, antibiosis, or some other form of biological or chemical control (Huber and Schneider 1982; Schneider 1982).

Greenhouse soils are usually so heavily amended that much of the work with field soils reviewed by Hornby (1983) and Schneider (1982), among many others, is hardly applicable. However, since greenhouse soils are usually sterilized or are replaced with various more inert media, they frequently begin with a biological vacuum, which can very quickly become occupied by dangerous populations of pathogens. There is an excellent opportunity, on the other hand, to make these soils suppressive to expected pathogens (Cook 1982).

Newly sterilized groundbed soil can contain 10^4 colony-forming units of *Fusarium oxysporum* f. sp. *radicis-lycopersici* per gram within a week. This population explosion is initiated by airborne microconidia (Rowe and Coplin 1976; Rowe et al 1977). It can be countered in soils amended with nonsterile peat or with green manure, especially lettuce residues (Jarvis 1977b; Jarvis and Thorpe 1981). The amended soils are suppressive, and levels of Fusarium crown and root rot are greatly reduced in tomato crops grown in them. The suppression of the disease by lettuce residues has been attributed to allelopathy (Jarvis and Thorpe 1981) and specifically to catecholic siderophores in the lettuce, especially in tissues challenged by the pathogen (Kasenberg and Traquair 1987, 1988, 1989). *F. oxysporum* f. sp. *radicis-lycopersici* is also susceptible to antagonism sensu lato from *Trichoderma harzianum, Aspergillus ochraceus*, and *Penicillium funiculosum* (Marois and Mitchell 1981a,b). The pathogen is a poor competitor in nonsterile soil, evidently, and is easily suppressed. Suppression probably results from at least three mechanisms: allelopathy, iron deprivation in the presence of siderophores, and competition from antagonists.

Other fungi may be similarly suppressed in greenhouse media. Tahvonen (1982a) found that light-colored sphagnum peat reduced the within-crop spread of *F. oxysporum* f. sp. *lycopersici*, an effect he attributed to streptomycetes (Tahvonen 1982b). Isolated, these organisms controlled damping-off by *Alternaria brassicicola* (Schwein.) Wiltshire and *Rhizoctonia solani* in cauliflower seedlings in sterilized or fresh peat. Wolffhechel (1989) also found organisms in peat that suppressed damping-off of cucumber by *Pythium ultimum* in sphagnum peat; they included *Trichoderma harzianum*, *Chrysosporium* spp., *Penicillium* spp., *Aspergillus* spp., and actinomycetes. *Gliocladium virens*, a hyperparasite, was also isolated from peat but inhibited the growth of cucumber seedlings when it was reintroduced into heat-treated peat.

Bolton (1977) noted that the suppression of root rot caused by *Pythium splendens* H. Braun in geranium cuttings differed in efficacy in different soilless mixes. He attributed the suppression to the activity of a *Trichoderma* sp., a *Penicillium* sp., and *Streptomyces* spp. These organisms were antagonistic to *P. splendens* in nonsterile soil–sand–peat, less antagonistic in autoclaved soil, and least effective in a mixture of terra-cotta pellets and vermiculite and a mixture of peat and terra-cotta pellets.

Soils suppressive to damping-off by *Rhizoctonia* spp. are also well known. In Hawaiian soils, suppressiveness was associated with populations of bacteria about four times higher than in conducive soils (Cheru and Ko 1988). Hyphae of *R. solani* lysed faster in suppressive soil, and the fungus did not survive as long as it did in conducive soil. Cheru and Ko concluded that suppressiveness to *R. solani* in radish monocropping resulted from enhanced competition from the bacteria.

Wijetunga and Baker (1979) also analyzed suppressiveness in the same damping-off disease of radish in successive monocrops. The soil became suppressive to *R. solani* if small hyphal fragments but not large fragments were added to it, and the disease incidence increased more rapidly when host tissue was reincorporated into the soil after each radish crop than when none was added. The ED_{50}, indicated by the ratio of inoculum density to disease incidence, was 8.0 propagules per gram in conducive soil and 21.5 in suppressive soil. Wijetunga and Baker deduced, from a series of dilutions of suppressive soil with increasing amounts of conducive soil, that the suppression could possibly result from hyperparasitism, interpreted as a "rhizosphere" effect adjacent to the thallus of *R. solani.*

A number of fungi isolated by Bourbos (1987) from the rhizosphere of tomato plants affected by *F. oxysporum* f. sp. *lycopersici* suppressed Fusarium wilt in sterile soil. These fungi included *Aspergillus alutaceus* Berk. & M. A. Curtis, *Paeciliomyces lilacinus, Penicillium chrysogenum* Thom and other *Penicillium* spp., and *Trichoderma viride*. Soil containing them suppressed the disease when it was added in a 1:10 ratio (w/w) to a naturally infested soil in a tomato greenhouse.

Siderophores

Suppressiveness in soils has been attributed to a number of mechanisms (Rovira 1982). They include hyperparasitism, the production of a specific antibiotic (as in the *Agrobacterium radiobacter* system), competition for infection sites on the root, the role of soil minerals in favoring the proliferation of saprophytic *Fusarium* spp. in the so-called Fusarium wilt–suppressive soils, and, frequently, the intervention of siderophores (Schroth and Hancock 1982).

Siderophores are extracellular, low-molecular-weight compounds with a high affinity for ferric iron (Leong 1986; Neilands and Leong 1986). The differential ability of organisms to sequester iron provides a competitive advantage and may well prevent the germination of infective propagules for which iron is essential. One example of this has already been noted, in soils amended with lettuce residues and suppressing Fusarium crown and root rot of tomato, caused by *F. oxysporum* f. sp. *radicis-lycopersici* (Kasenberg and Traquair 1987, 1988, 1989). In that case a catecholic siderophore comes from a higher plant; the pathogen has a hydroxamate siderophore with a lower affinity for iron.

Many studies have shown that conducive soils can be made suppressive by the introduction of microorganisms that produce extracellular siderophores complexing ferric iron needed by the pathogen (Kloepper et al 1980; Liao 1989; Paulitz et al 1989; Scher and Baker 1980, 1982; Simeoni et al 1987). In most cases, the siderophores are produced extracellularly from fluorescent *Pseudomonas* spp.

Siderophores have been implicated in Fusarium wilt–suppressive soils (C.-S. Park et al 1988; Scher and Baker 1982), often produced from the fluorescent pseudomonad *P. putida* and acting with nonpathogenic competitive forms of *F. oxysporum* against pathogenic forms. The siderophores of fluorescent pseudomonads compete for the ferric iron needed by the pathogen. *P. putida* together with nonpathogenic *F. oxysporum* suppressed cucumber wilt caused by *F. oxysporum* f. sp. *cucurbitacearum* at pH 6.7–8.1; separately they were not suppressive (Park et al 1988). Strains of *P. putida* reduced the germination of chlamydospores of both the pathogen and the nonpathogenic isolates of *F. oxysporum* in the rhizosphere, and the pseudomonad population in the rhizosphere increased in the presence of the nonpathogens at pH 8.1. It seemed to Park et al that populations of fluorescent pseudomonads and siderophore production are enhanced by cucumber root exudates induced by relatively high populations of nonpathogenic *F. oxysporum*.

Rhizosphere competence is indeed an essential attribute of organisms mediating suppressiveness. Park et al (1988) found that two strains of fluorescent pseudomonads differed in their control of cucumber wilt, even though they showed no differences in siderophore activity or antagonism

SUMMARY

Suppressive Soils

Some field soils are naturally suppressive: although a susceptible host and its pathogen are present, disease development is suppressed by fungistasis, poor competitive saprophytic ability of the pathogen, antibiosis, or some other form of biological or chemical control, often poorly understood. Greenhouse soils are frequently sterilized and lack the biological factors that may make them suppressive. Because they are biologically a near-vacuum, such soils can be made suppressive by the addition of appropriate microorganisms. It is hardly practicable to add vast amounts of inoculum to sterilized soils, but various amendments can introduce and maintain mixed populations of antagonistic organisms. Such amendments include peat, green manures, and composts. Fungi useful in imparting suppressiveness to soils include *Trichoderma* spp., *Aspergillus alutaceus, Paeciliomyces lilacinus, Penicillium chrysogenum*, and other *Penicillium* spp.

Suppressiveness in many soils has also been attributed to various pseudomonad bacteria that produce siderophores, particularly *Pseudomonas putida.* Siderophores are extracellular, low-molecular-weight compounds with a high affinity for ferric iron. The differential ability of organisms to sequester iron provides a competitive advantage to pseudomonads and a disadvantage to pathogens that require iron for germination and infection. Formae speciales of *Fusarium oxysporum* are particularly susceptible to this form of biological control. Siderophore interactions also occur on the phylloplane.

against *F. oxysporum* f. sp. *cucumerinum* in vitro and no differences in the inhibition of chlamydospore germination in the rhizosphere. *P. putida* strain N1R had greater rhizosphere competence and produced siderophores at higher iron levels than strain A12 of the same species, which provided inferior biological control.

The iron level in the soil is, indeed, a critical component of soil suppressiveness. Simeoni et al (1987) found that chlamydospore germination in *F. oxysporum* f. sp. *cucumerinum* was significantly reduced in the presence of a siderophore-producing strain of *P. putida*, compared with germination in the presence of a strain that did not produce siderophores, in soil containing 10^{-22} to 10^{-23} M Fe_3^+. Microbial activity reduced the amount of Fe_3^+ available for chlamydospore germination to 6 $\mu g \cdot kg^{-1}$, but toxic trace elements could decrease bacterial survival unless they were bound by a chelator. Simeoni et al suggested that the critical level of Fe_3^+ was 10^{-19} to 10^{-22} M, below which chlamydospore germination was suppressed, with optimum suppression occurring at concentrations between 10^{-22} and 10^{-27} M.

Siderophores on the phylloplane. Since bacteria are a normal component of the phylloplane and play a significant part in antagonism and competition against pathogens (Blakeman and Fokkema 1982), it is not unexpected that siderophore interactions occur there. Vedie and Le Normand (1984) described the iron-mediated competition between *Botrytis* spp., *Erwinia herbicola* (Löhnis) Dye, and *Pseudomonas fluorescens* (Trevisan) Migula. In culture, *P. fluorescens* inhibited the growth of *E. herbicola*, which in turn inhibited the growth of germ tubes of *B. cinerea* and *B. fabae* Sardiña. Ferric sulfate at 10^{-4} M inhibited necrosis by *B. cinerea* and *B. fabae* on leaves of *Vicia faba* L. and enhanced the population of *E. herbicola* on leaves at the expense of *P. fluorescens*. *E. herbicola* reversed the inhibition of the *Botrytis* spp. imposed by iron, which limited the formation of necrotic spots. The evident competition for iron by which *P. fluorescens* suppressed *E. herbicola* was relieved by supplying iron. Vedie and Le Normand believed *P. fluorescens* to produce a siderophore and suggested that if this facility could be genetically transferred to *E. herbicola*, its potential as a biocontrol agent would be considerably enhanced.

Delivery Systems

R. Baker (1986) and Scher and Castagno (1986), in a pair of complementary papers, outlined the practicalities of applied biological control. Baker remarked that in spite of decades of intensive research, few biological controls had then reached the market, and he advocated enhancement in the classic areas of biological control: antibiosis, exploitation of predation or hyperparasitism, and competition. Looking at the problem from an industrial viewpoint, Scher and Castagno said that for a biological control product to be successfully commercially developed, it should work well and reliably; should be needed by end users or others in the commercial distribution chain; should be technically feasible, economically feasible, and competitively attractive in comparison to other controls; should be acceptable to environmentalists and regulators; and, finally, should be compatible with the scientist's interests and with his or her agency's interests and activities.

Central to solving these practical problems is the selection of biological agents that are reliable in performance, survive in the field, are nonpathogenic to crops, animals, and people, and do not damage the environment (Forsyth 1990).

In some cases, simply enhancing natural populations of biological control agents may shortcut some of the problems. This can be done by changing the environment, as in the control of Fusarium wilts by supplying an appropriate nitrogen source together with lime and sometimes benomyl to soil to enhance populations of antagonists (J. P. Jones et al 1989). In the case of cucumber powdery mildew, naturally occurring populations of antagonists (Jarvis and Berry 1986) were enhanced by spraying leaves with microbial nutrients. The natural antagonists in peat (Tahvonen 1982a,b) also fit into this category, requiring no special delivery systems or registration by regulatory authorities.

The application of biological control agents has mostly followed methods developed for conventional chemical pesticides (Papavizas 1973; Ricard 1981). Microorganisms for the phylloplane are sprayed as suspensions of spores or bacteria in water, sometimes with conventional stickers and spreaders and with glycerol as a hygroscopic material (Spencer and Ebben 1981; Jarvis and Berry 1986). Adding spores in that manner, however, does not suffice for soil and rhizosphere application; activity is improved by

SUMMARY

Delivery Systems

Despite a very large volume of research, few biological controls have reached the market. For a product to be marketable, it has to be needed by growers and others in the commercial distribution chain; it should be technically and economically feasible and competitively attractive over other controls; it should be acceptable to environmentalists and regulators; and it should be at least as effective as other controls, reliable in performance, and persistent in its activity. Sometimes relatively simple modifications of the environment or cultural practices may suffice to enhance populations of biological control agents, but mostly it is necessary to deliver the agents to the plant or soil. Delivery systems usually duplicate those for conventional pesticides: sprays, drenches, root dips, soil amendments, and seed pelleting. Formulations for delivery therefore have to be compatible with existing technology and capable of being stored for long periods without special facilities.

providing an additional nutritious substrate that favors both competitive saprophytic ability and inoculum potential. R. Baker et al (1984) and Hadar et al (1979) found colonized peat–bran and wheat bran mixtures, respectively, to be suitable delivery substrates for *Trichoderma harzianum*, and Lewis and Papavizas (1987) used alginate pellets. Alginate pellets also proved suitable for the storage of high concentrations of inoculum of species of *Trichoderma, Cladosporium, Myrothecium, Chaetomium*, and *Drechslera* (Douville and Boland 1987). Viability was better after storage at 4°C than at 23°C and better at higher inoculum concentrations. Backman and Rodríguez-Kábana (1975) used diatomaceous earth granules impregnated with molasses and mineral salts for the culture and delivery of *T. harzianum*. They were stored dry at 5°C and were mixed with dry sterile granules and 10% molasses prior to delivery in the soil. The preparation does not necessarily have to be aseptic (Davet et al 1981).

The survival of biological controls is an important feature for commercial products. Cabanillas et al (1989) compared the survival of *Paeciliomyces lilacinus* in alginate pellets, diatomaceous earth granules, wheat grains, soil, and soil amended with chitin. Fungal viability was high in wheat and diatomaceous earth granules but poor in soil, amended or not. *P. lilacinus* protected tomatoes against *Meloidogyne incognita* best when delivered in pellets, wheat grains, or granules, both in the year of application and carrying over into the second year. Colony-forming units at the end of the second year were 10 times more numerous than before treatment.

The delivery of biological control agents in hydroponic systems presents greater problems, because of the rapid dilution and movement of the inoculum. Consequently, control has not proved consistent (van Peer et al 1988; van Steekelenburg and van der Sar 1987).

Seed can be pelleted with microorganisms as with chemicals (Gordon-Lennox et al 1987). Harman and Taylor (1988) and Harman et al (1989) obtained improved biological control and overall seedling performance from seeds treated with *Enterobacter cloacae* (Jordan) Hormaeche & Edwards and *Trichoderma* spp. if the seeds were osmotically primed at pH 4.1 before sowing.

For the treatment of pruning wounds with biological control agents, the method of Grosclaude et al (1973) is applicable to greenhouse crops in some situations, such as taking cuttings. Secateurs are fitted with a small pump that ejects a spore suspension onto the blades each time a cut is made.

Integrated Disease Management

CHAPTER 10

It is evident that greenhouse crops are protected only from the weather outside—and sometimes imperfectly at that—and scarcely at all from diseases. The structural design, orientation, topographical siting, and covering material have a permanent influence on the climate inside, and ventilation, cooling, and heating systems enable the grower to modify the climate daily to various extents. Increasing use of computers and advanced engineering permit finer and finer degrees of control over climate, irrigation, and fertilization. Notwithstanding, the climate inside the greenhouse is far from uniform. There are steep vertical and horizontal gradients in temperature and hence also in vapor pressure deficit and relative humidity. There are also pockets and corners of still air, which increase the likelihood of dew formation and impede the evaporation of inoculum drops. The problem is compounded by dense cropping and often by poor bench design.

The primary purpose of all greenhouse control systems, whether of climate, fertilization, irrigation, or mechanization, is to produce the maximum perfect-quality yield per unit of area within a rigidly delimited time, especially in florist crops. Plant protection, as such, often receives scant attention, and if crop management happens to produce a disease-free crop, it is sometimes more by luck than by conscious disease-escape management.

Grainger (1979), discussing the scientific basis for economic decisions for farmers, remarked that a piece of scientific fallacy is the tacit assumption that host plants are invariably receptive to pathogens. This assumption, he said, is incorrect, but a host plant can be made receptive at any time by bad management. This is often loosely characterized as stress: a crop pushed into high productivity by advanced horticultural technology can be said to be stressed in phytopathological terms (Jarvis 1989). Thus, *Penicillium oxalicum* Currie & Thom has emerged as an important pathogen of cucumber in high-technology production (Jarvis et al 1990), and *Plasmopara lactucae-radicis* M. E. Stanghellini & R. L. Gilbertson as a pathogen of lettuce roots in hydroponic production (Stanghellini et al 1990). Dhanvantari (1990) remarked that a new bacterial stem necrosis in tomato, caused by a *Pseudomonas* sp., was a function of stress in mature plants, associated with such factors as unbalanced nutrition, excessive humidity, or the onset of fruiting. It is these factors that require definition, in answer to Grainger's call for an equal study of host, pathogen, and environment (Grainger 1979).

In parentheses, it is interesting to note that experienced extension pathologists and horticulturalists rate bad management and human error quite highly as stress factors—errors such as miscalculating fertilizer and pesticide rates, using the same sprayer for herbicides and fungicides, buying bargain surplus transplants from suspect sources, or saving seed from a virus-infected crop. These sorts of problems may account for something on the order of 25% of an extension agent's caseload.

Self-diagnosis of diseases and, worse, ignorant self-prescription of remedies also contribute to bad-management stress, and it may well be that expert diagnostic systems might be incorporated into the grower's computer menu. However, diagnosis of disorders of complex etiology is one thing (Wallace 1978), and analyzing their base causes and prescribing remedies by computer is another (Latin et al 1987; Wahl 1989). At present, there is no substitute for the experience of the pathologist, the extension agent, and the grower working together in problem solving.

Consider the examples described by K. F. Baker et al (1954) and Dimock and Baker (1951) in analyzing the complex etiology of a disorder of column stock (*Matthiola incana* (L.) R. Br.) and that of rust of snapdragon (*Antirrhinum majus* L.), respectively. In the first case, salinity injury, climate, fungicide application, and *Botrytis cinerea* Pers.:Fr. all combined to kill column stock. In the second, climate, the source and age of spores, and the use of Bordeaux mixture (which kills *Fusarium roseum* Link:Fr., a natural antagonist of *Puccinia antirrhini* Dietel & Holw.) exacerbated snapdragon rust. More recently, Thompson and Jenkins (1985) analyzed the roles of temperature, leaf wetness, high relative humidity, prolonged high temperature, irrigation, rainfall, and the age of leaves in affecting the size of anthracnose lesions on cucumber and the production of conidia of *Colletotrichum lagenarium* (Pass.) Ellis & Halst. Only about 20% of statistical variation was explained by climatic factors, and only about 50% was explained when the time since inoculation was taken into account. Predictive accuracy remained elusive for this disease. Such examples would be difficult, but not impossible, to describe in an expert system.

The relatively simple microclimate factors that determine sporulation, spore dispersal, the imminence of dew for inoculum drops, soil temperature, water stress, insect vector activity, and many other biological functions in infection and pathogenesis should be relatively easy, however, to incorporate into expert systems for many diseases. The major problem would be to construct a general system for all diseases. The best that can be done, perhaps, is to construct compromise systems that would confer escape from two or three of the major diseases of a particular crop in a particular locality. Gray mold, whose biology is fairly well known, is a universal disease, for which an expert system might be constructed for greenhouse crops.

Integrated disease management, therefore, implies climate control for disease escape compatible with climate control for optimum crop yield and quality, along with measures for eliminating inoculum (sterilizing soil or using soilless media and obtaining disease-free planting material), measures for limiting disease spread, biological and pesticidal control, and, last but most important, the use of resistant germ plasm.

Not only is the manipulation of host germ plasm a primary role of plant breeders (Trenbath 1984), but genetic manipulation is a valuable tool for microbiologists in improving the antibiotic activity, sensu lato, of biological control organisms, including entomopathogens (Gutterson et al 1986; Heale 1988); improving rhizosphere and phyllosphere competence (Papavizas 1987); introducing avirulent mutants of a pathogen in biological control (Trigalet and Trigalet-Demery 1990); and improving the pesticide tolerance of biological control organisms in integrated programs (Papavizas et al 1982) and their survival in effective populations (May 1985).

Plant breeders run on a treadmill, because of the plasticity of pathogens, and host resistance will always have to be augmented by other means in order to keep inoculum pressures to a minimum. The possible emergence of new diseases or new races of old pathogens demands the constant vigilance of the grower. Fortunately, most greenhouse crops require daily attention for routine cultural operations, and aberrations are readily detected. Early warning of trouble and the establishment of economic thresholds of tolerable damage are essential in the decision to mobilize therapeutic measures (Young et al 1978; Zadoks 1985).

Stern et al (1959), referring rather to insect pests, defined the economic injury level as the lowest population at which a pest causes economic damage, and they defined the economic threshold as the pest population at which control measures should be introduced to prevent it from increasing to the economic injury level. In terms of disease and pathogen thresholds these definitions require reinterpretation depending not only on pathogen populations, such as sporulating sites or airborne spores, but also on host receptivity sensu Grainger (1979). For petal flecking by *B. cinerea* in florist crops, the disease threshold approaches zero; for stem lesions in tomato caused by the same pathogen, the tolerance level is much higher unless lesions girdle the stems. No threshold figures, however, appear to have been established for most greenhouse crops.

Economic decisions in closed systems require complex assessments of the costs and benefits of integrated disease and insect pest control, crop management inputs, and projected returns in temporal relation to disease and pest infestations (Edens and Haynes 1982; van Lenteren and Woets 1988). In greenhouses, for example, attempts to save energy by restricting ventilation are offset by increased risks of gray mold in tomato and florist crops or downy mildew in lettuce (Augsburger and Powell 1986; W. M. Morgan 1983a, 1984a,b, 1985; Morgan and Molyneux 1981). Since the value of the crop changes over several weeks or months, the grower must consider the timing of the onset of a disease and whether its effect is transitory (as in petal flecking) or more permanent (as in root rot). In the latter case, the only recourse may be to pull the crop out and replant.

It may well be possible to model aids to this decision making at policy, strategic, and tactical levels (Conway 1984a–c), but no modeling, save for production, has been done for greenhouse crop protection (Kranz et al 1984).

Different control strategies are appropriate for different sorts of pathogens. Conway (1984b) has argued that two groups of pests and pathogens can be distinguished: *r*-pathogens, which are essentially opportunistic, with high potential rates of population increase and very effective means of dispersal and host-finding ability, and *K*-pathogens, which occupy more specialized niches and have low potential rates of increase but show greater competitive ability; *r* and *K* are population growth parameters (Boyce 1984). The *r*-pathogens are the more difficult to control, with frequent invasions and massive, damaging outbreaks, which require fast and flexible response. *B. cinerea* is an *r*-pathogen. *K*-pathogens, on the other hand, are more easily controlled by reducing or eliminating the population or eliminating the pathogen niche (for example, by switching to a monogenically resistant cultivar). Formae speciales and races of *Fusarium oxysporum* are *K*-pathogens. Appropriate control strategies have been identified for *r*- and *K*-pathogens and intermediate pathogens (Table 35).

The need to consider integrated control of different pathogens in different ways has been reviewed by Ebben (1987). She compared and contrasted the integrated control of some diseases of greenhouse crops and pointed out where modifications of management practices could help to tilt the biological control mechanism in a more effective direction. Attempts to control the carnation rust fungus, *Uromyces dianthi* (Pers.:Pers.) Niessl, by *Verticillium lecanii* (A. Zimmerm.) Viégas (Spencer 1980; Spencer and Atkey 1981) fail because the establishment of *V. lecanii* on the leaf surface is inadequate to cope with the prolific sporulation of the rust. Ebben suggested that a more humid leaf microclimate could improve the activity of *V. lecanii*—which also requires a low vapor pressure deficit to parasitize aphids (Lisansky and Hall 1983). That is impracticable in carnation culture, however, and so a better-adapted mycoparasite might be sought.

Ebben (1987) was more optimistic about the integrated control of soilborne diseases, not least because the alternatives—sterilization or soil substitutes—are expensive and can be justified only in high-return crops. Lettuce growers,

Table 35. Principal control techniques appropriate for different pathogen strategies[a]

	r-Pathogens	Intermediate pathogens	*K*-pathogens
Pesticides	Early, wide-scale applications based on forecasting	Selective pesticides	Precisely targeted applications based on monitoring
Biological control		Introduction or enhancement of natural enemies	
Cultural control	Timing, cultivation, sanitation	→ ←	Changes in agronomic practice, destruction of alternate hosts
Resistance	General, polygenic resistance	→ ←	Specific, monogenic resistance

[a] Adapted from Conway (1984b).

for example, with low economic input and returns, still rely mostly on soil, so that if white mold (caused by *Sclerotinia sclerotiorum* (Lib.) de Bary) or lettuce drop (caused by *S. minor* Jagger) becomes a problem, soil sterilization may prove too expensive. Soilborne pathogens tend to build up in situ and spread slowly, allowing the simultaneous buildup of selected hyperparasites and antagonists. Ebben considered that soilborne pathogens are spread mostly by mismanagement, not least by the unsafe disposal of crop debris.

At the tactical level of formulating control measures are modeling in order to improve the precision of disease management by forecasting (Waggoner 1984) and its corollary, monitoring (Kranz et al 1984). Modeling for forecasting has not been done for greenhouse crop diseases, but studies such as those of Marois et al (1988) on gray mold in roses are a beginning. Monitoring to provide data for forecasting based on appropriate sampling techniques has not been done either, but daily inspection of greenhouse crops is commonplace.

The special case of policy with respect to pesticides has been discussed by Regev (1984), who argued that we have become addicted to pesticide use and have entered a perpetual cycle of pests–pesticides–more pests–more pesticides. Pesticides are designed to reduce crop losses, but we still incur losses, largely because of pest resistance and the suppression of natural enemies. There comes a point, therefore, when there has to be a retrenchment in pesticide application. Integrated control enables other control mechanisms to operate in the gap, even with enhanced effectiveness if natural enemies are no longer suppressed.

For virus diseases, Dunez (1987) compiled a checklist of integrated measures:

- Production of virus-free plants
- Diagnosis
- Maintenance of virus-free stocks
- Chemotherapy
- Breeding for resistance
- Use of viral genes or viral information in breeding for resistance

Thermotherapy and micropropagation may be added to that list.

At the tactical level, a number of successful case histories illustrate the need for adopting integrated control: it has been successful against Phialophora wilt of carnation (R. Baker 1980), Fusarium foot rot of Douglas fir (Sinclair et al 1975), downy mildew of lettuce (Crute 1984; W. M. Morgan 1984b), and gray mold of tomato (W. M. Morgan 1985).

The program described by Cho et al (1989) epitomizes the integrated approach, the principles of which are generally applicable to a wide variety of diseases, with the additional advantage of having control of the environment in the greenhouse. A multidisciplinary team was assembled in Hawaii to combat tomato spotted wilt virus, which has some 200 hosts among both crops and weeds. The team engaged in the following studies:

- Identification and detection of the virus
- Relationships between the virus and its thrips vector *Franklinella occidentalis* (Pergande)
- Chemical control of the vector
- A search for resistant germ plasm
- Management strategies
  - Precrop phase: rotation, placement (crop separation), control of alternate hosts and vector hosts
  - Crop phase: virus-free seedlings, insecticides, reduced cultivation
  - Postcrop phase: fallow for 3–4 wk to allow thrips to emerge and disperse, soil fumigation

Management was not completely effective if the virus and vector levels were high, and the cooperation of growers was essential for concerted management and placement of crops.

Greenhouse growers have a very large menu of available control measures and the opportunity to put them into practice in the controllable greenhouse environment. Major natural disease threats can be tackled by national and international regulatory policy, and concerted regional action can place a plant-protection umbrella over specialized horticultural industries. The integration of production and marketing tactics, strategy, and policy must take in the protection of plants from diseases and pests if horticulture is to continue to provide food and beauty for us all. As Horne (1989) pointed out, the groundwork for decision is a symphony of parts and processes brought together by people who share a variety of roles and responsibilities. The recommendation system for plant disease control is structured from scientific concepts and knowledge of crop production to form a disease control strategy. It represents a partnership between the scientific community and producers.

References

Abdel-Salem, M. M. 1934. Botrytis disease of lettuce. J. Pomol. 12:15-35.

Abiko, K., and Kishi, J. 1979. Influence of temperature and humidity on the outbreak of cucumber powdery mildew. Bull. Veg. Ornamental Crops Res. Stn. (Jpn.) Ser. A 5:167-176.

Adams, A. J., and Palmer, A. 1989. Air-assisted electrostatic application to glasshouse tomatoes: Droplet distribution and its effect upon whiteflies (*Trialeurodes vaporariorum*) in the presence of *Encarsia formosa*. Crop Prot. 8:40-48.

Adams, R. P., and Robinson, I. 1979. Treatment of irrigation water by ultra-violet radiation. Pages 91-97 in: Plant Pathogens. D. W. Lovelock, ed. Academic Press, London.

Adatia, M. H., and Besford, R. T. 1986. The effects of silicon on cucumber plants grown in recirculating nutrient solution. Ann. Bot. (London) 58:343-351.

Airhart, D. L. 1984. Overcoming horticultural problems in solar greenhouses. Acta Hortic. 148:785-790.

Aitken, E. A. B., Newbury, H. J., and Callow, J. A. 1989. Races of rust (*Puccinia antirrhini*) of *Antirrhinum majus* and the inheritance of host resistance. Plant Pathol. 38:169-175.

Aitken-Christie, J., and Davies, H. E. 1988. Development of a semi-automated micropropagation system. Acta Hortic. 230:81-87.

Albright, L. D. 1984. Modeling thermal mass effects in commercial greenhouses. Acta Hortic. 148:359-368.

Aldrich, R. A., and Nichols, L. P. 1969. Temperatures of soil mixtures during steam treatment. Pages 1-2, 9, 14 in: Pa. Flower Grow. Bull. 221.

Aldrich, R. A., White, J. W., and Manbeck, H. B. 1967. Transmission of solar energy through rigid plastic greenhouses. Pages 79-91 in: Proc. Natl. Agric. Plast. Conf. 7th.

Ali, M. K., Lepoivre, P., and Semal, J. 1988. In vitro selection of *Phytophthora citrophthora* isolates resistant to phosphorous acid and fosetyl-Al. Meded. Fac. Landbouwwet. Rijksuniv. Gent 53(2b):597-604.

Allen, W. R., and Matteoni, J. A. 1988. Cyclamen ringspot: Epidemics in Ontario greenhouses caused by tomato spotted wilt virus. Can. J. Plant Pathol. 10:41-46.

Allmaras, R. R., Kraft, J. M., and Miller, D. E. 1988. Effects of soil compaction and incorporated crop residue on root health. Annu. Rev. Phytopathol. 26:219-243.

Altman, J. 1970. Increased and decreased plant growth responses resulting from soil fumigation. Pages 216-222 in: Root Diseases and Soil-Borne Pathogens. T. A. Toussoun, R. V. Bega, and P. E. Nelson, eds. University of California Press, Berkeley.

Amdurskey, V. 1980. Computer models of short term plant growth and short wave radiation in greenhouses. Acta Hortic. 106:147-148.

Amsen, M. G. 1980. Is leaf temperature a suitable parameter for greenhouse environment control? Acta Hortic. 106:137-138.

Anderson, D. B. 1936. Relative humidity or vapor pressure deficit. Ecology 17:277-282.

Anderson, R. M. 1979. Parasite pathogenicity and the depression of host population equilibria. Nature (London) 279:150-152.

Anonymous. 1967. Tomato, heated: CO_2 enrichment of an early crop. Pages 86-88 in: Rep. Efford Exp. Hortic. Stn. 1967.

———. 1971. Hot-water treatment of plant material. Minist. Agric. Fish. Food (G.B.) Bull. 201.
———. 1973. Tomato mosaic control by inoculation. Minist. Agric. Fish. Food (G.B.) Short Term Leafl. 152.
———. 1976. Fogging under glass—Special review. Grower (London) 86:191-198.
———. 1983. Cucumbers. Grower Guide 15. Grower Books, London.
———. 1988a. Computers in greenhouses. Introducing a decision-management unit. Annu. Rep. Ohio Agric. Res. Dev. Cent.
———. 1988b. Crown and root rot. Grower (London) 109(14):56-57.
———. 1989. Tomatoes in bags of jelly. Grower (London) 111(8):19-21.
———. 1990. Targets for 1990. Grower (London) 113(2):HN12-15.
Ark, P. A. 1944. Studies on bacterial canker of tomato. Phytopathology 34:394-400.
Arnold, A. C. 1980. Developments in sampling and measuring techniques for monitoring small droplets. Pages 233-240 in: Spraying Systems for the 1980s. Monogr. 24. BCPC Publications, Croydon, England.
Atilano, R. A., and van Gundy, S. D. 1979. Systemic activity of oxamyl to *Meloidogyne javanica* at the root surface of tomato plants. Pages 339-349 in: The Soil-Root Interface. J. L. Harley and R. S. Russell, eds. Academic Press, London.
Augsburger, N. D., and Powell, C. C. 1986. Correct greenhouse ventilation: The basics of excessive humidity control. Pages 6-8 in: Ohio Florists Assoc. Bull. 675.
Avgelis, A. D., and Manios, V. I. 1989. Elimination of tomato mosaic virus by composting tomato residues. Neth. J. Plant Pathol. 95:167-170.
Backman, P. A., and Rodríguez-Kábana, R. 1975. A system for the growth and delivery of biological control agents to the soil. Phytopathology 65:819-821.
Baeten, S., Verlodt, H., El Gahem, S., and Harboui, Y. 1985. Visualization of temperature distribution in PE greenhouses with static aeration. Acta Hortic. 170:173-184.
Bailey, B. J. 1985. Wind dependent control of greenhouse temperature. Acta Hortic. 174:381-384.
Bainbridge, A., and Legg, B. J. 1976. Release of barley-mildew conidia from shaken leaves. Trans. Br. Mycol. Soc. 66:495-498.
Baker, E. A. 1974. The influence of environment on leaf surface wax development in *Brassica oleracea* var. *gemmifera*. New Phytol. 73:955-966.
Baker, K. F. 1952. A problem of seedsmen and flower growers—Seedborne parasites. Seed World 70(11):33-47.
———. 1956. Development and production of pathogen-free seed of three ornamental plants. Pages 68-71 in: Plant Dis. Rep. Suppl. 238.
———. 1962a. Principles of heat treatment of soil and planting material. J. Aust. Inst. Agric. Sci. 28:118-126.
———. 1962b. Thermotherapy of planting material. Phytopathology 52:1244-1255.
———. 1970. Selective killing of soil microorganisms by aerated steam. Pages 234-239 in: Root Diseases and Soil-Borne Pathogens. T. A. Toussoun, R. V. Bega, and P. E. Nelson, eds. University of California Press, Berkeley.
———. 1987. Evolving concepts of biological control of plant pathogens. Annu. Rev. Phytopathol. 25:67-85.
Baker, K. F., and Chandler, P. A. 1956. Development and production of pathogen-free propagative material of foliage and succulent plants. Pages 88-90 in: Plant Dis. Rep. Suppl. 238.
———. 1957. Development and maintenance of healthy planting stock. Pages 217-236 in: The U.C. System for Producing Healthy Container-Grown Plants. K. F. Baker, ed. Calif. Agric. Exp. Stn. Man. 23.
Baker, K. F., and Cook, R. J. 1974. Biological Control of Plant Pathogens. W. H. Freeman, San Francisco.
Baker, K. F., and Linderman, R. G. 1979. Unique features of the pathology of ornamental plants. Annu. Rev. Phytopathol. 17:253-277.
Baker, K. F., and Olsen, C. M. 1960. Aerated steam for soil treatment. (Abstr.) Phytopathology 50:82.
Baker, K. F., and Roistacher, C. N. 1957a. Principles of heat treatment of soil. Pages 138-161 in: The U.C. System for Producing Healthy Container-Grown Plants. K. F. Baker, ed. Calif. Agric. Exp. Stn. Man. 23.

_____. 1957b. Equipment for heat treatment of soil. Pages 162-196 in: The U.C. System for Producing Healthy Container-Grown Plants. K. F. Baker, ed. Calif. Agric. Exp. Stn. Man. 23.

Baker, K. F., Matkin, O. A., and Davis, L. H. 1954. Interaction of salinity injury, leaf age, fungicide application, climate, and *Botrytis cinerea* in a disease complex of column stock. Phytopathology 44:39-42.

Baker, R. 1965. The dynamics of inoculum. Pages 395-403 in: Ecology of Soil-Borne Plant Pathogens: Prelude to Biological Control. K. F. Baker and W. C. Snyder, eds. University of California Press, Berkeley.

_____. 1980. Measures to control Fusarium and Phialophora wilt pathogens of carnations. Plant Dis. 64:743-749.

_____. 1986. Biological control: An overview. Can. J. Plant Pathol. 8:218-221.

Baker, R., and Phillips, D. J. 1962. Obtaining pathogen-free stock by shoot tip culture. Phytopathology 52:1242-1244.

Baker, R., and Scher, F. M. 1987. Enhancing the activity of biological control agents. Pages 1-17 in: Innovative Approaches to Plant Disease Control. I. Chet, ed. John Wiley & Sons, New York.

Baker, R., Elad, Y., and Chet, I. 1984. The controlled experiment in the scientific method with special emphasis on biological control. Phytopathology 74:1019-1021.

Bakker, J. C. 1984a. Effects of changes in ventilation on cucumber. Acta Hortic. 148:519-524.

_____. 1984b. Physiological disorders in cucumber under high humidity conditions and low ventilation rates in greenhouses. Acta Hortic. 156:252-264.

_____. 1989. The effects of air humidity on growth and fruit production of sweet pepper (*Capsicum annuum* L.). J. Hortic. Sci. 64:41-46.

Bakker, J. C., and van de Vooren, J. 1984a. Water vapour transport from a greenhouse by ventilation. I. Effects on climate. Acta Hortic. 148:535-542.

_____. 1984b. Plant densities and training systems in greenhouse cucumber. Acta Hortic. 156:43-48.

Bakker, J. C., Welles, G. W. H., and van Uffelen, J. A. M. 1987. The effects of day and night humidity on yield and quality of glasshouse cucumbers. J. Hortic. Sci. 62:363-370.

Bald, J. G. 1956. Development and production of pathogen-free gladiolus cormels. Pages 81-84 in: Plant Dis. Rep. Suppl. 238.

Bals, E. J. 1978a. Reasons for CDA—Controlled drop application. Pages 659-666 in: Proc. Br. Crop Prot. Conf. Pests Dis. 1978.

_____. 1978b. Reduction of active ingredient dosage by selecting appropriate droplet size for the target. Pages 101-106 in: Symposium on Controlled Drop Application. Monogr. 22. BCPC Publications, Croydon, England.

Bant, J. H., and Storey, L. F. 1952. Hot-water treatment of celery seed in Lancashire. Plant Pathol. 3:21-25.

Bar-Joseph, M., Segev, D., Twizer, S., and Rosner, A. 1985. Detection of avocado sunblotch viroid by hybridization with synthetic oligonucleotide probes. J. Virol. Methods 10:69-73.

Bark, L. D., and Carpenter, W. J. 1969. Understanding the air circulation patterns produced by the horizontal unit heater/convection tube systems. Florists' Rev. 143(3713):18-19, 60-61.

Barkai-Golan, R., Lavy-Meir, G., and Kopeliovitch, E. 1989. Stimulation of fruit ethylene production by wounding and by *Botrytis cinerea* and *Geotrichum candidum* infection in normal and non-ripening tomatoes. J. Phytopathol. (Berlin) 125:148-156.

Bartok, J. W., and Aldrich, R. A. 1984. Low cost solar collectors for greenhouse water heating. Acta Hortic. 148:771-774.

Bashi, E., and Aylor, D. E. 1983. Survival of detached sporangia of *Peronospora destructor* and *Peronospora tabacina.* Phytopathology 73:1135-1139.

Bateman, D. F. 1961. The effect of soil moisture upon development of poinsettia root rots. Phytopathology 51:445-451.

_____. 1962. Relation of soil pH to development of poinsettia root rots. Phytopathology 52:559-566.

Bateman, D. F., and Dimock, A. W. 1959. The influence of temperature on root rots of poinsettia caused by *Thielaviopsis basicola, Rhizoctonia solani,* and *Pythium ultimum.* Phytopathology 49:641-647.

Bates, M. L., and Stanghellini, M. E. 1984. Root rot of hydroponically grown spinach caused by *Pythium aphanidermatum* and *P. dissotocum*. Plant Dis. 68:989-991.

Beach, W. S. 1949. The effects of excess solutes, temperature and moisture upon damping-off. Pa. Agric. Exp. Stn. Bull. 509.

Beaumont, A., Dillon Weston, W. A. R., and Wallace, E. R. 1936. Tulip fire. Ann. Appl. Biol. 23:57-88.

Beck, G. E., and Vaughn, J. R. 1949. Botrytis leaf and blossom blight of *Saintpaulia*. Phytopathology 39:1054-1056.

Beech, M. 1989. Healthy plants for a healthy industry. Grower (London) 112(13):55-57.

Bender, C. L., and Coyier, D. L. 1984. Isolation and identification of races of *Sphaerotheca pannosa* var. *rosae*. Phytopathology 74:100-103.

Benhamou, N., Lafontaine, J. G., Joly, J. R., and Ouellette, G. B. 1985. Ultrastructural localization in host tissues of a toxic glycopeptide produced by *Ophiostoma ulmi* monoclonal antibodies. Can. J. Bot. 63:1185-1195.

Benn, W. 1987. Technologischer Fortschritt bei der Kaltvernebelung von Pflanzenshutzmitteln in Gewächshäusen. Nachrichtenbl. Pflanzenschutzdienst DDR 41:241-246.

Benoit, F., and Ceustermans, N. 1989. Recommendations for the commercial production of butterhead lettuce in NFT. Soilless Cult. 5:1-12.

Bergman, H. F. 1959. Oxygen deficiency as a cause of disease in plants. Bot. Rev. 25:418-485.

Bergstrom, G. C., Johnson, M. C., and Kuć, J. 1982. Effects of local infection of cucumber by *Colletotrichum lagenarium, Pseudomonas lachrymans,* or tobacco necrosis virus on systemic resistance to cucumber mosaic virus. Phytopathology 72:922-926.

Bernstein, L. 1964. Salt tolerance of plants. U.S. Dep. Agric. Agric. Inf. Bull. 283.

Besri, M. 1982. Solar heating (solarization) of tomato supports for control of *Didymella lycopersici* Kleb. stem canker. (Abstr.) Phytopathology 72:939.

Bewley, W. F. 1923. Diseases of Glasshouse Plants. Benn, London.

Bhatt, D. D. 1962. The role of associated fungi in prevention of Botrytis rot of strawberries. (Abstr.) Phytopathology 52:359.

Biemond, T. 1989. A glasshouse climate model as part of a bio-economic model. Acta Hortic. 260:275-280.

Bilgrami, K. S. 1963. Role of moisture in spreading the infection caused by some species of *Phyllosticta*. Sci. Cult. 29:453-454.

Bird, G. W. 1969. Depth of migration of *Meloidogyne incognita* (Nematodea) associated with greenhouse tomato and cucumber roots. Can. J. Plant Sci. 49:132-134.

Biro, R. L., and Jaffe, M. J. 1984. Thigmomorphogenesis: Ethylene evolution and its role in the changes observed in mechanically perturbed bean plants. Physiol. Plant. 62:289-296.

Blackman, V. H., and Welsford, E. J. 1916. Studies in the physiology of parasitism. II. Infection by *Botrytis cinerea*. Ann. Bot. (London) 30:389-398.

Blad, B. L., Steadman, J. R., and Weiss, A. 1978. Canopy structure and irrigation influence white mold disease and microclimate of dry edible beans. Phytopathology 68:1431-1437.

Blair, I. D. 1943. Behaviour of the fungus *Rhizoctonia solani* Kühn in the soil. Ann. Appl. Biol. 30:118-127.

Blakeman, J. P. 1972. Effect of plant age on inhibition of *Botrytis cinerea* spores by bacteria on beet-root leaves. Physiol. Plant Pathol. 2:143-152.

———. 1978. Microbial competition for nutrients and germination of fungal spores. Ann. Appl. Biol. 89:151-155.

———. 1980. Behavior of conidia on aerial plant surfaces. Pages 115-151 in: The Biology of *Botrytis*. J. R. Coley-Smith, K. Verhoeff, and W. R. Jarvis, eds. Academic Press, London.

———. 1985. Ecological succession of leaf surface microorganisms in relation to biological control. Pages 6-30 in: Biological Control on the Phylloplane. C. E. Windels and S. E. Lindow, eds. American Phytopathological Society, St. Paul, MN.

Blakeman, J. P., and Brodie, I. D. S. 1977. Competition for nutrients between epiphytic organisms and germination of spores of plant pathogens on beetroot leaves. Physiol. Plant Pathol. 10:29-42.

Blakeman, J. P., and Fokkema, N. J. 1982. Potential for biological control of plant diseases on the phylloplane. Annu. Rev. Phytopathol. 20:167-192.

Blakeman, J. P., and Fraser, A. K. 1971. Inhibition of *Botrytis cinerea* spores by bacteria on the surface of chrysanthemum leaves. Physiol. Plant Pathol. 1:45-54.

Blaker, N. S., and MacDonald, J. D. 1981. Predisposing effects of soil moisture extremes on the susceptibility of rhododendron to Phytophthora root and crown rot. Phytopathology 71:831-834.

Bloomberg, W. J. 1979. Model simulations of infection of Douglas-fir seedlings by *Fusarium oxysporum*. Phytopathology 69:1072-1077.

Bogemann, B. 1980. Economic aspects of growth regulation by climate control. Acta Hortic. 106:191-192.

Boland, C. J., and Inglis, G. D. 1988. Antagonism of white mold (*Sclerotinia sclerotiorum*) of bean by fungi from bean and rapeseed flowers. Can. J. Bot. 67:1775-1781.

Bollen, G. J. 1969a. Effects of steam sterilization on the biological characteristics of soil. Meded. Dir. Tuinbouw (Neth.) 32:475-480.

_____. 1969b. The selective effect of heat treatment on the microflora of a greenhouse soil. Neth. J. Plant Pathol. 75:157-163.

_____. 1985. The fate of plant pathogens during composting of crop residues. Pages 282-287 in: Composting of Agricultural and Other Wastes. J. K. R. Gasser, ed. Elsevier Applied Science, London.

Bollen, G. J., and Scholten, G. 1971. Acquired resistance to benomyl and some other systemic fungicides in a strain of *Botrytis cinerea* in cyclamen. Neth. J. Plant Pathol. 77:83-90.

Bolton, A. T. 1976. Fungicide resistance in *Botrytis cinerea*, the result of selective pressure on resistant strains already present in nature. Can. J. Plant Sci. 56:861-864.

_____. 1977. The severity of root rot and persistence of *Pythium splendens* in geranium cuttings grown in soilless mixes. Can. J. Plant Sci. 57:87-92.

Bolton, A. T., and Svejda, F. J. 1979. A new race of *Diplocarpon rosae* capable of causing severe black spot on *Rosa rugosa* hybrids. Can. Plant Dis. Surv. 59:38-41.

Bonde, R., and Schultz, E. S. 1943. Potato cull piles as a source of late-blight infection. Am. Potato J. 20:112-118.

Boote, K. J., Jones, J. W., Mishoe, J. W., and Berger, R. D. 1983. Coupling pests to crop growth simulators to predict yield reductions. Phytopathology 73:1581-1587.

Bos, L., van Dorst, H. J. M., Huttinga, H., and Maat, D. Z. 1984. Further characterization of melon necrotic spot virus causing severe disease in glasshouse cucumbers in the Netherlands and its control. Neth. J. Plant Pathol. 90:55-69.

Bot, G. P. A. 1980. Validation of a dynamical model of greenhouse climate. Acta Hortic. 106:149-158.

_____. 1983. Greenhouse climate: From physical processes to a dynamic model. Ph.D. thesis, Agricultural University, Wageningen, Netherlands.

Bouhot, D., and Smith, I. M. 1988. *Pythium* spp. Pages 200-202 in: European Handbook of Plant Diseases. I. M. Smith, J. Dunez, R. A. Lelliott, D. H. Phillips, and S. A. Archer, eds. Blackwell, Oxford.

Bourbos, V. A. 1987. Prospect for biological control of *Fusarium oxysporum* f. sp. *lycopersici* (Sacc.) Snyder and Hansen in glasshouse tomatoes. Pages 223-232 in: Integrated and Biological Control in Protected Crops. R. Cavalloro, ed. A. A. Balkema, Rotterdam.

Bowman, G. E. 1972. The influence of greenhouse covering, propagating bench design and pot type on environmental temperature. Agric. Meteorol. 10:211-223.

Boyce, M. S. 1984. Restitution of *r*- and *K*-selection as a model of density-dependent natural selection. Annu. Rev. Ecol. Syst. 15:427-447.

Bradbury, J. F. 1967. *Pseudomonas lachrymans*. Descriptions of Pathogenic Fungi and Bacteria, No. 124. Commonwealth Mycological Institute, Kew, England.

Bradfield, E. G., and Guttridge, C. G. 1984. Effects of night-time humidity and nutrient solution concentration on the calcium content of tomato fruit. Sci. Hortic. (Amsterdam) 22:207-217.

Bravenboer, L. 1974. Pest and disease control in glasshouses in north-west Europe. Outlook Agric. 8:95-99.

Bravenboer, L., and Pet, G. 1962. Control of soil-borne diseases in tomatoes by grafting on resistant rootstocks. Pages 317-324 in: Proc. Int. Hortic. Congr. 16th.

Bravenboer, L., and Strijbosch, T. 1975. Humidity conditions and the development of fungus diseases in glasshouses. Acta Hortic. 51:333-335.

Brazelton, R. W. 1968. Sterilizing soil mixes with aerated steam. Agric. Eng. 49:400-401.

Brian, P. W. 1949. Studies on the biological activity of griseofulvin. Ann. Bot. (London) 13:59-77.

Brierley, P. 1952. Exceptional heat tolerance and some other properties of chrysanthemum stunt virus. Plant Dis. Rep. 36:243-244.

Broadbent, L. 1960. The epidemiology of tomato mosaic: A review of the literature. Pages 96-116 in: Annu. Rep. Glasshouse Crops Res. Inst. 1960.

——. 1962. The epidemiology of tomato mosaic. II. Smoking tobacco as a source of virus. Ann. Appl. Biol. 50:461-466.

——. 1963. The epidemiology of tomato mosaic. III. Cleaning virus from hands and tools. Ann. Appl. Biol. 52:225-232.

——. 1965a. The epidemiology of tomato mosaic. VIII. Virus infection through tomato roots. Ann. Appl. Biol. 55:57-66.

——. 1965b. The epidemiology of tomato mosaic. IX. Transmission of TMV by birds. Ann. Appl. Biol. 55:67-69.

——. 1976. Epidemiology and control of tomato mosaic virus. Annu. Rev. Phytopathol. 14:75-96.

Broadbent, L., and Fletcher, J. T. 1963. The epidemiology of tomato mosaic. IV. Persistence of virus on clothing and glasshouse structures. Ann. Appl. Biol. 52:233-241.

——. 1966. The epidemiology of tomato mosaic. XII. Sources of TMV in commercial tomato crops under glass. Ann. Appl. Biol. 57:113-120.

Broadbent, L., Read, W. H., and Last, F. T. 1965. The epidemiology of tomato mosaic. X. Persistence of TMV-infected debris in soil, and the effects of soil partial sterilization. Ann. Appl. Biol. 55:471-483.

Broadbent, P., Baker, K. F., and Waterworth, Y. 1971. Bacteria and actinomycetes antagonistic to fungal root pathogens in Australian soils. Aust. J. Biol. Sci. 24:925-944.

Brodie, I. D. S., and Blakeman, J. P. 1975. Competition for carbon compounds by a leaf surface bacterium and conidia of *Botrytis cinerea.* Physiol. Plant Pathol. 6:125-135.

Brooks, F. T. 1908. Observations on the biology of *Botrytis cinerea.* Ann. Bot. (London) 29:479-487.

Brown, G. E., and Kennedy, B. W. 1966. Effect of oxygen concentration on Pythium seed rot of soybean. Phytopathology 56:407-411.

Brown, J. K., and Stanghellini, M. E. 1988. Lettuce infectious yellows virus in hydroponically grown lettuce in Pennsylvania. Plant Dis. 72:453.

Brown, M. E., and Beringer, J. E. 1983. The potential of antagonists for fungal control. Agric. Ecosyst. Environ. 10:127-141.

Brown, W. 1922. Studies in the physiology of parasitism. VIII. On the exosmosis of nutrient substances from the host into the infection drop. Ann. Bot. (London) 36:101-119.

Brown, W., and Montgomery, N. 1948. Problems in the cultivation of winter lettuce. Ann. Appl. Biol. 35:161-180.

Bruehl, G. W. 1989. Integrated control of soil-borne plant pathogens. Can. J. Plant Pathol. 11:153-157.

Bruehl, G. W., and Lai, P. 1966. Prior-colonization as a factor in the saprophytic survival of several fungi in wheat straw. Phytopathology 56:766-768.

Bugbee, W. M., and Anderson, N. A. 1963. Whitefly transmission of *Xanthomonas pelargonii* and histological examination of leafspots of *Pelargonium hortorum.* Phytopathology 53:177-178.

Bunt, A. C. 1961. Some physical properties of pot plant composts and their effect on plant growth. III. Compaction. Plant Soil 15:228-242.

——. 1976. Modern Potting Composts. George Allen and Unwin, London.

Bunt, A. C., and Kulweic, Z. J. 1970. The effect of container porosity on root environment and plant growth. I. Temperature. Plant Soil 32:65-80.

Burdon, J. J., and Chilvers, G. A. 1975a. Epidemiology of damping-off disease (*Pythium irregulare*) in relation to density of *Lepidium sativum* seedlings. Ann. Appl. Biol. 81:135-143.

——. 1975b. A comparison between host density and inoculum density effects on the frequency of primary infection foci in *Pythium*-induced damping-off disease. Aust. J. Bot. 23:899-904.

——. 1982. Host density as a factor in plant disease ecology. Annu. Rev. Phytopathol. 20:143-166.

Burdon, J. J., Jarosz, A. M., and Kirby, G. C. 1989. Pattern and patchiness in plant-pathogen

interactions—Causes and consequences. Annu. Rev. Ecol. Syst. 20:119-136.
Burrage, S. W. 1971. The microclimate at the leaf surface. Pages 92-101 in: Ecology of Leaf Surface Micro-organisms. T. F. Preece and C. H. Dickinson, eds. Academic Press, London.
Businger, J. A. 1963. The glasshouse climate. Pages 277-318 in: Physics of Plant Environment. N. R. van Wijk, ed. North-Holland, Amsterdam.
Butt, D. J. 1978. Epidemiology of powdery mildews. Pages 51-81 in: The Powdery Mildews. D. M. Spencer, ed. Academic Press, London.
Byther, R. 1965. Ecology of plant pathogens in soil. V. Inorganic nitrogen utilization as a factor of competitive saprophytic ability of *Fusarium roseum* and *F. solani*. Phytopathology 55:852-858.
Cabanillas, E., Barker, K. R., and Nelson, L. A. 1989. Survival of *Paeciliomyces lilacinus* in selected carriers and related effects on *Meloidogyne incognita* on tomato. J. Nematol. 21:121-130.
Calvert, A., and Hand, D. W. 1975. CO_2 enrichment is still important. Grower (London) 84:617-620.
Campbell, G. S. 1977. An Introduction to Environmental Biophysics. Springer-Verlag, Berlin.
Campbell, T. A. 1984. Inheritance of seedling resistance to gray mold in kenaf. Crop Sci. 24:733-735.
Caron, M., Fortin, J. A., and Richard, C. 1985a. Effect of vesicular-arbuscular mycorrhizae on Fusarium crown and root rot of tomato as influenced by phosphorus concentrations. (Abstr.) Can. J. Plant Pathol. 7:443.
_____. 1985b. Influence of substrate on the interaction of *Glomus intraradices* and *Fusarium oxysporum* f. sp. *radicis-lycopersici* on tomatoes. Plant Soil 87:233-239.
Caron, M., Richard, C., and Fortin, J. A. 1986. Effect of preinfestation of the soil by a vesicular arbuscular mycorrhizal fungus, *Glomus intraradices,* on Fusarium crown and root rot of tomatoes. Phytoprotection 67:15-19.
Carpenter, W. J. 1969. Temperature patterns and greenhouse heating. Pages 6-8 in: Pa. Flower Grow. Bull. 221.
Carpenter, W. J., and Bark, L. D. 1967a. Temperature patterns in greenhouse heating. I. Introduction. Florists' Rev. 139(3609):43-45.
_____. 1967b. Temperature patterns in greenhouse heating. II. Temperature patterns and steam coils. Florists' Rev. 139(3611):21-22.
Carpenter, W. J., and Nautiyal, J. P. 1969. Light intensity and air movement effects on leaf temperatures and growth of shade-requiring greenhouse crops. J. Am. Soc. Hortic. Sci. 94:212-214.
Carpenter, W. J., and Willis, W. W. 1957. Comparisons of low pressure mist, atomized fog, and evaporative-fan-and-pad systems for greenhouse and plant response. Proc. Am. Soc. Hortic. Sci. 70:490-500.
Caruso, F. L., and Kuć, J. 1979. Induced resistance of cucumber to anthracnose and angular leaf spot by *Pseudomonas lachrymans* and *Colletotrichum lagenarium*. Physiol. Plant Pathol. 14:191-201.
Castanho, B., and Butler, E. E. 1978a. Rhizoctonia decline: A degenerative disease of *Rhizoctonia solani*. Phytopathology 68:1505-1510.
_____. 1978b. Rhizoctonia decline: Studies on hypovirulence and potential use in biological control. Phytopathology 68:1511-1514.
Castanho, B., Butler, E. E., and Shepherd, R. J. 1978. The association of double-stranded RNA with Rhizoctonia decline. Phytopathology 68:1515-1519.
Cayley, G. R., Etheridge, P., Griffiths, G. C., Phillips, F. T., Pye, B., and Scott, G. C. 1984. A review of the performance of electrostatically charged rotary atomisers on different crops. Ann. Appl. Biol. 105:379-386.
Challa, H., and van de Vooren, J. 1980. A strategy for climate control in greenhouses in early winter production. Acta Hortic. 106:159-164.
Challa, H., Bot, G. P. A., Nederhoff, E. M., and van de Braak, N. J. 1988. Greenhouse climate control in the nineties. Acta Hortic. 230:459-470.
Chandra, P. 1982. Thermal radiation exchange in a greenhouse with a transmitting cover. J. Agric. Eng. Res. 27:261-265.
Channon, A. G., Cheffins, N. J., Hitchon, G. M., and Baker, J. 1978. The effect of inoculation with an attenuated mutant strain of tobacco mosaic virus on the growth and yield of early

glasshouse tomato crops. Ann. Appl. Biol. 88:121-129.

Charles-Edwards, D. A. 1982. Physiological Determinants of Crop Growth. Academic Press, London.

Chase, A. R. 1988. Effect of fertilizer rate on growth of *Ficus lyrata* and susceptibility to *Pseudomonas cichorii*. HortScience 23:151-152.

_____. 1989. Effect of fertilizer level on severity of Xanthomonas leaf spot of *Pilea spruceana*. J. Environ. Hortic. 7:47-49.

_____. 1991. Greenhouse ornamental crops—Pest management systems for disease. Pages 709-728 in: Handbook of Pest Management in Agriculture. D. Pimental, ed. CRC Press, Boca Raton, FL.

Chase, A. R., and Jones, J. B. 1986. Effects of host nutrition, leaf age, and preinoculation light levels on severity of leaf spot of dwarf schefflera caused by *Pseudomonas cichorii*. Plant Dis. 70:561-563.

Chase, A. R., and Osborne, L. S. 1983. Influence of an insecticidal soap on several foliar diseases of foliage plants. Plant Dis. 67:1021-1023.

Chase, A. R., and Poole, R. T. 1984. Influence of foliar applications of micronutrients and fungicides on foliar necrosis and leaf spot disease of *Chrysalidocarpus lutescens*. Plant Dis. 68:195-197.

_____. 1985a. Host nutrition and severity of Myrothecium leaf spot of *Dieffenbachia maculata* 'Perfection.' Sci. Hortic. (Amsterdam) 25:85-92.

_____. 1985b. Root rot: Is steam treatment the answer? Greenhouse Manager (June), pp. 142-149.

Chaumont, J. P., and Simeray, J. 1985. Propriétés allélopathiques de 114 extraits de carpophores de champignons sur la germination de semences de radis. Rev. Ecol. Biol. Sol 22:331-339.

Chérif, M., Benhamou, N., Chamberland, H., and Dostaler, D. 1989. Isolation of fungi from various peats and selection of those showing antagonistic properties against *Fusarium oxysporum* f. sp. *radicis-lycopersici*: Cytochemical investigation of dual cultures. (Abstr.) Can. J. Plant Pathol. 11:188.

Cheru, L. L., and Ko, W. H. 1988. Characteristics of inhibition of suppressive soil created by monoculture with radish in the presence of *Rhizoctonia solani*. J. Phytopathol. (Berlin) 126:237-245.

Chet, I. 1987. *Trichoderma*—Application, mode of action, and potential as a biocontrol agent of soilborne plant pathogenic fungi. Pages 137-160 in: Innovative Approaches to Plant Disease Control. I. Chet, ed. John Wiley & Sons, New York.

Chet, I., and Baker, R. 1981. Isolation and biocontrol potential of *Trichoderma hamatum* from soil naturally suppressive to *Rhizoctonia solani*. Phytopathology 71:286-290.

Chet, I., Harman, G. E., and Baker, R. 1981. *Trichoderma hamatum:* Its interaction with *Rhizoctonia solani* and *Pythium* spp. Microb. Ecol. 7:29-38.

Cho, J. J., Mau, R. F. L., German, T. L., Hartmann, R. W., Yudin, L. S., Gonsalves, D., and Provvidenti, R. 1989. A multidisciplinary approach to management of tomato spotted wilt virus in Hawaii. Plant Dis. 73:375-383.

Cirulli, M., and Ciccarese, F. 1975. Interactions between TMV isolates, temperature, allelic condition and combination of the *Tm* resistance genes in tomato. Phytopathol. Mediterr. 14:100-105.

Clark, M. F. 1981. Immunosorbent assays in plant pathology. Annu. Rev. Phytopathol. 19:83-106.

Clarkson, D. T., and Hanson, J. B. 1980. The mineral nutrition of higher plants. Annu. Rev. Plant Physiol. 31:239-298.

Clayton, E. E. 1923a. The relation of temperature to the Fusarium wilt of tomato. Am. J. Bot. 10:71-88.

_____. 1923b. The relation of soil moisture to the Fusarium wilt of tomatoes. Am. J. Bot. 10:133-147.

_____. 1936. Water soaking of leaves in relation to development of the wildfire disease of tobacco. J. Agric. Res. (Washington, D.C.) 52:239-260.

Clerjeau, M., Moreau, C., and Pigameau, B. 1984. Effectiveness of fosetyl Al against strains of *Plasmopara viticola* and *Phytophthora infestans* that have developed resistance to anilide fungicides. Pages 497-502 in: Proc. Br. Crop Prot. Conf. Pests Dis. 1984.

Cline, M. N., Chastagner, G. A., Aragaki, M., Baker, R., Daughtrey, M. L., Lawson, R. H., MacDonald, J. D., Tammen, J. F., and Worf, G. L. 1988. Current and future research directions of ornamental pathology. Plant Dis. 72:926-934.

Cline, S., and Neely, D. 1984. Relationship between juvenile-leaf resistance to anthracnose and the presence of juglone and hydrojuglone glucoside in black walnut. Phytopathology 74:185-188.

Cobb, G. S., Hanan, J. J., and Baker, R. 1978. Environmental factors affecting rose powdery mildew in greenhouses. HortScience 13:464-466.

Cockshull, K. E. 1985. Greenhouse climate and crop response. Acta Hortic. 174:285-292.

Coffee, R. A. 1979. Electrodynamic energy—A new approach to pesticide application. Pages 777-789 in: Proc. Br. Crop Prot. Conf. Pests Dis. 1979.

_____. 1980. Electrodynamic spraying. Pages 95-107 in: Spraying Systems for the 1980s. Monogr. 24. BCPC Publications, Croydon, England.

Cohen, D. 1977. Thermotherapy and meristem-tip culture of some ornamental plants. Acta Hortic. 78:381-388.

Cohen, H., Nitzany, F. A., and Vilda, T. 1961. Tomato yellow top virus in Israel. Hassadeh 42:139-140.

Cohen, Y. 1981. Downy mildew of cucurbits. Pages 341-354 in: The Downy Mildews. D. M. Spencer, ed. Academic Press, London.

Cole, J. S., and Zvenyika, Z. 1988. Integrated control of *Rhizoctonia solani* and *Fusarium solani* in tobacco transplants with *Trichoderma harzianum* and triadimenol. Plant Pathol. 37:271-277.

Coley-Smith, J. R. 1980. Sclerotia and other structures in survival. Pages 85-114 in: The Biology of *Botrytis*. J. R. Coley-Smith, K. Verhoeff, and W. R. Jarvis, eds. Academic Press, London.

Colhoun, J. 1973. Effects of environmental factors on plant disease. Annu. Rev. Phytopathol. 11:343-364.

_____. 1979. Predisposition by the environment. Pages 75-96 in: Plant Disease: An Advanced Treatise. Vol. 4. J. G. Horsfall and E. B. Cowling, eds. Academic Press, New York.

Collier, G. F., and Tibbits, T. W. 1984. Effects of relative humidity and root temperature on calcium concentration and tip burn development in lettuce. J. Am. Soc. Hortic. Sci. 109:128-131.

Conway, G. R. 1984a. Introduction. Pages 1-11 in: Pest and Pathogen Control: Strategic, Tactical, and Policy Models. G. R. Conway, ed. John Wiley & Sons, Chichester.

_____. 1984b. Strategic models. Pages 15-28 in: Pest and Pathogen Control: Strategic, Tactical, and Policy Models. G. R. Conway, ed. John Wiley & Sons, Chichester.

_____. 1984c. Policy models. Pages 397-405 in: Pest and Pathogen Control: Strategic, Tactical, and Policy Models. G. R. Conway, ed. John Wiley & Sons, Chichester.

Cook, R. J. 1982. Use of pathogen-suppressive soils for disease control. Pages 51-65 in: Suppressive Soils and Plant Disease. R. W. Schneider, ed. American Phytopathological Society, St. Paul, MN.

Cook, R. J., and Baker, K. F. 1983. The Nature and Practice of Biological Control of Plant Pathogens. American Phytopathological Society, St. Paul, MN.

Cook, R. J., and Flentje, N. T. 1967. Chlamydospore germination and germling survival of *Fusarium solani* f. *pisi* in soil as affected by soil water and pea seed exudation. Phytopathology 57:178-182.

Cook, R. J., and Papendick, R. I. 1972. Influence of water potential of soils and plants on root disease. Annu. Rev. Phytopathol. 10:349-374.

Cooper, A. 1979. The ABC of NFT. Grower Books, London.

Copeman, R. J., Mauza, B. E., and Eaton, G. W. 1988. Increased yields from cross-protected, Fusarium crown and root rot infected tomato plants grown in sawdust culture. (Abstr.) Can. J. Plant Pathol. 10:362.

Corden, M. E. 1965. Influence of calcium nutrition on Fusarium wilt of tomato and polygalacturonase activity. Phytopathology 55:222-224.

Couteaudier, Y., Alabouvette, C., and Soulas, M.-L. 1985. Nécrose du collet et pourriture des racines de tomate. Rev. Hortic. 254:39-42.

Cox, R. S., and Winfree, J. P. 1957. Observations on the effect of fungicides on gray mold and leaf spot and on the chemical composition of strawberry plant tissues. Plant Dis.

Rep. 41:755-759.
Cremer, M. C., and Schenk, P. K. 1967. Notched leaf in *Gladiolus* spp., caused by viruses of the rattle virus group. Neth. J. Plant Pathol. 73:33-48.
Critten, D. L. 1983. A computer model to calculate the daily light integral and transmissivity of a greenhouse. J. Agric. Eng. Res. 28:61-76.
———. 1984. Improving glasshouse light transmission. Pages 18-19 in: Fruit, Vegetables and Science. Agriculture and Food Research Council, London.
Crute, I. R. 1984. The integrated use of genetic and chemical methods for control of lettuce downy mildew (*Bremia lactucae*). Crop Prot. 3:223-242.
Crute, I. R., and Dixon, G. R. 1981. Downy mildew diseases caused by the genus *Bremia*. Pages 421-460 in: The Downy Mildews. D. M. Spencer, ed. Academic Press, London.
Crute, I. R., Norwood, J. M., and Gordon, P. L. 1987. The occurrence, characteristics and distribution in the United Kingdom of resistance to phenylamide fungicides in *Bremia lactucae* (lettuce downy mildew). Plant Pathol. 36:297-315.
Curtis, L. C. 1943. Deleterious effects of guttated fluid on foliage. Am. J. Bot. 30:778-781.
Curtis, O. F. 1936. Leaf temperature and the cooling of leaves by radiation. Plant Physiol. 11:343-364.
Daie, J. 1985. Carbohydrate partitioning and metabolism in crops. Hortic. Rev. 7:69-108.
Dale, A. C., Puri, V. M., and Hammer, P. A. 1984. A special heat conserving glasshouse with solar heated soil for crop (tomato) production. Acta Hortic. 148:731-738.
Damagnez, J. 1981. Formation du microclimat dans les serres. Acta Hortic. 107:33-41.
Darrow, G. M., and Waldo, C. F. 1932. Effect of fertilizers on plant growth, yield and decay of strawberries in North Carolina. Proc. Am. Soc. Hortic. Sci. 29:318-324.
Daughtrey, M. L., and Schippers, P. A. 1980. Root death and associated problems. Acta Hortic. 98:283-291.
Davet, P. 1971. Recherches sur le *Colletotrichum coccodes*. II. Étude des sources d'infection au Liban. Phytopathol. Mediterr. 10:159-163.
Davet, P., Artigues, M., and Martin, C. 1981. Production en conditions non aseptiques d'inoculum de *Trichoderma harzianum* Rifai pour des essais de lutte biologique. Agronomie (Paris) 1:933-936.
Davidson, O. W. 1953. High-pressure sprays for greenhouse air-conditioning. N.J. Plant Flower Grow. Assoc. 3(4):1-7.
Davies, J. M. L. 1980. Disease in NFT. Acta Hortic. 98:299-305.
Davis, D. 1967. Cross-protection in Fusarium wilt diseases. Phytopathology 57:311-314.
———. 1968. Partial control of Fusarium wilt in tomato by formae of *Fusarium oxysporum*. Phytopathology 58:121-122.
Davis, E. F. 1928. The toxic principle of *Juglans nigra* as identified with synthetic juglone, and its toxic effects on tomato and alfalfa plants. Am. J. Bot. 15:620.
Davis, R. P., and Dennis, C. 1981. Studies on the survival and infective ability of dicarboximide-resistant strains of *Botrytis cinerea*. Ann. Appl. Biol. 98:395-402.
Dawson, J. R., Johnson, R. A. H., Adams, P., and Last, F. T. 1965. Influence of steam/air mixtures, when used for heating soil, on biological and chemical properties that affect seedling growth. Ann. Appl. Biol. 56:243-251.
Dawson, J. R., Kilby, A. A. T., Ebben, M. H., and Last, F. T. 1967. The use of steam/air mixtures for partially sterilizing soils infested with cucumber root pathogens. Ann. Appl. Biol. 60:215-222.
Dean, R. A., and Kuć, J. 1986. Induced systemic protection in cucumber: Time of production and movement of the signal. Phytopathology 76:966-970.
de Boodt, M., and Verdonck, O. 1972. The physical properties of the substrates in horticulture. Acta Hortic. 26:37-44.
de Brouwer, W. M. T. J., and van Dorst, H. J. M. 1979. The relationship between the *Aphis gossypii* group and cucumber mosaic virus in autumn cucumbers. Neth. J. Agric. Sci. 23:269-278.
de Halleux, D., Nijskens, J., Deltour, J., Coutisse, S., and Nisen, A. 1985. Effet de la condensation sur les transferts thermiques des couvertures de serres: Cas du verre et du polyéthylène. Plasticulture 66:19-25.
de Jong, J., and Honma, S. 1976. Evaluation of screening techniques and determination of criteria for assessing resistance to *Clavibacter michiganense* in tomato. Euphytica 25:405-

414.
Dekker, J. 1976. Acquired resistance to fungicides. Annu. Rev. Phytopathol. 14:405-428.
_____. 1987. Build-up and persistence of fungicide resistance. Pages 153-168 in: Rational Pesticide Use. K. J. Brent and R. K. Atkin, eds. Cambridge University Press, Cambridge.
de Koning, A. N. M. 1988a. The effect of different day/night temperature regimes in growth, development and yield of glasshouse tomatoes. J. Hortic. Sci. 63:465-471.
_____. 1988b. An algorithm for controlling the average 24-hour temperature in glasshouses. J. Hortic. Sci. 63:473-477.
De Long, R. E., and Powell, C. C. 1988. Development of a deterministic model for greenhouse rose powdery mildew using modified leaf wetness sensors. (Abstr.) Phytopathology 78:1520.
Delp, C. J. 1954. Effect of temperature and humidity on the grape powdery mildew fungus. Phytopathology 44:615-626.
Demmink, J. F., Baayen, R. P., and Sparnaaij, L. D. 1989. Evaluation of the virulence of races 1, 2 and 4 of *Fusarium oxysporum* f. sp. *dianthi* in carnation. Euphytica 42:55-63.
de Tempe, J. 1979. The fungicide-treated seed in the soil. Pages 205-230 in: Soil Disinfestation. D. Mulder, ed. Elsevier, Amsterdam.
Devergne, J.-C., and Cardin, L. 1970. Étude sérologique comparative de plusieurs isolats du virus de la mosaique du concombre (CMV): Relations sérologiques au niveau du virus et de l'antigène soluble. Ann. Phytopathol. 2:639-661.
de Waard, M. A. 1984. Negatively correlated cross-resistance and synergism as strategies in coping with fungicide resistance. Pages 573-584 in: Proc. Br. Crop. Prot. Conf. Pests Dis. 1984.
Dhanvantari, B. N. 1990. Stem necrosis of greenhouse tomato caused by a novel *Pseudomonas* sp. Plant Dis. 74:124-127.
Dhanvantari, B. N., and Dirks, V. A. 1987. Bacterial stem rot of greenhouse tomato: Etiology, spatial distribution, and the effect of high humidity. Phytopathology 77:1457-1463.
Diachun, S., Valleau, W. D., and Johnson, E. M. 1942. Relation of moisture to invasion of tobacco leaves by *Bacterium tabacum* and *Bacterium angulatum*. Phytopathology 32:379-387.
Dickinson, C. H., and Dooley, M. J. 1967. The microbiology of cut-away peat. I. Descriptive ecology. Plant Soil 27:172-196.
Digat, B. 1987. Méthodologie de la detection des bactéries pathogènes du pelargonium et organisation de la certification sanitaire de cette culture en France. EPPO Bull. 17:281-286.
Dimock, A. W. 1956. Production of chrysanthemum propagating material free from certain pathogens. Pages 59-62 in: Plant Dis. Rep. Suppl. 238.
_____. 1962. Obtaining pathogen-free stock by cultured cutting techniques. Phytopathology 52:1239-1241.
Dimock, A. W., and Baker, K. F. 1951. Effect of climate on disease development, injuriousness, and fungicidal control, as exemplified by snapdragon rust. Phytopathology 41:536-552.
Dimock, A. W., and Tammen, J. 1969. Powdery mildew of roses. Pages 163-171 in: Roses: A Manual on the Culture, Management, Diseases, Insects, Economics and Breeding of Greenhouse Roses. J. W. Mastalerz and R. W. Langhans, eds. Pennsylvania Flower Growers, New York State Flower Growers Association, and Roses Inc., Haslett, MI.
Diop-Bruckler, M., and Molot, P. M. 1987. Intérêt de quelques hyperparasites dans la lutte contre *Leveillula taurica* et *Sphaerotheca fuliginea*. EPPO Bull. 17:593-600.
Diprose, M. F., and Evans, G. H. 1988. The use of dielectric heating for controlling soilborne diseases. (Abstr.) Page 7 in: Br. Soc. Plant Pathol. Symp. Novel Unusual Methods Dis. Control.
Dirks, V. A., and Jarvis, W. R. 1975. A basal stem and root rot of greenhouse tomatoes caused by *Fusarium oxysporum* and its spread in crops. (Abstr.) Proc. Am. Phytopathol. Soc. 2:36.
Ditner, J. L., Lindsay, S. C., Brundrett, E., and Jewett, T. J. 1985. Development of a microprocessor based greenhouse environment controller. Acta Hortic. 174:497-504.
Dodds, J. A., Morris, T. J., and Jordan, R. L. 1984. Plant viral double-stranded RNA. Annu. Rev. Phytopathol. 22:151-168.
Domsch, K. H. 1964. Soil fungicides. Annu. Rev. Phytopathol. 2:293-320.

Doolittle, S. P., and Walker, M. N. 1926. Control of cucumber mosaic by eradication of wild host plants. U.S. Dep. Agric. Bull. 1461.

Douville, Y., and Boland, G. J. 1989. Influence of propagule concentration and temperature on storage of potential biological control agents in alginate pellets. (Abstr.) Can. J. Plant Pathol. 11:188.

Drew, M. C., and Lynch, J. M. 1980. Soil anaerobiosis, microorganisms, and root function. Annu. Rev. Phytopathol. 18:37-66.

Duffus, J. E. 1965. Beet pseudo-yellows virus, transmitted by the greenhouse whitefly (*Trialeurodes vaporariorum*). Phytopathology 55:450-453.

Duinveld, T. L. J., and Beijersbergen, J. C. M. 1975. On the resistance to benomyl of fungi isolated from bulbs and corms. Acta Hortic. 47:143-148.

Duncan, G. A., and Walker, J. N. 1973. Poly-tube heating ventilation systems and equipment. Ky. Agric. Exp. Stn. AEN-7.

Dunez, J. 1987. Perspectives in the control of plant viruses. Pages 297-324 in: Innovative Approaches to Plant Disease Control. I. Chet, ed. John Wiley & Sons, New York.

Duniway, J. M. 1976. Movement of zoospores of *Phytophthora cryptogea* in soils of various textures and matric potentials. Phytopathology 66:877-882.

———. 1977. Predisposing effect of water stress on the severity of Phytophthora root rot in safflower. Phytopathology 67:884-889.

———. 1979. Water relations of water molds. Annu. Rev. Phytopathol. 17:431-460.

Durkin, D., and Janick, J. 1966. The effect of plant density on greenhouse carnation production. Proc. Am. Soc. Hortic. Sci. 89:609-614.

Duvdevani, S., Reichert, I., and Palti, J. 1946. The development of downy and powdery mildew of cucumbers as related to dew and other environmental factors. Palest. J. Bot. Rehovot Ser. 5:127-151.

Eastin, J. D., and Sullivan, C. Y. 1984. Environmental stress influences on plant persistence, physiology, and production. Pages 201-236 in: Physiological Basis of Crop Growth and Development. M. B. Tesar, ed. American Society of Agronomy and Crop Science Society of America, Madison, WI.

Eastin, J. D., Haskins, F. A., Sullivan, C. Y., and van Bavel, C. H. M., eds. 1969. Physiological Aspects of Crop Yield. American Society of Agronomy, Madison, WI.

Ebben, M. H. 1968. Root pathogens of glasshouse crops. Pages 95-96 in: Annu. Rep. Glasshouse Crops Res. Inst. 1968.

———. 1971. Tomato brown root rot: The build-up of soil inoculum and its control by fumigation. Pages 243-250 in: Proc. Br. Insectic. Fungic. Conf. 6th.

———. 1987. Observations on the role of biological control methods within integrated system, with reference to three contrasting diseases of protected crops. Pages 197-208 in: Integrated and Biological Control in Protected Crops. R. Cavalloro, ed. A. A. Balkema, Rotterdam.

Ebben, M. H., and Last, F. T. 1975. Incidence of root rots—Their prediction and relation to yield losses: A glasshouse study. Pages 6-10 in: Biology and Control of Soil-Borne Plant Pathogens. G. W. Bruehl, ed. American Phytopathological Society, St. Paul, MN.

Ebben, M. H., and Spencer, D. M. 1976. Fusarium wilt of carnation caused by *Fusarium oxysporum* f. sp. *dianthi:* Relation of soil inoculum level to disease severity and control. Pages 113-115 in: Annu. Rep. Glasshouse Crops Res. Inst. 1976.

———. 1978. The use of antagonistic organisms for the control of black root rot of cucumber. Ann. Appl. Biol. 89:103-106.

Edens, T. C., and Haynes, D. L. 1982. Closed system agriculture: Resource constraints, management options, and design alternatives. Annu. Rev. Phytopathol. 20:363-395.

Edgington, L. V., and Peterson, C. A. 1977. Systemic fungicides: Theory, uptake and translocation. Pages 51-89 in: Antifungal Compounds. Vol. 2. M. R. Siegel and H. D. Sisler, eds. Marcel Dekker, New York.

Edgington, L. V., Corden, M. E., and Dimond, A. E. 1961. The role of pectic substances in chemically induced resistance to Fusarium wilt of tomato. Phytopathology 51:179-182.

Elad, Y. 1988a. Involvement of ethylene in the disease caused by *Botrytis cinerea* on rose and the possibility of control. Ann. Appl. Biol. 113:589-598.

———. 1988b. Latent infection of *Botrytis cinerea* in rose flowers and combined chemical and physiological control of the disease. Crop Prot. 7:361-366.

Elad, Y., and Chet, I. 1987. Possible role of competition for nutrients in biocontrol of Pythium

damping-off by bacteria. Phytopathology 77:190-195.
Engelhard, A. W., ed. 1989. Soilborne Plant Pathogens: Management of Diseases with Macro- and Microelements. American Phytopathological Society, St. Paul, MN.
English, J. T., Thomas, C. S., Marois, J. J., and Gubler, W. D. 1989. Microclimates of grapevine canopies associated with leaf removal and control of Botrytis bunch rot. Phytopathology 79:395-401.
Ercolani, G. L., and Casolari, A. 1966. Recherche di microflora in pomodori sani. Ind. Conserve 41:15-22.
Erwin, D. C., Bartnicki-Garcia, S., and Tsao, P. H., eds. 1983. *Phytophthora:* Its Biology, Taxonomy, Ecology, and Pathology. American Phytopathological Society, St. Paul, MN.
Ewart, J. M., and Chrimes, J. R. 1980. Effects of chlorine and ultraviolet light in disease control in NFT. Acta Hortic. 98:317-323.
Farrara, B. F., Ilott, T. W., and Michelmore, R. W. 1987. Analysis of genetic factors for resistance to downy mildew (*Bremia lactucae*) in species of lettuce (*Lactuca sativa* and *L. serriola*). Plant Pathol. 36:499-514.
Faulwetter, R. F. 1917. Wind-blown rain, a factor in disease dissemination. J. Agric. Res. (Washington, D.C.) 10:639-648.
Favrin, R. J., Rahe, J. E., and Mauza, B. 1988. *Pythium* spp. associated with crown rot of cucumbers in British Columbia greenhouses. Plant Dis. 72:683-687.
Ferare, J., and Goldsberry, K. L. 1984a. Plant responses to different plastic greenhouse covers. Acta Hortic. 148:619-626.
_____. 1984b. Environmental conditions created by plastic greenhouse covers. Acta Hortic. 148:675-682.
Fery, R. L., and Janick, J. 1970. Response of the tomato to population pressure. J. Am. Soc. Hortic. Sci. 95:614-624.
Flaherty, A. 1989. Wind it out, roll it up. Grower (London) 111(8):23.
Fletcher, J. T. 1969. Studies on the overwintering of tomato mosaic in root debris. Plant Pathol. 18:97-108.
_____. 1984. Diseases of Greenhouse Plants. Longman, London.
Fletcher, J. T., and Butler, D. 1975. Strain changes in populations of tobacco mosaic virus from tomato crops. Ann. Appl. Biol. 81:409-412.
Fletcher, J. T., and Rowe, J. M. 1975. Observations and experiments on the use of an avirulent mutant strain of tobacco mosaic virus as a means of controlling tomato mosaic. Ann. Appl. Biol. 81:171-179.
Flor, H. H. 1942. Inheritance of pathogenicity in *Melampsora lini.* Phytopathology 32:653-669.
_____. 1971. Current status of the gene-for-gene concept. Annu. Rev. Phytopathol. 9:275-296.
Fokkema, N. J., Riphagen, I., Poot, R. J., and de Jong, C. 1983. Aphid honeydew, a potential stimulant of *Cochliobolus sativus* and *Septoria nodorum* on wheat leaves, and the competitive role of the saprophytic mycoflora. Trans. Br. Mycol. Soc. 81:355-363.
Follett, B. K. M, and Follett, D. E., eds. 1981. Biological Clocks in Seasonal Reproductive Cycles. Scientechnica, Bristol, England.
Forsyth, S. F. 1990. Regulatory issues and approaches for plant disease biocontrol. Can. J. Plant Pathol. 12:318-321.
Forsythe, W. E., ed. 1964. Smithsonian Physical Tables. Smithsonian Institution, Washington, D.C.
Foster, R. L., and Walker, J. C. 1947. Predisposition of tomato to Fusarium wilt. J. Agric. Res. (Washington, D.C.) 74:165-185.
Frampton, V. L., and Longrée, K. 1941. The vapor pressure gradient above a transpiring leaf. Phytopathology 31:1040-1042.
Franc, G. D., Harrison, M. D., and Lahman, L. K. 1988. A simple day-degree model for initiating chemical control of potato early blight in Colorado. Plant Dis. 72:851-854.
Frappell, B. D. 1979. Competition in vegetable crop communities. J. Aust. Inst. Agric. Sci. 45:211-217.
Fraser, R. P. 1958. The fluid kinetics of application of pesticidal chemicals. Adv. Pest Control Res. 2:1-106.
Fravel, D. R. 1988. Role of antibiosis in the biocontrol of plant diseases. Annu. Rev.

Phytopathol. 26:75-91.
Frick, E. L. 1970. The effects of volume, drop size and concentration, and their interaction on the control of apple powdery mildew by dinocap. Pages 23-33 in: Br. Crop Prot. Counc. Monogr. 2.
Frinkling, H. D., and Scholte, B. 1983. Dissemination of mildew spores in a glasshouse. Philos. Trans. R. Soc. London B 302:575-582.
Fulton, J. P. 1967. Dual transmission of tobacco ringspot virus and tomato ringspot virus by *Xiphinema americanum.* Phytopathology 57:535-537.
Fulton, J. P., Gergerich, R. C., and Scott, H. A. 1987. Beetle transmission of plant viruses. Annu. Rev. Phytopathol. 25:111-123.
Fulton, R. W. 1986. Practices and precautions in the use of cross protection for plant virus disease control. Annu. Rev. Phytopathol. 24:67-81.
Funck-Jensen, D., and Hockenhull, D. 1983a. The influence of some factors on the severity of Pythium root rot of lettuce in soilless (hydroponic) growing systems. Acta Hortic. 133:129-136.
_____. 1983b. Is damping-off, caused by *Pythium,* less of a problem in hydroponics than in traditional growing systems? Acta Hortic. 133:137-145.
Galea, V. J., and Price, T. V. 1988. Survival of the lettuce anthracnose fungus (*Microdochium panattonianum*) in Victoria. Plant Pathol. 37:54-63.
Gardiner, D. C., Horst, R. K., and Nelson, P. E. 1987. Symptom enhancement of Fusarium wilt of chrysanthemum by high temperatures. Plant Dis. 71:1106-1109.
_____. 1989. Influence of night temperature on disease development in Fusarium wilt of chrysanthemum. Plant Dis. 73:34-37.
Gardiner, R. B., Jarvis, W. R., and Shipp, J. L. 1990. Ingestion of *Pythium* spp. by larvae of the fungus gnat *Bradysia impatiens* (Diptera: Sciaridae). Ann. Appl. Biol. 116:205-212.
Garibaldi, A., and Tamietti, G. 1984. Attempts to use soil solarization in closed greenhouses in northern Italy for controlling corky root of tomato. Acta Hortic. 152:237-243.
Garrett, S. D. 1970. Pathogenic Root-Infecting Fungi. Cambridge University Press, Cambridge.
Garzoli, K. 1985. A simple greenhouse climate model. Acta Hortic. 174:393-400.
Garzoli, K., and Blackwell, J. 1973. The response of a glasshouse to high solar radiation and ambient temperature. J. Agric. Eng. Res. 18:205-216.
Garzoli, K., and Shell, G. S. G. 1984. Performance and cost analysis of an Australian solar greenhouse. Acta Hortic. 148:723-730.
Gasiorkiewicz, E. C., and Olsen, C. J. 1956. Progress in the development and production of virus-free carnation varieties. Pages 77-80 in: Plant Dis. Rep. Suppl. 238.
Gates, D. A. 1980. Biophysical Ecology. Springer-Verlag, New York.
Gäumann, A. E. 1951. Pflanzliche Infektionslehre. Verlag Birkhäuser, Basel.
Geiger, R. 1966. The Climate near the Ground. Harvard University Press, Cambridge.
Geissler, T. 1979. Gemuseproduktion unter Glas und Plasten Produktionsverfahren. VEB Deutscher Landwirtschaftsverlag, Berlin.
Georgopoulos, S. G. 1977. A new class of carboxin-resistant mutants of *Ustilago maydis.* Neth. J. Plant Pathol. 83(suppl. 1):235-242.
Germing, G. H. 1985. Greenhouse design and cladding materials. Acta Hortic. 170:253-257.
Gessler, C., and Kuć, J. 1982. Induction of resistance to Fusarium wilt in cucumber by root and foliar pathogens. Phytopathology 72:1439-1441.
Ghabrial, S. A., and Shepherd, R. J. 1980. A sensitive radioimmunoabsorbent assay for the detection of plant viruses. J. Gen. Virol. 48:311-317.
Giacomelli, G. A., Giniger, M. S., Krass, A. E., and Mears, D. R. 1985. Improved methods of greenhouse evaporative cooling. Acta Hortic. 174:49-55.
Gillard, A., and van den Brande, J. 1955. Quelques problèmes concernant les nématodes des racines (*Meloidogyne* spp.) en Belgique, particulièrement la désinfection des tubercules de *Begonia multiflora* par traitement à l'eau chaude. Parasitica 11:74-80.
Gindrat, D., van der Hoeven, E., and Moody, A. R. 1977. Control of *Phomopsis sclerotioides* with *Gliocladium roseum* or *Trichoderma.* Neth. J. Plant Pathol. 83(suppl. 1):429-438.
Giniger, M. S., Stine, C. B., Giacomelli, G. A., and Mears, D. R. 1985. Developing control strategies for a large thermal mass. Acta Hortic. 174:401-406.
Goedhart, M., Nederhoff, E. M., Udink ten Cate, A. J., and Bot, G. P. A. 1984. Methods and instruments for ventilation rate measurements. Acta Hortic. 148:393-400.

Goisque, M. J., Louvet, H., Marten, C., Lagier, J., Davet, P., Couteaudier, Y., and Louvet, J. 1984. La désinfection soliare des sols sous serre. Plasticulture 64:32-38.

Gold, S. E., and Stanghellini, M. E. 1985. Effects of temperature on Pythium root rot of spinach grown under hydroponic conditions. Phytopathology 75:333-337.

Goldsberry, K. L., and Wolnik, D. 1966. Evaluation of greenhouse shading compounds. Colo. Flower Grow. Assoc. Bull. 190.

Gonzalez, A. G. 1977. Lactuceae—A chemical review. Pages 1082-1095 in: The Biology and Chemistry of the Compositae. Vol. II. J. B. Harborne and B. L. Turner, eds. Academic Press, New York.

Gonzalez, D., and Rawlins, N. A. 1969. Relation of aphid populations to field spread of lettuce mosaic virus in New York. J. Econ. Entomol. 62:1109-1114.

Good, J. M., and Carter, R. L. 1965. Nitrification lag following soil fumigation. Phytopathology 55:1147-1150.

Goode, M. J., and Sasser, M. 1980. Prevention—The key to controlling bacterial spot and bacterial speck of tomato. Plant Dis. 64:831-834.

Gordon-Lennox, G., Walther, D., and Gindrat, D. 1987. Utilisation d'antagonistes pour l'enrobage des semences: Efficacité et mode d'action contre les agents de la fonte des semis. EPPO Bull. 17:631-637.

Goring, C. A. I. 1962. Theory and principles of soil fumigation. Adv. Pest Control Res. 5:47-84.

Grainger, J. 1962. The host plant as a habitat for fungal and bacterial parasites. Phytopathology 52:140-150.

_____. 1968. C_p/R_s and the disease potential of plants. Hortic. Res. 8:1-40.

_____. 1979. Scientific proportion and economic decisions for farmers. Annu. Rev. Phytopathol. 17:223-252.

Grainger, J., and Kennedy, R. R. 1964. Temperature distribution and performance in balloon-sheet soil steaming. Hortic. Res. 4:25-41.

Grange, R. I., and Hand, D. W. 1987. A review of the effects of atmospheric humidity on growth of horticultural crops. J. Hortic. Sci. 62:125-134.

Graves, C. J. 1983. The nutrient film technique. Hortic. Rev. 5:1-44.

Gregory, P. H., Guthrie, E. J., and Bunce, M. E. 1959. Experiments on splash dispersal of fungus spores. J. Gen. Microbiol. 20:328-354.

Griffin, D. M. 1963a. Soil physical factors and the ecology of fungi. II. Behaviour of *Pythium ultimum* at small water suctions. Trans. Br. Mycol. Soc. 46:368-372.

_____. 1963b. Soil physical factors and the ecology of fungi. III. Activity of fungi in relatively dry soil. Trans. Br. Mycol. Soc. 46:373-377.

_____. 1968. A theoretical study relating the concentration and diffusion of oxygen to the ecology of organisms in soil. New Phytol. 67:561-577.

_____. 1972. Ecology of Soil Fungi. Chapman and Hall, London.

Griffiths, E. 1981. Iatrogenic plant diseases. Annu. Rev. Phytopathol. 19:69-82.

Grosclaude, C., Ricard, J., and Dubos, B. 1973. Inoculation of *Trichoderma viride* spores via pruning shears for biological control of *Stereum purpureum* on plum tree wounds. Plant Dis. Rep. 57:25-28.

Gutterson, N. I., Layton, T. J., Ziegle, J. S., and Warren, G. J. 1986. Molecular cloning of genetic determinants for inhibition of fungal growth by a fluorescent pseudomonad. J. Bacteriol. 165:696-703.

Hadar, Y., and Mandelbaum, R. 1986. Suppression of Pythium aphanidermatum damping-off in container media containing composted liquorice roots. Crop Prot. 5:88-92.

Hadar, Y., Chet, I., and Henis, Y. 1979. Biological control of Rhizoctonia solani damping-off with wheat bran culture of *Trichoderma harzianum*. Phytopathology 69:64-68.

Haggett, B. G. D. 1989. Sensors and instrumentation in horticulture. (Abstr.) Biologist 36:87.

Hakkart, F. A., and Versluijs, J. M. 1988. Virus elimination by meristem-tip culture from a range of *Alstromeria* varieties. Neth. J. Plant Pathol. 94:49-56.

Halber, M. 1963. *Botrytis* sp. in Douglas fir seedlings. Plant Dis. Rep. 47:556.

Hall, R. A., Gillespie, A. T., and Burges, H. D. 1982. Use of fungi as microbial insecticides. Pages 98-103 in: Annu. Rep. Glasshouse Crops Res. Inst. 1982.

Hall, T. J., and Bowes, S. A. 1980. Screening for disease resistance in tomato. Pages 157-170 in: Annu. Rep. Glasshouse Crops Res. Inst. 1980.

Hames, B. D., and Higgins, S. J., eds. 1985. Nucleic Acid Hybridization, a Practical Approach. IRL Press, Oxford.

Hammarlund, C. 1925. Zur Genetik, Biologie, und Physiologie einiger Erysiphaceen. Hereditas (Lund, Swed.) 6:1-126.

Hammett, K. R. W., and Manners, J. G. 1973. Conidium liberation in *Erysiphe graminis.* II. Conidial chain and pustule structure. Trans. Br. Mycol. Soc. 61:121-133.

Hanan, J. J. 1958. Air movement and temperature control. Colo. Flower Grow. Assoc. Res. Bull. 98:1-2.

_____. 1965. Preliminary measurements of flower temperature. Colo. Flower Grow. Assoc. Res. Bull. 188:1-3.

_____. 1970a. Statistical analysis of flower temperatures in the carnation. J. Am. Soc. Hortic. Sci. 95:68-73.

_____. 1970b. Some observations on radiation in greenhouses. Colo. Flower Grow. Assoc. Res. Bull. 239:1-4.

Hanan, J. J., Holley, W. D., and Goldsberry, K. L. 1978. Greenhouse Management. Springer-Verlag, Berlin.

Hancock, J. G. 1981. Longevity of *Pythium ultimum* in moist soils. Phytopathology 71:1033-1037.

Hannsler, G., and Hermanns, M. 1981. *Verticillium lecanii* as a parasite on cysts of *Heterodera schachtii.* Z. Pflanzenkrankh. Pflanzenschutz 80:678-681.

Hanson, W. J., and Nex, R. W. 1953. Diffusion of ethylene dibromide in soils. Soil Sci. 76:209-214.

Harazono, Y., Taenaka, T., and Yabuki, K. 1984. Optimizing environmental control of greenhouse climate for cucumber growth by means of "learning control method." Acta Hortic. 148:259-265.

Harazono, Y., Kamiya, H., and Yabuki, K. 1988. A control method based on an artificial intelligence technique and its application for controlling plant environment. Acta Hortic. 230:209-213.

Hardwick, R. C., Cole, R. A., and Fyfield, T. P. 1984. Injury and death of cabbage (*Brassica oleracea*) seedlings caused by vapours of dibutyl phthalate emitted from certain plastics. Ann. Appl. Biol. 105:97-105.

Harling, R., Taylor, G. S., Matthews, P., and Arthur, A. E. 1988. The effect of temperature on symptom expression and colonization in resistant and susceptible carnation cultivars infected with *Fusarium oxysporum* f. sp. *dianthi.* J. Phytopathol. (Berlin) 121:103-117.

Harman, G. E., and Taylor, A. G. 1988. Improved seedling performance by integration of biological control agents at favorable pH levels with solid matrix priming. Phytopathology 78:520-525.

Harman, G. E., Taylor, A. G., and Stasz, T. E. 1989. Combining effective strains of *Trichoderma harzianum* and solid matrix priming to improve biological seed treatments. Plant Dis. 73:631-637.

Harris, K. F., and Maramorosch, K. 1980. Vectors of Plant Pathogens. Academic Press, New York.

Harrison, B. D. 1974. Potato mop-top virus. Descriptions of Plant Viruses, No. 138. Commonwealth Mycological Institute and Association of Applied Biologists, Kew, England.

Hashimoto, Y., Morimoto, T., and Funada, S. 1981. Computer processing of speaking plant for climate control and computer aided plantation (computer aided cultivation). Acta Hortic. 115:317-325.

Hashimoto, Y., Strain, B. R., Morimoto, T., and Fukuyama, T. 1984. System identification of plant responses in energy conservative greenhouses. Acta Hortic. 148:287-295.

Hausbeck, M. K., and Pennypacker, S. P. 1987. Effects of grower activity and environmental modification on *Botrytis cinerea* conidial concentration in a greenhouse. (Abstr.) Phytopathology 77:1733.

_____. 1988. Variation in resistance of cutting-propagated geraniums to leaf blight caused by *Botrytis cinerea.* (Abstr.) Phytopathology 78:1508.

_____. 1991. Influence of grower activity and disease incidence on concentrations of airborne conidia of *Botrytis cinerea* among geranium stock plants. Plant Dis. 75:798-803.

Hawker, L. 1950. Physiology of Fungi. University Press, London.

Hayward, A. C. 1974. Latent infections by bacteria. Annu. Rev. Phytopathol. 12:87-97.

Heale, J. B. 1988. The potential impact of fungal genetics and molecular biology on biological control, with particular reference to entomopathogens. Pages 211-234 in: Fungi in Biological Control Systems. M. N. Burge, ed. University of Manchester Press, Manchester.

Heale, J. B., and Stringer-Calvert, A. 1974. Invertase levels and induced resistance in tissue cultures of *Daucus carota* L. invaded by fungi. Cytobios 10:167-180.

Healy, M. J. R. 1955. Statistical techniques for inspection sampling. Trop. Agric. (Trinidad) 32:10-19.

Hearn, A. R., Fitter, D. J., and Adams, P. 1981. An automated system for sampling flowing nutrient solutions. Pages 160-173 in: Annu. Rep. Glasshouse Crops Res. Inst. 1981.

Hege, H., and Ross, H. 1972. Das Dämpfen von Boden und Erden. Kuratorium für Technik und Bauwesen in der Landwirtschaft, Frankfurt am Main.

Heijna, B. J. 1966. Improved soil steam sterilization. Pages 81-101 in: Jaarb. Tuinbouwtech. Inst. Tuinbouwtechniek.

Heine, C. G. 1980. A review of pesticide application systems. Pages 75-83 in: Spraying Systems for the 1980s. Monogr. 24. BCPC Publications, Croydon, England.

Hemwall, J. B. 1960. Theoretical considerations of several factors influencing the effectivity of soil fumigants under field conditions. Soil Sci. 90:157-168.

Hendrix, F. F., Jr., and Campbell, W. A. 1973. Pythiums as plant pathogens. Annu. Rev. Phytopathol. 11:77-98.

Henis, Y., and Chet, I. 1975. Microbial control of plant pathogens. Adv. Appl. Microbiol. 19:85-111.

Henis, Y., and Katan, J. 1975. Effect of inorganic amendments and soil reaction on soil-borne plant diseases. Pages 100-106 in: Biology and Control of Soil-Borne Plant Pathogens. G. W. Bruehl, ed. American Phytopathological Society, St. Paul, MN.

Henis, Y., Ghaffar, A., and Baker, R. 1978. Integrated control of Rhizoctonia solani damping-off of radish: Effect of successive plantings, PCNB, and *Trichoderma harzianum* on pathogen and disease. Phytopathology 68:900-907.

Hennebert, G. L., and Gilles, G. L. 1958. Épidémiologie de *Botrytis cinerea* Pers. sur les fraisiers. Meded. Landbouwhogesch. Opzoekingsstn. Staat Gent 23:864-888.

Hesketh, T., Skilton, R. A., and Studman, C. J. 1986. Advanced digital control for New Zealand greenhouses. J. Agric. Eng. Res. 34:207-218.

Hijwegen, T. 1986. Biological control of cucumber powdery mildew by *Tilletiopsis minor*. Neth. J. Plant Pathol. 92:93-95.

______. 1988. Effect of seventeen fungicolous fungi on sporulation of cucumber powdery mildew. Neth. J. Plant Pathol. 94:185-190.

Hijwegen, T., and Buchenauer, H. 1984. Isolation and identification of hyperparasitic fungi associated with Erysiphaceae. Neth. J. Plant Pathol. 90:79-84.

Hirano, S. S., and Upper, C. D. 1983. Ecology and epidemiology of foliar bacterial plant pathogens. Annu. Rev. Phytopathol. 21:243-269.

Hirst, J. M. 1959. Spore liberation and dispersal. Pages 529-538 in: Plant Pathology: Problems and Progress 1908-1958. C. S. Holton, G. W. Fischer, R. W. Fulton, H. Hart, and S. E. A. McCallan, eds. University of Wisconsin Press, Madison.

Hislop, E. C., and Baines, C. R. 1980. An analysis of some spray factors affecting the protection of foliage by fungicides. Pages 23-33 in: Spraying Systems for the 1980s. Monogr. 24. BCPC Publications, Croydon, England.

Hite, R. E. 1973. The effect of irradiation on the growth and asexual reproduction of *Botrytis cinerea*. Plant Dis. Rep. 57:131-135.

Hobbs, E. L., and Waters, W. E. 1964. Influence of nitrogen and potassium on susceptibility of *Chrysanthemum morifolium* to *Botrytis cinerea*. Phytopathology 54:674-676.

Hoch, H. C., and Provvidenti, R. 1979. Mycoparasitic relationships: Cytology of the *Sphaerotheca fuliginea-Tilletiopsis* sp. interaction. Phytopathology 69:359-362.

Hoffmann, G. M., and Malkomes, H.-P. 1974. Bromide residues in vegetable crops after soil fumigation with methyl bromide. Agric. Environ. 1:321-328.

______. 1979. The fate of fumigants. Pages 291-335 in: Soil Disinfestation. D. Mulder, ed. Elsevier, Amsterdam.

Hoitink, H. A. J. 1980. Composted bark, a lightweight growth medium with fungicidal properties. Plant Dis. 64:142-147.

Hoitink, H. A. J., and Fahy, P. C. 1986. Basis for the control of soilborne plant pathogens with composts. Annu. Rev. Phytopathol. 24:93-114.

Hoitink, H. A. J., Herr, L. J., and Schmitthenner, A. F. 1976. Survival of some plant pathogens during composting of hardwood tree bark. Phytopathology 66:1369-1372.

Hoitink, H. A. J., Nelson, E. B., and Gordon, D. T. 1982. Composted bark controls soil pathogens of plants. Ohio Rep. 67:7-10.

Holden, W. S. 1970. Water Treatment and Examination. J. & A. Churchill, London.

Holliday, P., and Mulder, J. L. 1976. *Fulvia fulva.* Descriptions of Pathogenic Fungi and Bacteria, No. 487. Commonwealth Mycological Institute, Kew, England.

Hollings, M. 1962. Heat treatment in the production of virus-free ornamental plants. NAAS Q. Rev. 57:31-34.

——. 1965. Disease control through virus-free stock. Annu. Rev. Phytopathol. 3:367-396.

Homma, Y., Kubo, C., Ishir, M., and Ohata, K. 1979a. Effect of organic amendments to soil on suppression of disease severity of tomato Fusarium-wilt. Pages 89-101 in: Bull. Shikoku Agric. Exp. Stn. 34.

——. 1979b. Mechanism of suppression of tomato Fusarium-wilt by amendment of chicken manure to soil. Pages 103-121 in: Bull. Shikoku Agric. Exp. Stn. 34.

Honda, Y., and Yunoki, T. 1977. Control of Sclerotinia disease of greenhouse eggplant and cucumber by inhibition of development of apothecia. Plant Dis. Rep. 61:1036-1040.

Hondelmann, W., and Richter, E. 1973. Über die Anfälligkeit von Erdbeerklonen gegen *Botrytis cinerea* Pers. in Abhängigkeit von Pektinquantität und -qualität der Früchte. Gartenbauwissenschaft 38:311-314.

Hornby, D. 1983. Suppressive soils. Annu. Rev. Phytopathol. 21:65-85.

Horne, C. W. 1989. Groundwork for decision: Developing recommendations for plant disease control. Plant Dis. 73:943-948.

Horsfall, J. G., and Dimond, A. E. 1957. Interactions of tissue sugar, growth substances, and disease susceptibility. Z. Pflanzenkrankh. Pflanzenpathol. Pflanzenschutz 64:415-421.

Hoshi, T., and Kozai, T. 1984. Knowledge-based and hierarchically distributed online system for greenhouse management. Acta Hortic. 148:301-303.

Howard, E. L., and Horsfall, J. G. 1959. Therapy. Pages 563-604 in: Plant Pathology: An Advanced Treatise. J. G. Horsfall and A. E. Dimond, eds. Academic Press, New York.

Howell, C. R. 1982. Effect of *Gliocladium virens* on *Pythium ultimum, Rhizoctonia solani,* and damping-off of cotton seedlings. Phytopathology 72:496-498.

Hsaio, T. C. 1973. Plant responses to water stress. Annu. Rev. Plant Physiol. 24:519-576.

Huang, H. C., and Hoes, J. A. 1980. Importance of plant spacing and sclerotial position to development of Sclerotinia wilt of sunflower. Plant Dis. 64:81-84.

Huber, D. M., and Schneider, R. W. 1982. The description and occurrence of suppressive soils. Pages 1-7 in: Suppressive Soils and Plant Disease. R. W. Schneider, ed. American Phytopathological Society, St. Paul, MN.

Huber, D. M., and Watson, R. D. 1974. Nitrogen form and plant disease. Annu. Rev. Phytopathol. 12:139-165.

Huisman, O. C. 1982. Interrelations of root growth dynamics to epidemiology of root-invading fungi. Annu. Rev. Phytopathol. 20:303-327.

Hurd, R. G., and Graves, C. J. 1984. The influence of different temperature patterns having the same integral on the earliness and yield of tomatoes. Acta Hortic. 148:547-554.

Hurd, R. G., and Sheard, G. 1981. Fuel Savings in Greenhouses: The Biological Approach. Grower Guide 20. Grower Books, London.

Hussey, N. W., Read, H. W., and Hesling, J. J. 1967. The Pests of Protected Cultivation. Elsevier, New York.

Ingold, C. T. 1971. Fungal Spores: Their Liberation and Dispersal. Clarendon Press, Oxford.

Ingram, D. L. 1985. Modeling high temperature and exposure time interactions on *Pittosporum tobira* root cell membrane thermostability. J. Am. Soc. Hortic. Sci. 110:470-473.

Ishiba, C., Tani, T., and Murata, M. 1981. Protection of cucumber against anthracnose by a hypovirulent strain of *Fusarium oxysporum* f. sp. *cucumerinum.* Ann. Phytopathol. Soc. Jpn. 47:352-359.

Jackson, M. B. 1980. Aeration in the nutrient film technique of glasshouse crop production and the importance of oxygen, ethylene and carbon dioxide. Acta Hortic. 98:61-78.

Jackson, R. M. 1965. Antibiosis and fungistasis of soil microorganisms. Pages 363-369 in: Ecology of Soil-Borne Pathogens. K. F. Baker and W. C. Snyder, eds. University of California

Press, Berkeley.

Jaffe, M. J., and Biro, R. 1979. Thigmomorphogenesis: The effect of mechanical perturbation on the growth of plants, with special reference to anatomical changes, the role of ethylene, and interaction with other environmental stresses. Pages 25-59 in: Stress Physiology in Crop Plants. H. Mussell and R. C. Staples, eds. John Wiley & Sons, New York.

Jamart, G., de Prest, G., and Kamoen, O. 1988. Control of *Pythium* spp. on ornamental plants in a nutrient-film system. Meded. Fac. Landbouwwet. Rijksuniv. Gent 53(2b):625-634.

Janick, J., and Durkin, D. 1968. The effect of plant density on greenhouse chrysanthemum quality. Proc. Am. Soc. Hortic. Sci. 93:583-588.

Jarrett, P., Burges, H. D., and Matthews, G. A. 1978. Penetration of controlled drop spray of *Bacillus thuringiensis* into chrysanthemum beds compared to high volume spray and thermal fog. Pages 75-81 in: Symposium on Controlled Drop Application. Monogr. 22. BCPC Publications, Croydon, England.

Jarvis, W. R. 1960. An apparatus for studying hygroscopic responses in fungal conidiophores. Trans. Br. Mycol. Soc. 43:357-364.

———. 1962a. The dispersal of *Botrytis cinerea* in a raspberry plantation. Trans. Br. Mycol. Soc. 45:549-559.

———. 1962b. The epidemiology of *Botrytis cinerea* in strawberries. Pages 258-262 in: Proc. Int. Hortic. Congr. 16th.

———. 1962c. Splash dispersal of spores of *Botrytis cinerea.* Nature (London) 193:599.

———. 1962d. The infection of strawberry and raspberry fruits by *Botrytis cinerea* Pers. Ann. Appl. Biol. 50:569-575.

———. 1964. Thermal and translocated induction of endophytic mycelium in two powdery mildews. Nature (London) 203:895.

———. 1965. The biological basis for the design of control measures in Botrytis diseases. Pages 108-115 in: Proc. Br. Insectic. Fungic. Conf. 1965.

———. 1977a. *Botryotinia* and *Botrytis* species: Taxonomy, physiology and pathogenicity. Can. Dep. Agric. Monogr. 15.

———. 1977b. Biological control of *Fusarium.* Can. Agric. 22:28-30.

———. 1980a. Taxonomy. Pages 1-17 in: The Biology of *Botrytis.* J. R. Coley-Smith, K. Verhoeff, and W. R. Jarvis, eds. Academic Press, London.

———. 1980b. Epidemiology. Pages 219-250 in: The Biology of *Botrytis.* J. R. Coley-Smith, K. Verhoeff, and W. R. Jarvis, eds. Academic Press, London.

———. 1985. Problems of latency in *Botrytis* diseases. (Abstr.) Can. J. Plant Pathol. 7:445.

———. 1988. Fusarium crown and root rot of tomatoes. Phytoprotection 69:49-64.

———. 1989. Managing diseases in greenhouse crops. Plant Dis. 73:190-194.

Jarvis, W. R., and Barrie, S. D. 1988. Stem rot of greenhouse cucumbers caused by *Penicillium crustosum.* Plant Dis. 72:363.

Jarvis, W. R., and Berry, J. W. 1986. Integrated control of cucumber powdery mildew in the greenhouse. (Abstr.) Can. J. Plant Pathol. 8:351.

Jarvis, W. R., and Nuttall, V. W. 1981. Cucumber diseases. Can. Dep. Agric. Publ. 1684.

Jarvis, W. R., and Shoemaker, R. A. 1978. Taxonomic status of *Fusarium oxysporum* causing foot and root rot of tomato. Phytopathology 68:1679-1680.

Jarvis, W. R., and Slingsby, K. 1975. Tolerance of *Botrytis cinerea* and rose powdery mildew to benomyl. Can. Plant Dis. Surv. 55:44.

———.1977. The control of powdery mildew of greenhouse cucumber by water sprays and *Ampelomyces quisqualis.* Plant Dis. Rep. 61:728-730.

Jarvis, W. R., and Thorpe, H. J. 1981. Control of Fusarium foot and root rot of tomatoes by soil amendment with lettuce residues. Can. J. Plant Pathol. 3:159-162.

Jarvis, W. R., Thorpe, H. J., and Meloche, R. B. 1983. Survey of greenhouse management practices in Essex County, Ontario, in relation to Fusarium foot and root rot of tomato. Plant Dis. 67:38-40.

Jarvis, W. R., Shaw, L. A., and Traquair, J. A. 1989. Factors affecting antagonism of cucumber powdery mildew by *Stephanoascus flocculosus* and *S. rugulosus.* Mycol. Res. 92:162-165.

Jarvis, W. R., Barrie, S. D., Traquair, J. A., and Stoessl, A. 1990. Morphological and chemical studies of *Penicillium oxalicum,* newly identified as a pathogen on greenhouse cucumber. Can. J. Bot. 68:21-25.

Jenkins, B. M., Sachs, R. M., and Forister, C. W. 1988. A comparison of bench top and perimeter heating of greenhouses. Calif. Agric. 42(1):13-15.

Jenkins, S. F., Jr. 1981. Use of chlorine to suppress root infecting pathogens of vegetables growing in recirculating hydroponic systems. (Abstr.) Phytopathology 71:883.

Jenkins, S. F., Jr., and Averre, C. W. 1983. Root diseases of vegetables in hydroponic culture systems in North Carolina greenhouses. Plant Dis. 67:968-970.

Jenkins, S. F., Jr., Winstead, N. W., and McCombs, C. L. 1964. Pathogenic comparisons of three new and four previously described races of *Glomerella cingulata* var. *orbiculare*. Plant Dis. Rep. 48:619-622.

Jenns, A. E., and Kuć, J. 1979. Graft transmission of systemic resistance of cucumber to anthracnose induced by *Colletotrichum lagenarium* and tobacco necrosis virus. Phytopathology 69:753-756.

______. 1980. Characteristics of anthracnose resistance induced by localized infection of cucumber with tobacco necrosis virus. Physiol. Plant Pathol. 7:81-91.

Jensen, M. H., and Collins, W. L. 1985. Hydroponic vegetable production. Hortic. Rev. 7:483-558.

Jewett, T. J., Jackson, H. A., Bruno, L., Shaw, M. B., and McGugan, C. A. 1984. Evaluation of an air conditioning water chiller/heat pump system with heat storage for solar heating of greenhouses. Acta Hortic. 148:691-698.

Johnson, J. 1914. The control of damping-off disease in plant beds. Pages 29-61 in: Wis. Agric. Exp. Stn. Res. Bull. 31.

______. 1936. Relation of root pressure to plant disease. Science (Washington, D.C.) 84:135-136.

Johnson, J., and Ogden, W. B. 1929. The overwintering of tobacco mosaic virus. Wash. Agric. Exp. Stn. Bull 308.

Johnson, P. W. 1975. Effects of rate and depth of application of nematicide on nematode vertical distribution and tomato production in a sandy loam greenhouse soil. Can. J. Plant Sci. 55:1017-1021.

Johnson, P. W., and McKeen, C. D. 1973. Vertical movement and distribution of *Meloidogyne incognita* (Nematodea) under tomato in a sandy loam greenhouse soil. Can. J. Plant Sci. 53:837-841.

Jones, D., and Watson, D. 1969. Parasitism and lysis by soil fungi of *Sclerotinia sclerotiorum* (Lib.) de Bary, a phytopathogenic fungus. Nature (London) 224:287-288.

Jones, J. P., and Woltz, S. S. 1969. Fusarium wilt (race 2) of tomato: Calcium, pH, and micronutrient effects on disease development. Plant Dis. Rep. 53:276-279.

Jones, J. P., Crill, P., and Volin, R. B. 1975. Effect of light duration on Verticillium wilt of tomato. Phytopathology 65:647-648.

Jones, J. P., Engelhard, A. W., and Woltz, S. S. 1989. Management of Fusarium wilt of vegetables and ornamentals by macro- and microelement nutrition. Pages 18-32 in: Soilborne Plant Pathogens: Management of Diseases with Macro- and Microelements. A. W. Engelhard, ed. American Phytopathological Society, St. Paul, MN.

Jones, P., Jacobson, B. K., and Jones, J. W. 1988. Applying expert systems concepts to real-time greenhouse controls. Acta Hortic. 230:201-208.

Jordan, V. W. L., and Richmond, D. V. 1972. The effects of glass cloche and coloured polyethylene tunnels on microclimate, growth, yield and disease severity of strawberry plants. J. Hortic. Sci. 47:419-426.

Kamerman, W. 1975. Biology and control of *Xanthomonas hyacinthi* in hyacinths. Acta Hortic. 47:99-105.

Kano, A., and Shimaji, H. 1988. Greenhouse environmental control system with a crop model and an expert system. Acta Hortic. 230:229-236.

Kasenberg, T. R., and Traquair, J. A. 1987. Allelopathic biocontrol of Fusarium crown and root rot in greenhouse tomatoes. (Abstr.) Can. J. Plant Pathol. 9:280.

______. 1988. Effects of phenolics on the growth of *Fusarium oxysporum* f. sp. *radicis-lycopersici* in vitro. Can. J. Bot. 66:1174-1177.

______. 1989. Lettuce siderophores and biocontrol of Fusarium rot in greenhouse tomatoes. (Abstr.) Can. J. Plant Pathol. 11:192.

Kassaby, F. Y. 1985. Solar-heating soil for control of damping-off diseases. Soil Biol. Biochem. 17:429-434.

Kassanis, B. 1954. Heat therapy of virus-infected plants. Ann. Appl. Biol. 41:470-474.
Katan, J. 1971. Symptomless carriers of the tomato Fusarium wilt pathogen. Phytopathology 61:1213-1217.
_____. 1979. Solar heating of the soil and other economical environmentally safe methods of controlling soilborne pathogens, weeds and pests. (Abstr.) Phytopathology 69:1033-1034.
_____. 1980. Solar pasteurization of soils for disease control: Status and prospects. Plant Dis. 64:450-454.
_____. 1981. Solar heating (solarization) of soil for control of soilborne pests. Annu. Rev. Phytopathol. 19:211-236.
_____. 1989. Soil temperature interactions with the biotic components of vascular wilt diseases. Pages 353-366 in: Vascular Wilt Diseases of Plants: Basic Studies and Control. E. C. Tjamos and C. H. Beckman, eds. NATO ASI and Springer-Verlag, Berlin.
Katan, J., Greenberger, A., Alon, H., and Grinstein, A. 1976. Solar heating by polyethylene mulching for the control of diseases caused by soil-borne pathogens. Phytopathology 66:683-688.
Katan, T., and Ovadia, S. 1985. Effect of chlorothalonil on resistance of *Botrytis cinerea* to dicarboximides in cucumber glasshouses. EPPO Bull. 15:365-369.
Kato, T., Suzuki, K., Takahashi, J., and Kamoshita, K. 1984. Negatively correlated cross-resistance between benzimidazole fungicides and methyl *N*-(3,5-dichlorophenyl)carbamate. J. Pestic. Sci. 9:489-495.
Kavanagh, T., and Herlihy, M. 1975. Microbiological aspects. Pages 39-49 in: Peat in Horticulture. D. W. Robinson and J. G. D. Lamb, eds. Academic Press, London.
Kempton, R. G., and Maw, G. A. 1972. Soil fumigation with methyl bromide: Bromide accumulation in lettuce plants. Ann. Appl. Biol. 72:71-79.
_____. 1973. Soil fumigation with methyl bromide: The uptake and distribution of inorganic bromide in tomato plants. Ann. Appl. Biol. 74:91-98.
_____. 1974. Soil fumigation with methyl bromide: The phytotoxicity of inorganic bromine to carnation plants. Ann. Appl. Biol. 76:217-229.
Kendrick, J. B., and Walker, J. C. 1948. Predisposition of tomato to bacterial canker. J. Agric. Res. (Washington, D.C.) 77:169-186.
Kennedy, J. S., Day, M. F., and Eastop, V. F. 1962. A Conspectus of Aphids as Vectors of Plant Viruses. Commonwealth Institute of Entomology, London.
Kerling, L. C. P. 1964. Fungi in the phyllosphere of leaves of rye and strawberry. Meded. Landbouwhogesch. Opzoekingsstn. Staat Gent 29:885-895.
Kerr, A. 1964. The influence of soil moisture on infection of peas by *Pythium ultimum*. Aust. J. Biol. Sci. 17:676-685.
_____. 1980. Biological control of crown gall through production of agrocin 84. Plant Dis. 64:25-30.
Kim, S. H., Forer, L. B., and Longenecker, J. L. 1975. Recovery of plant pathogens from commercial peat-products. (Abstr.) Proc. Am. Phytopathol. Soc. 2:124.
King, E. 1963. Zur Klimatologie des thermischen Wirkungsgrads von Mattenkühlungen in Gewächshäusern. Gartenbauwissenschaft 28:133-136.
Kirkby, E. A. 1969. Ion uptake and ionic balance in plants in relation to the form of nitrogen nutrition. Pages 215-235 in: Ecological Aspects of the Mineral Nutrition of Plants. I. H. Rorison, ed. Blackwell, Oxford.
Kitchen, J. T. 1953. The heat unit theory for vegetables. Hortic. News 34(5):1-4.
Kloepper, J. W., Leong, J., Teintze, M., and Schroth, M. N. 1980. *Pseudomonas* siderophores: A mechanism explaining disease-suppressive soils. Curr. Microbiol. 4:317-320.
Knies, P., van der Braak, N. J., and Breuer, J. J. G. 1984. Infrared heating in greenhouses. Acta Hortic. 148:73-80.
Knight, D. E., and Keyworth, W. G. 1960. Didymella stem-rot of outdoor tomatoes. I. Studies on sources of infection and their elimination. Ann. Appl. Biol. 48:245-258.
Kodama, T., and Fukai, T. 1982. Solar heating in closed plastic house for control of soilborne diseases. V. Application for control of Fusarium wilt of strawberry. Ann. Phytopathol. Soc. Jpn. 48:570-577.
Köhl, J., and Schlösser, E. 1988. Specificity in decay of sclerotia of *Botrytis cinerea* by species and strains of *Trichoderma*. Meded. Fac. Landbouwwet. Rijksuniv. Gent 53(2a):339-346.
_____. 1989. Decay of sclerotia of *Botrytis cinerea* by *Trichoderma* spp. at low temperatures.

J. Phytopathol. (Berlin) 125:320-326.

Kontaxis, D. G. 1962. Leaf trichomes as avenues for infection by *Corynebacterium michiganense.* Phytopathology 52:1306-1307.

Kooistra, E. 1968. Powdery mildew resistance in cucumber. Euphytica 17:236-244.

Koths, J. S. 1983. HAF keeps plants dry. Pages 3-4 in: Conn. Greenhouse Newsl. 117.

Koths, J. S., and Bartok, J. W. 1985. Horizontal air flow. Conn. Greenhouse Newsl. 125.

Kozai, T. 1976. Optimizing control of plant growth using a digital computer. J. Agric. Meteorol. Soc. Jpn. 32:41-49.

Kozel, P. C., and Tukey, H. B. 1968. Loss of gibberellins by leaching from stems and foliage of *Chrysanthemum morifolium* 'Princess Anne.' Am. J. Bot. 55:1184-1189.

Kramer, R. J. 1939. Effect of drops of water on leaf temperatures. Am. J. Bot. 26:12-14.

Kranz, J., Hau, B., and Aust, H. J. 1984. Monitoring in crop protection modeling. Pages 233-241 in: Pest and Pathogen Control: Strategic, Tactical, and Policy Models. G. R. Conway, ed. John Wiley & Sons, Chichester.

Krauss, A. 1969. Einfluss der Ernährung der Pflanzen mit Mineralstoffen und den Befall mit parasitären Krankheiten und Schädlingen. Z. Pflanzenernaehr. Bodenkd. 124:129-149.

_____. 1971. Einfluss der Ernährung des Salats mit Massennährstoffen auf den Befall mit *Botrytis cinerea.* Z. Pflanzenernaehr. Bodenkd. 128:12-23.

Kretchman, D. W. 1968. A preliminary report on several aspects of fruit setting of greenhouse tomatoes. Pages 5-8 in: Res. Summ. Ohio Agric. Res. Dev. Cent. 26.

Kreutzer, W. A. 1963. Selective toxicity of chemicals to soil microorganisms. Annu. Rev. Phytopathol. 1:101-126.

Krikun, J., Nachmias, A., Offenbach, R., and Ucko, O. 1983. Establishment of the crown and root rot pathogen of tomato in the Negev region. Phytoparasitica 11:3-4.

Krug, H. 1963. Ein Beitrag zur Klimatisierung von Gewächshäusern durch Verdungstungkühlung. Gartenbauwissenschaft 28:123-132.

Krug, H., and Liebig, H. P. 1980. A plant production model to optimize temperature in greenhouses on the base of ecological and economical data. Acta Hortic. 106:179-182.

Kuć, J. 1987. Plant immunization and its applicability for disease control. Pages 255-274 in: Innovative Approaches to Plant Disease Control. I. Chet, ed. John Wiley & Sons, New York.

Kuć, J., and Richmond, S. 1977. Aspects of the protection of cucumber against *Colletotrichum lagenarium* by *Colletotrichum lagenarium.* Phytopathology 67:533-536.

Kuć, J., Shockley, G., and Kearney, K. 1975. Protection of cucumber against *Colletotrichum lagenarium.* Physiol. Plant Pathol. 7:195-199.

Kuniyasu, K., and Yamakawa, K. 1983. Control of Fusarium wilt of tomato caused by *Fusarium oxysporum* f. sp. *lycopersici* race J3 by grafting to KNVF and KVF rootstocks of the interspecific hybrids between *Lycopersicon esculentum* × *L. hirsutum.* Ann. Phytopathol. Soc. Jpn. 49:581-586.

Kurata, K. 1988. Greenhouse control by machine learning. Acta Hortic. 230:195-200.

Kuter, G. A., and Hoitink, H. A. J. 1985. Use of combinations of microbial antagonists to suppress Rhizoctonia and Pythium damping-off in compost-amended container media. (Abstr.) Phytopathology 75:1344.

Langerak, C. J. 1977. The role of antagonists in the chemical control of *Fusarium oxysporum* f. sp. *narcissi.* Neth. J. Plant Pathol. 83(suppl. 1):365-381.

Lankow, R. K. 1971. Growth responses of strains of *Botrytis cinerea* tolerant and susceptible to 2,6-dichloro-4-nitroaniline. Phytopathology 61:900.

Last, F. T., and Ebben, M. H. 1966. The epidemiology of tomato brown root rot. Ann. Appl. Biol. 57:95-112.

Last, F. T., Ebben, M. H., Hoare, R. C., Turner, E. A., and Carter, A. R. 1969. Build-up of tomato brown rot caused by *Pyrenochaeta lycopersici* Schneider and Gerlach. Ann. Appl. Biol. 64:449-459.

Latin, R. X., Miles, G. E., and Rettinger, J. C. 1987. Expert systems in plant pathology. Plant Dis. 71:866-872.

Law, S. E. 1980. Droplet charging and electrostatic deposition of pesticide sprays—Research and development in the U.S.A. Pages 85-94 in: Spraying Systems for the 1980s. Monogr. 24. BCPC Publications, Croydon, England.

Leach, C. M. 1962. Sporulation of diverse species of fungi under near-ultraviolet radiation.

Can. J. Bot. 40:151-161.
_____. 1985. Effect of still and moving moisture-saturated air on sporulation of *Drechslera* and *Peronospora*. Trans. Br. Mycol. Soc. 84:179-183.
Leach, J. G. 1964. Observations on cucumber beetles as vectors of cucurbit wilt. Phytopathology 54:606-607.
Leach, L. D. 1947. Growth rates of host and pathogen as factors determining the severity of preemergence damping-off. J. Agric. Res. (Washington, D.C.) 75:161-179.
Leben, C. 1964. Influence of bacteria isolated from healthy cucumber leaves on two leaf diseases of cucumber. Phytopathology 54:405-408.
_____. 1965a. Epiphytic microorganisms in relation to plant disease. Annu. Rev. Phytopathol. 3:209-230.
_____. 1965b. Influence of humidity on the migration of bacteria on cucumber seedlings. Can. J. Microbiol. 11:671-676.
_____. 1981. How plant-pathogenic bacteria survive. Plant Dis. 65:633-637.
Leben, C., and Daft, G. C. 1965. Influence of an epiphytic bacterium on cucumber anthracnose, early blight of tomato, and northern leaf blight of corn. Phytopathology 55:760-762.
Lecoq, H. 1988. Cucumber green mottle mosaic virus (CGMMV). Page 52 in: European Handbook of Plant Diseases. I. M. Smith, J. Dunez, R. A. Lelliott, D. H. Phillips, and S. A. Archer, eds. Blackwell, Oxford.
Lecoq, H., Cohen, S., Pitrat, M., and Labonne, G. 1979. Resistance to cucumber mosaic virus transmission by aphids in *Cucumis melo*. Phytopathology 69:1223-1225.
Leistra, M., Smelt, J. H., and Nollen, H. M. 1974. Concentration-time relationship for methyl isothiocyanate in soil after injection of metham-sodium. Pestic. Sci. 5:409-417.
Lelliott, R. A. 1984. Cost-benefit analysis as used in the United Kingdom for eradication-campaigns against alien pests and diseases. EPPO Bull. 14:337-341.
_____. 1988a. *Erwinia chrysanthemi* pv. *dianthicola* (Hellmers) Dickey. Pages 192-193 in: European Handbook of Plant Diseases. I. M. Smith, J. Dunez, R. A. Lelliott, D. H. Phillips, and S. A. Archer, eds. Blackwell, Oxford.
_____. 1988b. *Pseudomonas caryophylli* (Burkholder) Starr & Burkholder. Pages 138-139 in: European Handbook of Plant Diseases. I. M. Smith, J. Dunez, R. A. Lelliott, D. H. Phillips, and S. A. Archer, eds. Blackwell, Oxford.
_____. 1988c. *Xanthomonas campestris* pv. *pelargonii* (Brown) Dye. Pages 164-165 in: European Handbook of Plant Diseases. I. M. Smith, J. Dunez, R. A. Lelliott, D. H. Phillips, and S. A. Archer, eds. Blackwell, Oxford.
Lenz, J., Paulus, A. O., and Bald, J. G. 1971. Systemic fungicides for control of some diseases of Easter lilies. Calif. Agric. 25(3):4-5.
Leong, J. 1986. Siderophores: Their biochemistry and possible role in the biocontrol of plant pathogens. Annu. Rev. Phytopathol. 24:187-209.
Levitt, J. 1980. Responses of Plants to Environmental Stresses. Academic Press, New York.
Lewis, J. A., and Papavizas, G. C. 1970. Evolution of volatile sulfur-containing compounds from decomposition of crucifers in soil. Soil Biol. Biochem. 2:239-246.
_____. 1971. Effect of sulfur-containing volatile compounds and vapors from cabbage decomposition on *Aphanomyces euteiches*. Phytopathology 61:208-214.
_____. 1974. Effect of volatiles from decomposing plant tissues on pigmentation, growth, and survival of *Rhizoctonia solani*. Soil Sci. 118:156-163.
_____. 1987. Application of *Trichoderma* and *Gliocladium* in alginate pellets for control of Rhizoctonia damping-off. Plant Pathol. 36:438-446.
Liao, C. H. 1989. Antagonism of *Pseudomonas putida* strain PP22 to phytopathogenic bacteria and its potential use as a biocontrol agent. Plant Dis. 73:223-226.
Linderman, R. G. 1989. Organic amendments and soil-borne diseases. Can. J. Plant Pathol. 11:180-183.
Linderman, R. G., Moore, L. W., Baker, K. F., and Cooksey, D. A. 1983. Strategies for detecting and characterizing systems for biological control of soilborne plant pathogens. Plant Dis. 67:1058-1064.
Lindquist, R. K., and Powell, C. C. 1980. Evaluation of low-volume pesticide applicators for use on glasshouse crops. Pages 271-281 in: Spraying Systems for the 1980s. Monogr. 24. BCPC Publications, Croydon, England.
Linfield, C. A. 1987. Cobweb disease of carnations—A new disease of imported flowers.

Plant Pathol. 36:222-223.
Lisansky, S., and Hall, R. A. 1983. Use of fungi as microbial insecticides. Pages 327-345 in: The Filamentous Fungi. Vol. 4, Fungal Technology. J. E. Smith, D. R. Berry, and B. Kristiansen, eds. Edward Arnold, London.
Locke, J. C., Marois, J. J., and Papavizas, G. C. 1985. Biological control of Fusarium wilt of greenhouse-grown chrysanthemums. Plant Dis. 69:167-169.
Lockhart, C. L., and Forsyth, F. R. 1964. Influence of fungicides on the tomato and growth of *Botrytis cinerea* Pers. Nature (London) 204:1107-1108.
Lockwood, J. L. 1988. Evolution of concepts associated with soilborne plant pathogens. Annu. Rev. Phytopathol. 26:93-121.
Longrée, K. 1939. The effect of temperature and relative humidity on the powdery mildew of roses. Cornell Univ. Agric. Exp. Stn. Mem. 223.
Lopez-Real, J., and Foster, M. 1985. Plant pathogen survival during the composting of agricultural wastes. Pages 291-300 in: Composting of Agricultural and Other Wastes. J. K. R. Gasser, ed. Elsevier Applied Science, London.
Lorbeer, J. W. 1980. Variation in *Botrytis* and *Botryotinia*. Pages 19-39 in: The Biology of *Botrytis*. J. R. Coley-Smith, K. Verhoeff, and W. R. Jarvis, eds. Academic Press, London.
Lorenz, D. H., and Eichorn, K. W. 1982. *Botrytis cinerea* and its resistance to dicarboximide fungicides. EPPO Bull. 12:125-129.
Lot, H. 1988. Lettuce mosaic virus (LMV). Pages 40-41 in: European Handbook of Plant Diseases. I. M. Smith, J. Dunez, R. A. Lelliott, D. H. Phillips, and S. A. Archer, eds. Blackwell, Oxford.
Lumsden, R. D., and Locke, J. C. 1989. Biological control of damping-off caused by *Pythium ultimum* and *Rhizoctonia solani* with *Gliocladium virens* in soilless mix. Phytopathology 79:361-366.
Lyle, E. W. 1938. The black spot disease of roses, and its control under greenhouse conditions. Cornell Univ. Agric. Exp. Stn. Bull. 690.
_____. 1956. Development of rose propagative material free from black spot. Pages 91-92 in: Plant Dis. Rep. Suppl. 238.
Lynch, J. M. 1979. Straw residues as substrates for growth and product formation by soil microorganisms. Pages 105-112 in: Straw Decay and Its Effect on Disposal and Utilization. E. Grossbard, ed. John Wiley & Sons, Chichester.
Mabbet, T. 1986. Fogging, a versatile method for propagation and management. Agric. Int. 38:78-80.
MacDonald, J. D. 1982. Effect of salinity stress on the development of Phytophthora root rot of chrysanthemum. Phytopathology 72:214-219.
MacWithey, H. S. 1967. Protective fungicide treatments for control of winter rhizome rot caused by *Botrytis cinerea*. Plant Dis. Rep. 51:83-86.
Magie, R. O. 1980. Fusarium disease of gladioli controlled by inoculation of corms with non-pathogenic fusaria. Proc. Fla. State Hortic. Soc. 93:172-175.
Maher, M. J., and O'Flaherty, T. 1973. An analysis of greenhouse climate. J. Agric. Eng. Res. 18:197-203.
Mahrer, Y., and Katan, J. 1981. Spatial soil temperature regime under transparent polyethylene mulch: Numerical and experimental studies. Soil Sci. 131:82-87.
Malathrakis, N. E. 1985. The fungus *Acremonium alternatum* Link:Fr., a hyperparasite of the cucurbit powdery mildew pathogen *Sphaerotheca fuliginea*. Z. Pflanzenkrankh. Pflanzenschutz 92:509-515.
_____. 1987. Parasitism of cucurbit powdery mildew pathogen *Sphaerotheca fuliginea* by *Acremonium alternatum*. Pages 193-196 in: Integrated and Biological Control in Protected Crops. R. Cavalloro, ed. A. A. Balkema, Rotterdam.
_____. 1989. Resistance of *Botrytis cinerea* to dichlofluanid in greenhouse vegetables. Plant Dis. 73:138-141.
Malathrakis, N. E., and Saris, G. E. 1983. Bromine residues in greenhouse cucumbers and the possibility of reducing them. Acta Hortic. 152:297-304.
Malathrakis, N. E., and Vakalounakis, D. J. 1983. Resistance to benzimidazole fungicides in the gummy stem blight pathogen in cucumber. Plant Pathol. 32:395-399.
Malathrakis, N. E., Kapetanakis, G. E., and Linardakis, D. S. 1983. Brown root rot of tomato, and its control in Crete. Ann. Appl. Biol. 102:251-256.

Maliga, P. 1984. Isolation and characterization of mutants in plant cell culture. Annu. Rev. Plant Physiol. 35:519-542.

Malkomes, H.-P. 1971. Untersuchungen verschiedener Einflüsse auf die Zersetzung von Methylbromid (Terabol) bei Begasung von Boden und Pflanzsubstraten. Z. Pflanzenkrankh. Pflanzenschutz 79:274-290.

_____. 1972. Der Einfluss von Bodenbegasungen mit Methylbromid (Terabol) auf gärtnerische Kulturpflanzen. II. Bromidaufnahme und -toleranz bei Gemüsepflanzen. Z. Pflanzenkrankh. Pflanzenschutz 79:321-338.

Manbeck, H. B., and Aldrich, R. A. 1967. Analytical determination of the direct visible solar energy transmitted by rigid plastic greenhouses. Trans. ASAE 10:564-567.

Marks, C. F., Elliott, J. M., and Tu, C. M. 1972. Effects of deep fumigation on *Pratylenchus penetrans,* flue-cured tobacco, and soil nitrate content. Can. J. Plant Sci. 52:425-430.

Marois, J. J., and Mitchell, D. J. 1981a. Effects of fumigation and fungal antagonists on the relationships of inoculum density to infection incidence and disease severity in Fusarium crown rot of tomato. Phytopathology 71:167-170.

_____. 1981b. Effects of fungal communities on the pathogenic and saprophytic activities of *Fusarium oxysporum* f. sp. *radicis-lycopersici.* Phytopathology 71:1251-1256.

Marois, J. J., Mitchell, D. J., and Sonoda, R. M. 1981. Biological control of Fusarium crown rot of tomato under field conditions. Phytopathology 71:1257-1260.

Marois, J. J., Redmond, J. C., and MacDonald, J. D. 1988. Quantification of the impact of environment on the susceptibility of *Rosa hybrida* flowers to *Botrytis cinerea.* J. Am. Soc. Hortic. Sci. 113:842-845.

Marrou, J., and Migliori, A. 1971. Essai de protection des cultures de tomate contre le virus de la mosaïque du tabac: Mise en évidence d'une spécificité étroite de la prémunition entre souches de ce virus. Ann. Phytopathol. 3:447-459.

Martin, F. N., and Hancock, J. G. 1982. The effects of Cl^- and *Pythium oligandrum* on the ecology of *Pythium ultimum.* (Abstr.) Phytopathology 72:996.

Martin, H. 1964. The Scientific Principles of Crop Protection. Edward Arnold, London.

Martin, J. P. 1966. Bromine. Pages 62-64 in: Diagnostic Criteria for Plants and Soils. H. D. Chapman, ed. University of California Press, Berkeley.

Marx, D. H. 1972. Ectomycorrhizae as biological deterrents to pathogenic root infections. Annu. Rev. Phytopathol. 10:429-454.

Massey, A. B. 1925. Antagonism of the walnuts (*Juglans nigra* L. and *J. cinerea* L.) in certain plant associations. Phytopathology 15:773-784.

Mastalerz, J. W. 1977. The Greenhouse Environment. John Wiley & Sons, New York.

Matteoni, J. A., and Broadbent, A. B. 1988. Wounds caused by *Liriomyza trifolii* (Diptera: Agromizidae) as sites for infection of chrysanthemum by *Pseudomonas cichorii.* Can. J. Plant Pathol. 10:47-52.

Matteoni, J. A., Allen, W. R., and Broadbent, A. B. 1988. Tomato spotted wilt virus in greenhouse crops in Ontario. Plant Dis. 72:801.

Matthews, G. A. 1989. Electrostatic spraying of pesticides: A review. Crop Prot. 8:3-15.

Mattson, R. H., and Maxwell, T. J. 1971. Studies of greenhouse orientation in a heliodon. HortScience 6:209-210.

Maude, R. B. 1964. Studies on *Septoria* on celery seed. Ann. Appl. Biol. 54:313-326.

Mauromicale, G., Cosentino, S., and Copani, V. 1988. Validity of thermal unit summations for purposes of prediction in *Phaseolus vulgaris* L. cropped in Mediterranean environment. Acta Hortic. 229:321-325.

May, R. M. 1985. Host-parasite associations: Their population biology and population genetics. Pages 243-262 in: Ecology and Genetics of Host-Parasite Interactions. D. Rollinson and R. M. Anderson, eds. Academic Press, London.

McClellan, W. D., Baker, K. F., and Gould, C. J. 1949a. Occurrence of the Botrytis disease of gladiolus in the United States in relation to temperature and humidity. Phytopathology 39:260-271.

McClellan, W. D., Christie, J. R., and Horn, N. L. 1949b. Efficacy of soil fumigants as affected by soil temperature and moisture. Phytopathology 39:272-283.

McGrady, J. J., and Cotter, D. J. 1989. Fresh conifer bark reduces root-knot nematode galling of greenhouse tomatoes. HortScience 24:973-975.

McKeen, C. D. 1954. Methyl bromide as a soil fumigant for controlling soil-borne pathogens

and certain other organisms in vegetable seedbeds. Can. J. Bot. 32:101-115.
———. 1973. Occurrence, epidemiology, and control of bacterial canker of tomato in southwest Ontario. Can. Plant Dis. Surv. 53:127-130.
McLaughlin, S. B., and Taylor, G. E. 1981. Relative humidity: Important modifier of pollutant activity by plants. Science (Washington, D.C.) 211:167-169.
McWhorter, F. P. 1939. Botrytis blight of *Antirrhinum* related to trichome disposition. Phytopathology 29:651-652.
Mee, T. R. 1977. Man-made fog for freeze protection and microclimate control. Pages 203-208 in: Proc. Int. Soc. Citric. 1977. Vol. 1.
Meinkoth, J., and Wahl, G. 1984. Hybridization of nucleic acids immobilized on solid supports. Anal. Biochem. 138:276-284.
Melchers, L. E. 1926. Botrytis blossom blight and leaf spot of geranium and its relation to the gray mold of lettuce. J. Agric. Res. (Washington, D.C.) 32:883-894.
Meneley, J. C., and Stanghellini, M. E. 1972. Occurrence and significance of soft-rotting bacteria in healthy vegetables. (Abstr.) Phytopathology 62:778.
Merriman, P. R., Samson, I. M., and Schippers, B. 1981. Stimulation of germination of sclerotia of *Sclerotium cepivorum* at different depths in soil by artificial onion oil. Neth. J. Plant Pathol. 87:45-53.
Methy, M., and Salager, J. L. 1989. A microcomputer-based fast data acquisition system for in-vivo measurements of stress effects in crop plants by chlorophyll fluorescence induction. Comput. Electron. Agric. 4:121-128.
Miller, D. E., and Burke, D. W. 1975. Effect of soil aeration on Fusarium root rot of beans. Phytopathology 65:519-523.
Miller, J. C. 1963. Ecological relationships among parasites and the practice of biological control. Environ. Entomol. 12:620-624.
Miller, P. C., and Stoner, W. A. 1979. Canopy structure and environmental interactions. Pages 428-458 in: Topics in Plant Population Biology. O. T. Solbrig, S. Iain, G. B. Johnson, and P. H. Raven, eds. Columbia University Press, New York.
Miller, S. A., and Martin, R. R. 1988. Molecular diagnosis of plant disease. Annu. Rev. Phytopathol. 26:409-432.
Molyneux, C. J. 1988. A Practical Guide to NFT. Nutriculture, Ormskirk, England.
Monsion, M., and Dunez, J. 1971. État des recherches poursuivies en France sur les maladies à virus du chrysanthème. Rev. Zool. Agric. Pathol. Veg. 70:95-103.
Monteith, J. L. 1972. Survey of Instruments for Micrometeorology. IBP Handb. 22. Blackwell, Oxford.
Moody, A. R., and Gindrat, D. 1977. Biological control of cucumber black root rot by *Gliocladium roseum.* Phytopathology 67:1159-1162.
Morgan, J. M. 1984. Osmoregulation and water stress in higher plants. Annu. Rev. Plant Physiol. 35:299-319.
Morgan, W. M. 1981. The distribution and persistence of iprodione applied by thermal fogging in a glasshouse tomato crop. Ann. Appl. Biol. 98:93-99.
———. 1983a. Energy saving and the environmental control of disease in glasshouses. Page 1118 in: Int. Congr. Plant Prot. Proc. Conf. 10th.
———. 1983b. Viability of *Bremia lactucae* oospores and stimulation of their germination by lettuce seedlings. Trans. Br. Mycol. Soc. 80:403-408.
———. 1984a. The effect of night temperature and glasshouse ventilation on the incidence of *Botrytis cinerea* in a late planted tomato crop. Crop Prot. 3:243-251.
———. 1984b. Integration of environmental and fungicidal control of *Bremia lactucae* in a glasshouse lettuce crop. Crop Prot. 3:349-361.
———. 1985. Influence of energy-saving night temperature regimes on *Botrytis cinerea* in an early-season glasshouse tomato crop. Crop Prot. 4:97-110.
Morgan, W. M., and Ledieu, M. S. 1979. Pests and diseases of glasshouse crops. Pages 7.1-7.81 in: Pest and Disease Control Handbook. N. Scopes and M. Ledieu, eds. BCPC Publications, Croydon, England.
Morgan, W. M., and Molyneux, S. A. 1981. The effect of the glasshouse environment on disease. Pages 131-133 in: Annu. Rep. Glasshouse Crops Res. Inst. 1981.
Morotchovski, S. F. M., and Vitas, K. I. 1939. Main results of scientific research work during 1937 of the Pan-Soviet Research Institute for the sugar industry. (In Russian.) Pages 257-

260 in: Pishch. Promst. Leningr.
Morris, L. G. 1957. The use of steam for soil sterilization. Ind. Heat. Eng. 19:3-6, 24, 38-41, 67-71.
_____. 1959. Principles of glasshouse heating. Agriculture (London) 66:403-407.
Morris, L. G., Trickett, E. S., Vanstone, F. H., and Wells, D. A. 1958. The limitation of maximum temperature in a glasshouse by the use of a water film on the roof. J. Agric. Eng. Res. 3:121-130.
Mossop, D. W., and Procter, C. H. 1975. Cross protection of greenhouse tomatoes against tobacco mosaic virus. N.Z. J. Exp. Agric. 3:343-348.
Munnecke, D. E. 1956. Development and production of pathogen-free geranium propagative material. Pages 93-95 in: Plant Dis. Rep. Suppl. 238.
_____. 1957. Chemical treatment of nursery soils. Pages 197-209 in: The U.C. System for Producing Healthy Container-Grown Plants. K. F. Baker, ed. Calif. Agric. Exp. Stn. Man. 23.
Munnecke, D. E., and Chandler, P. A. 1957. A leaf spot of *Philodendron* related to stomatal exudation and to temperature. Phytopathology 47:299-303.
Munnecke, D. E., and Ferguson, J. 1960. The effect of soil fungicides upon soil-borne plant pathogenic bacteria and soil nitrogen. Plant Dis. Rep. 44:552-555.
Munnecke, D. E., Moore, B. J., and Abu-El-Haj, F. 1971. Soil moisture effects on control of *Pythium ultimum* or *Rhizoctonia solani* with methyl bromide. Phytopathology 61:194-197.
Munnecke, D. E., Kolbezen, M. J., and Wilbur, W. D. 1973. Effect of methyl bromide or carbon disulfide on *Armillaria* and *Trichoderma* growing on agar medium and relation to survival of *Armillaria* in soil following fumigation. Phytopathology 63:1352-1357.
Musselman, R. C., Sterrett, J. L., and Granett, A. L. 1985. A portable fogging apparatus for field or greenhouse use. HortScience 20:1127-1129.
Nagy, G. S. 1976. Studies on the powdery mildews of cucurbits II. Life cycle and epidemiology of *Erysiphe cichoracearum* and *Sphaerotheca fuliginea*. Acta Phytopathol. Acad. Sci. Hung. 11:205-210.
Nederhoff, E. M. 1984. A method to determine ventilation in greenhouses. Acta Hortic. 148:345-350.
Nederpel, L. 1972. Sterilisation of glasshouse soil. Pages 99-100 in: Annu. Rep. Glasshouse Crops Res. Exp. Stn. Naaldwijk 1972.
_____. 1979. Soil sterilization and pasteurization. Pages 29-37 in: Soil Disinfestation. D. L. Mulder, ed. Elsevier, Amsterdam.
Neilands, J. B., and Leong, S. A. 1986. Siderophores in relation to plant growth and disease. Annu. Rev. Plant Physiol. 37:187-208.
Nelson, E. B., and Hoitink, H. A. J. 1982. Factors affecting suppression of *Rhizoctonia solani* in container media. Phytopathology 72:275-279.
Newhall, A. G. 1928. The relation of humidity and ventilation to the leaf mold disease of tomatoes. Pages 119-122 in: Ohio Agric. Exp. Stn. Bimon. Bull. 13.
_____. 1955. Disinfection of soil by heat, flooding and fumigation. Bot. Rev. 21:189-250.
Newhook, F. J. 1951a. Microbiological control of *Botrytis cinerea* Pers. I. The role of pH changes and bacterial antagonism. Ann. Appl. Biol. 38:169-184.
_____. 1951b. Microbiological control of *Botrytis cinerea* Pers. II. Antagonism by fungi and actinomycetes. Ann. Appl. Biol. 38:185-202.
_____. 1957. The relationship of saprophytic antagonism to control of *Botrytis cinerea* Pers. on tomatoes. N.Z. J. Sci. Technol. Sect. A 38:473-481.
Nichols, R. 1966. Ethylene production during the senescence of flowers. J. Hortic. Sci. 41:279-290.
Nijskens, J., Deltour, J., Nisen, A., and Coutisse, S. 1984. Agronomic and radiometric characterization of greenhouse materials. Acta Hortic. 148:663-673.
Nobécourt, P. 1928. Contribution à l'Étude de l'Immunité chez les Végétaux. Bosc Frères, Lyons.
Northover, J., and Matteoni, J. A. 1986. Resistance of *Botrytis cinerea* to benomyl and iprodione in vineyards and greenhouses after exposure to the fungicides alone or mixed with captan. Plant Dis. 70:398-402.
Norwood, J. M., and Crute, I. R. 1983. Infection of lettuce by oospores of *Bremia lactucae*.

Trans. Br. Mycol. Soc. 81:144-147.

Nyland, G., and Milbrath, J. A. 1962. Obtaining virus-free stock by index techniques. Phytopathology 52:1235-1239.

Ohata, K., Serizowa, S., and Shirata, A. 1982. Infection source of the bacterial rot of lettuce caused by *Pseudomonas cichorii.* Bull. Natl. Inst. Agric. Sci. (Tokyo) Ser. C 36:75-80.

Ohr, H. D., Munnecke, D. E., and Bricker, J. L. 1973. The interaction of *Armillaria mellea* and *Trichoderma* spp. as modified by methyl bromide. Phytopathology 63:965-973.

Okada, M., and Sameshima, R. 1986. Predicting environmental conditions for dew formation on the leaves of glasshouse crops. J. Agric. Meteorol. (Tokyo) 42:51-55.

Olsen, C. M., and Baker, K. F. 1968. Selective heat treatment of soil, and its effect on the inhibition of *Rhizoctonia solani* by *Bacillus subtilis.* Phytopathology 58:79-87.

Oostenbrink, M., Kuiper, K., and S'Jacob, J. J. 1957. Tagetes als Fiendpflanzen von *Pratylenchus*-Arten. Nematologica 2(suppl.):424S-433S.

Orlikowski, L. B. 1980. Persistence of *Phytophthora cryptogea* in greenhouse substrate used for gerbera growing. Pr. Inst. Sadow. Kwiaciarstwa Skierniewicach B5:131-140.

Orlikowski, L., Hetman, J., and Tjia, B. 1974. Control of seed-borne *Botrytis cinerea* (Pers. ex Fr.) on *Gerbera jamesonii.* HortScience 9:239-240.

Osborne, L. S., and Chase, A. R. 1984. Influence of acephate and oxamyl on *Alternaria panax* and Alternaria leaf spot of schefflera. Plant Dis. 68:870-872.

Oshima, N., Osawa, T., Morita, H., and Mori, K. 1978. A new attenuated virus L_{11A237}. Ann. Phytopathol. Soc. Jpn. 44:504-508.

Palti, J., and Cohen, Y. 1980. Downy mildew of cucurbits (*Pseudoperonospora cubensis*): The fungus and its hosts, distribution, epidemiology and control. Phytoparasitica 8:109-147.

Paludan, N. 1973. Tobak-mosaik-virus (TMV): Infektionsforsøg, krydsbeskyttelse, smitteds-punkt og udbytte med tomatlinier af TMV hos tomat. Tidsskr. Planteavl 77:494-515.

Panayotakou, M., and Malathrakis, N. E. 1983. Resistance of *Botrytis cinerea* to dicarboximide fungicides in protected crops. Ann. Appl. Biol. 102:293-299.

Papadopoulos, A. P., and Liburdi, N. 1989. The "Harrow Fertigation Manager"—A computerized multifertilizer injector. Acta Hortic. 260:255-265.

Papavizas, G. C. 1963. Microbial antagonism in bean rhizosphere as affected by oat straw and supplemental nitrogen. Phytopathology 53:1430-1435.

———. 1973. Status of applied biological control of soil-borne plant pathogens. Soil Biol. Biochem. 5:709-720.

———. 1985. *Trichoderma* and *Gliocladium:* Biology, ecology, and potential for biocontrol. Annu. Rev. Phytopathol. 23:23-54.

———. 1987. Genetic manipulation to improve the effectiveness of biocontrol fungi for plant disease control. Pages 193-212 in: Innovative Approaches to Plant Disease Control. I. Chet, ed. John Wiley & Sons, New York.

Papavizas, G. C., and Lewis, J. A. 1981. Induction of new biotypes of *Trichoderma harzianum* resistant to benomyl and other fungicides. (Abstr.) Phytopathology 71:247-248.

Papavizas, G. C., and Lumsden, R. D. 1980. Biological control of soilborne fungal propagules. Annu. Rev. Phytopathol. 18:389-413.

Papavizas, G. C., Lewis, J. A., and Abd-El Moity, T. H. 1982. Evaluation of new biotypes of *Trichoderma harzianum* for tolerance to benomyl and enhanced biocontrol capabilities. Phytopathology 72:126-132.

Papendick, R. I., and Campbell, G. S. 1975. Water potential in the rhizosphere and plant and methods of measurement and experimental control. Pages 39-49 in: Biology and Control of Soil-Borne Plant Pathogens. G. W. Bruehl, ed. American Phytopathological Society, St. Paul, MN.

Pappas, A. C. 1980. Effectiveness of metalaxyl and phosetyl-Al against *Pseudoperonospora cubensis* (Berk. & Curt.) Rostow isolates from cucumbers. Pages 146-148 in: Proc. Congr. Mediterr. Phytopathol. Union 5th.

Pares, R. D., and Gunn, L. V. 1989. The role of non-vectored soil transmission as a primary source of infection by pepper mild mottle and cucumber mosaic viruses in glasshouse-grown *Capsicum* in Australia. J. Phytopathol. (Berlin) 126:353-360.

Park, C.-S., Paulitz, T. C., and Baker, R. 1988. Biocontrol of Fusarium wilt of cucumber resulting from interactions between *Pseudomonas putida* and nonpathogenic isolates of

Fusarium oxysporum. Phytopathology 78:190-194.

Park, D. 1965. Survival of microorganisms in soil. Pages 82-98 in: Ecology of Soil-Borne Plant Pathogens: Prelude to Biological Control. K. F. Baker and W. C. Snyder, eds. University of California Press, Berkeley.

Parlitz, M. 1984. Control algorithms for climate control of greenhouses. Acta Hortic. 148:245-249.

Parry, K. E., and Wood, R. K. S. 1958. The adaptation of fungi to fungicides: Adaptation to copper and mercury salts. Ann. Appl. Biol. 46:446-456.

_____. 1959a. The adaptation of fungi to fungicides: Adaptation to thiram, ziram, ferbam, nabam and zineb. Ann. Appl. Biol. 47:10-16.

_____. 1959b. The adaptation of fungi to fungicides: Adaptation to captan. Ann. Appl. Biol. 47:118-135.

Pategas, K. G., Schuerger, A. C., and Wetter, C. 1989. Management of tomato mosaic virus in hydroponically grown pepper (*Capsicum annuum*). Plant Dis. 73:570-573.

Pathak, S., and Chorin, M. 1968. Effect of humidity and temperature conditions on germination of the conidia of *Sphaerotheca pannosa* (Wallr.) Lev. var. *rosae* Woron. on young and old leaves of three rose varieties. Phytopathol. Mediterr. 7:123-128.

Patrick, Z. A. 1986. Allelopathic mechanisms and their exploitation for biological control. Can. J. Plant Pathol. 8:225-228.

Patrick, Z. A., and Toussoun, T. A. 1965. Plant residues and organic amendments in relation to biological control. Pages 440-459 in: Ecology of Soil-Borne Plant Pathogens: Prelude to Biological Control. K. F. Baker and W. C. Snyder, eds. University of California Press, Berkeley.

Paul, W. R. C. 1929. A comparative morphological and physiological study of a number of strains of *Botrytis cinerea* Pers. with special reference to their virulence. Trans. Br. Mycol. Soc. 47:351-355.

Paulitz, T. C., Ishimaru, C. A., and Loper, J. E. 1989. The role of siderophores in the biological control of Pythium damping-off of cucumber by *Pseudomonas putida*. (Abstr.) Phytopathology 79:911.

Paulitz, T. C., Ahmad, J. S., and Baker, R. 1990. Integration of *Pythium nunn* and *Trichoderma harzianum* isolate T-95 for the biological control of Pythium damping-off of cucumber. Plant Soil 121:243-250.

Pennypacker, B. W. 1989. The role of mineral nutrition in the control of Verticillium wilt. Pages 33-45 in: Soilborne Plant Pathogens: Management of Diseases with Macro- and Microelements. A. W. Engelhard, ed. American Phytopathological Society, St. Paul, MN.

Pepin, H. S., and MacPherson, E. A. 1982. Strains of *Botrytis cinerea* resistant to benomyl and captan in the field. Plant Dis. 66:404-405.

Perera, R. G., and Wheeler, B. E. J. 1975. Effect of water droplets on the development of *Sphaerotheca pannosa* on rose leaves. Trans. Br. Mycol. Soc. 64:313-319.

Peries, O. S. 1962. Studies on strawberry mildew caused by *Sphaerotheca macularis* (Wallr. ex Fries) Jaczewski. I. Biology of the fungus. Ann. Appl. Biol. 50:211-214.

Pérombelon, M. C. M., and Kelman, A. 1980. Ecology of the soft rot erwinias. Annu. Rev. Phytopathol. 18:361-387.

Peterson, M. J., Sutherland, J. R., and Tuller, S. E. 1988. Greenhouse environment and epidemiology of grey mould of container-grown Douglas fir seedlings. Can. J. For. Res. 18:974-980.

Pfister, S. E., and Peterson, J. L. 1990. Biological control of Fusarium wilt in China aster with chitin and chitinoclastic bacteria. (Abstr.) Phytopathology 80:122.

Philipp, W.-D., and Crüger, G. 1979. Parasitismus von *Ampelomyces quisqualis* auf Echten Mehltaupilzen an Gurken und anderen Gemüsearten. Z. Pflanzenkrankh. Pflanzenschutz 86:129-142.

Philipp, W.-D., Beuther, E., and Grossman, F. 1982. Untersuchungen über den Einfluss von Fungiziden auf *Ampelomyces quisqualis* im Hinblick auf eine integrierte Bekämpfung von Gurkenmehltau unter Glas. Z. Pflanzenkrankh. Pflanzenschutz 89:575-581.

Philipp, W.-D., Grauer, U., and Grossman, F. 1984. Ergänzende Untersuchungen zur biologischen und integrierten Bekämpfung von Gurkenmehltau unter Glas durch *Ampelomyces quisqualis*. Z. Pflanzenkrankh. Pflanzenschutz 91:438-443.

Phillips, D. H. 1959. The destruction of *Didymella lycopersici* Kleb. in tomato haulm composts.

Ann. Appl. Biol. 47:240-253.

Picken, A. J. F. 1984. A review of pollination and fruit set in the tomato (*Lycopersicon esculentum* Mill.). J. Hortic. Sci. 59:1-13.

Pierik, R. L. M. 1987. In Vitro Culture of Higher Plants. Martinus Nijhoff, Dordrecht.

———. 1988. Handicaps for the large scale commercial application of micropropagation. Acta Hortic. 230:63-71.

Pinckard, J. A. 1942. The mechanism of spore dispersal in *Peronospora tabacina* and certain other downy mildew fungi. Phytopathology 32:505-511.

Plaut, J. L., and Berger, R. D. 1981. Infection rates in three pathosystem epidemics initiated with reduced disease severities. Phytopathology 71:917-921.

Ploetz, R. C., and Engelhard, A. W. 1979. The Botrytis blight of *Exacum affine*. Proc. Fla. State Hortic. Soc. 92:353-355.

Pommer, E.-H., and Lorenz, G. 1982. Resistance of *Botrytis cinerea* Pers. to dicarboximide fungicides—A literature review. Crop Prot. 1:221-230.

Porter, M. A., and Grodzinski, B. 1985. CO_2 enrichment of protected crops. Hortic. Rev. 7:345-398.

Powell, C. C., Jr. 1988. The safety and efficacy of fungicides for use in Rhizoctonia crown rot control of directly potted unrooted poinsettia cuttings. Plant Dis. 72:693-695.

———. 1990. Studies on the chemical and environmental control of powdery mildew on greenhouse roses. Roses Inc. Bull. (Sept.), pp. 51-66.

Powelson, R. L. 1960. Initiation of strawberry fruit rot caused by *Botrytis cinerea*. Phytopathology 50:491-494.

Preuss, K. P. 1983. Day-degree methods for pest management. Environ. Entomol. 12:613-619.

Price, D. 1970. Tulip fire caused by *Botrytis tulipae* (Lib.) Lind; the leaf-spotting phase. J. Hortic. Sci. 45:233-238.

———. 1980. Fungal flora of tomato roots in nutrient film culture. Acta Hortic. 98:269-275.

Price, D., and Bateson, M. 1976. Investigations on root microflora. Pages 117-118 in: Annu. Rep. Glasshouse Crops Res. Inst. 1976.

Price, D., and Briggs, J. B. 1974. The control of *Botrytis tulipae* (Lib.) Lind, the cause of tulip fire, by dipping. Exp. Hortic. 26:36-39.

Price, D., and Dickinson, A. 1980. Fungicides and the nutrient film technique. Acta Hortic. 98:277-282.

Price, T. V. 1970. Epidemiology and control of powdery mildew (*Sphaerotheca pannosa*) on rose. Ann. Appl. Biol. 65:231-248.

Price, T. V., and Fox, P. 1986. Studies on the behaviour of furalaxyl on pythiaceous fungi and cucumbers in recirculating hydroponic systems. Aust. J. Agric. Res. 37:65-77.

Price, T. V., and Maxwell, M. K. 1980. Studies of disease problems and their control in hydroponics in Australia. Acta Hortic. 98:307-316.

Price, T. V., and Nolan, P. D. 1984. Incidence and distribution of *Pythium, Phytophthora* and *Fusarium* spp. in recirculating nutrient film hydroponic systems. Pages 523-532 in: Proc. Int. Congr. Soilless Cult. 6th.

Proeseler, G. 1980. Piesmids. Pages 97-113 in: Vectors of Plant Pathogens. K. F. Harris and K. Maramorosch, eds. Academic Press, New York.

Pullman, G. S., DeVay, J. E., and Garber, R. H. 1981. Soil solarization and thermal death: A logarithmic relationship between time and temperature for four soilborne plant pathogens. Phytopathology 71:959-964.

Punithalingam, E. 1967. *Septoria chrysanthemella*. Descriptions of Pathogenic Fungi and Bacteria, No. 137. Commonwealth Mycological Institute, Kew, England.

———. 1968a. *Puccinia chrysanthemi*. Descriptions of Pathogenic Fungi and Bacteria, No. 175. Commonwealth Mycological Institute, Kew, England.

———. 1968b. *Puccinia horiana*. Descriptions of Pathogenic Fungi and Bacteria, No. 176. Commonwealth Mycological Institute, Kew, England.

———. 1980. *Didymella chrysanthemi*. Descriptions of Pathogenic Fungi and Bacteria, No. 662. Commonwealth Mycological Institute, Kew, England.

Purdy, L. H. 1967. Application and use of soil and seed-treatment fungicides. Pages 155-239 in: Fungicides: An Advanced Treatise. Vol. 1. D. C. Torgeson, ed. Academic Press, New York.

Pussard-Radulesco, E. 1930. Recherches biologiques et cytologiques sur quelques thysanoptères. Ann. Epiphyt. 3-4:179-180.

Putnam, A. R., and Duke, W. B. 1978. Allelopathy in agroecosystems. Annu. Rev. Phytopathol. 16:431-451.

Quiot, J. B., Devergne, J. C., Marchoux, G., Cardin, L., and Douine, L. 1979. Écologie et épidémiologie du virus de la mosaïque du concombre dans le sud-est de la France. VI. Conservation de deux types de populations virales dans les plantes sauvages. Ann. Phytopathol. 11:349-357.

Ragazzi, A. 1980. Sporulazione in serra di *Sphaerotheca pannosa* (Wallr.) Lev. var. *rosae* Woron. Riv. Patol. Veg. 17:23-33.

Ramirez-Villapudua, J., and Munnecke, D. E. 1987. Control of cabbage yellows (*Fusarium oxysporum* f. sp. *conglutinans*) by solar heating of field soils amended with dry cabbage residues. Plant Dis. 71:217-221.

Ramsey, G. B. 1941. *Botrytis* and *Sclerotinia* as potato tuber pathogens. Phytopathology 31:439-448.

Ramsey, J. A., Butler, C. G., and Sang, J. H. 1938. The humidity gradient at the surface of a transpiring leaf. J. Exp. Biol. 15:255-265.

Rand, F. V., and Enlows, E. M. A. 1916. Transmission and control of bacterial wilt of cucurbits. J. Agric. Res. (Washington, D.C.) 6:417-434.

Rast, A. T. B. 1972. MII-16, an artificial symptomless mutant of tobacco mosaic virus for seedling inoculation of tomato crops. Neth. J. Plant Pathol. 78:110-112.

_____. 1975. Variability of tobacco mosaic virus in relation to control of tomato mosaic in glasshouse tomato crops by resistance breeding and cross protection. Versl. Landbouwkd. Onderz. 834.

_____. 1979. Infection of tomato seed by different strains of tobacco mosaic virus with particular reference to the symptomless mutant, MII-16. Neth. J. Plant Pathol. 85:223-233.

Rattink, H. 1982. Disinfection of potting soil by means of gamma-radiation. Meded. Fac. Landbouwwet. Rijksuniv. Gent 47:869-873.

_____. 1988. Possibilities of cross-protection against Fusarium wilt by non-pathogenic isolates of *Fusarium oxysporum* f. sp. *dianthi*. Acta Hortic. 216:131-140.

_____. 1990. Epidemiology of *Fusarium oxysporum* f. sp. *cyclaminis* in an ebb and flow system. Neth. J. Plant Pathol. 96:171-177.

Reavill, M. J. 1954. The effect of certain chloronitrobenzenes on germination, growth and sporulation of some fungi. Ann. Appl. Biol. 41:448-460.

Regev, U. 1984. An economic analysis of man's addiction to pesticides. Pages 441-453 in: Pest and Pathogen Control: Strategic, Tactical, and Policy Models. G. R. Conway, ed. John Wiley & Sons, Chichester.

Renz, M., and Kurz, C. 1984. A colorimetric method for DNA hybridization. Nucleic Acids Res. 12:3435-3444.

Reuveni, R., and Rotem, J. 1974. Effect of humidity on epidemiological patterns of the powdery mildew (*Sphaerotheca fuliginea*) on squash. Phytoparasitica 2:25-33.

Reuveni, R., Raviv, M., and Bar, R. 1989. Sporulation of *Botrytis cinerea* as affected by photoselective sheets and filters. Ann. Appl. Biol. 115:417-424.

Rewal, N., and Grewal, J. S. 1989. Inheritance of resistance to *Botrytis cinerea* Pers. in *Cicer arietinum* L. Euphytica 44:61-63.

Ricard, J. C. 1981. Commercialization of a *Trichoderma* based mycofungicide: Some problems and solutions. Biocontrol News Inf. 2:95-98.

Rice, E. L. 1974. Allelopathy. Academic Press, New York.

Richardson, F. 1982. Interplanting cuts costs for tomatoes. Grower (London) 98(Dec.):15-17.

Richardson, M. J. 1979. An Annotated List of Seed-Borne Diseases. 3rd ed. Phytopathol. Pap. 23. Commonwealth Mycological Institute, Kew, England.

_____. 1981. Supplement 1 to An Annotated List of Seed-Borne Diseases. International Seed Testing Association, Zurich.

_____. 1983. Seed health testing. Pages 365-373 in: Plant Pathologist's Pocketbook. 2nd ed. A. Johnston and C. Booth, eds. Commonwealth Mycological Institute, Kew, England.

Roberts, P. A., Magyarosy, A. C., Matthews, W. C., and May, D. M. 1988. Effects of metam-sodium applied by drip irrigation on root-knot nematodes, *Pythium ultimum*, and *Fusarium* sp. in soil and on carrot and tomato roots. Plant Dis. 72:213-217.

Robertson, G. I. 1973. Occurrence of *Pythium* spp. in New Zealand soils, sands, pumices, and peat, and on roots of container-grown plants. N.Z. J. Agric. Res. 16:357-365.

Rodríguez-Kábana, R., Backman, P. A., and Curl, E. A. 1977. Control of seed and soilborne plant diseases. Pages 117-161 in: Antifungal Compounds. Vol. 1. M. R. Siegel and H. D. Sisler, eds. Marcel Dekker, New York.

Roebroeck, E. J. A., Jansen, M. J. W., and Mes, J. J. 1991. A mathematical model describing the combined effect of exposure time and temperature of hot water treatments on survival of gladiolus cormels. Ann. Appl. Biol. 119:89-96.

Roest, S., and Gilisson, L. J. W. 1989. Plant regeneration from protoplasts: A literature review. Acta Bot. Neerl. 38:1-33.

Rogers, M. N. 1957. Some effects of moisture and of leaf-temperature depression on the development of powdery mildew of roses. (Abstr.) Phytopathology 47:29.

———. 1959. Some effects of moisture and host plant susceptibility on the development of powdery mildew on roses caused by *Sphaerotheca pannosa* var. *rosae*. Cornell Univ. Agric. Exp. Stn. Mem. 363.

Roistacher, C. N., Baker, K. F., and Bald, J. G. 1957. Hot-water treatment of gladiolus cormels for the eradication of *Fusarium oxysporum* f. sp. *gladioli*. Hilgardia 26:659-684.

Roitman, J. N., Mahoney, N. E., Janisiewicz, W. J., and Benson, M. 1990. A new chlorinated phenylpyrrole antibiotic produced by the antifungal bacterium *Pseudomonas cepacia*. J. Agric. Food Chem. 38:529-537.

Rotem, J., and Cohen, Y. 1966. The relationship between mode of irrigation and severity of tomato foliage diseases in Israel. Plant Dis. Rep. 50:635-639.

Rotem, J., and Palti, J. 1969. Irrigation and plant diseases. Annu. Rev. Phytopathol. 7:267-288.

Rotem, J., Cohen, Y., and Bashi, E. 1978. Host and environmental influences on sporulation in vivo. Annu. Rev. Phytopathol. 16:83-101.

Rouse, D. I. 1988. Use of crop growth-models to predict the effects of disease. Annu. Rev. Phytopathol. 26:183-201.

Rouse, D. I., Nordheim, E. V., Hirano, S. S., and Upper, C. D. 1985. A model relating the probability of foliar disease incidence to the population frequencies of bacterial plant pathogens. Phytopathology 75:505-509.

Rovira, A. D. 1982. Organisms and mechanisms involved in some soils suppressive to soilborne plant diseases. Pages 23-33 in: Suppressive Soils and Plant Disease. R. W. Schneider, ed. American Phytopathological Society, St Paul, MN.

Rowe, R. C., and Coplin, D. L. 1976. Dispersal of *Fusarium oxysporum* in tomato greenhouses. (Abstr.) Proc. Am. Phytopathol. Soc. 3:221.

Rowe, R. C., and Farley, J. D. 1978. Control of Fusarium crown and root rot of greenhouse tomatoes by inhibiting recolonization of steam-disinfested soil with a captafol drench. Phytopathology 68:1221-1224.

Rowe, R. C., Farley, J. D., and Coplin, D. L. 1977. Airborne spore dispersal and recolonization of steamed soil by *Fusarium oxysporum* in tomato greenhouses. Phytopathology 67:1513-1517.

Runia, W. T. 1983. A recent development in steam sterilization. Acta Hortic. 152:195-200.

———. 1986. Disinfestation of substrates used in protected cultivation. Soilless Cult. 2:35-44.

———. 1987. Pesticides in rockwool crops, grown in open drainage system. Page 84 in: Annu. Rep. Glasshouse Crops Res. Exp. Stn. Naaldwijk 1987.

———. 1988. Elimination of plant pathogens in drainwater from soilless cultures. Pages 429-443 in: Proc. Int. Soc. Soilless Cult. 1988.

Runia, W. T., van Os, E. A., and Bollen, J. G. 1988. Disinfection of drainwater from soilless cultures by heat treatment. Neth. J. Agric. Sci. 36:231-238.

Rytter, J. L., Lukezic, F. L., Craig, R., and Moorman, G. W. 1989. Biological control of geranium rust by *Bacillus subtilis*. Phytopathology 79:367-370.

Saettler, A. W., Schaad, N. W., and Roth, D. A., eds. 1989. Detection of Bacteria in Seed and Other Planting Material. American Phytopathological Society, St Paul, MN.

Saffel, R. A. 1985. The control of temperature, humidity and carbon dioxide in experimental glasshouses. Acta Hortic. 174:443-447.

Salinas, J., Glandorf, D. C. M., Picavet, F. D., and Verhoeff, K. 1989. Effects of temperature, relative humidity and age of conidia on the incidence of spotting on gerbera flowers caused

by *Botrytis cinerea.* Neth. J. Plant Pathol. 95:51-64.
Salt, S. D., Pan, S. Q., and Kuć, J. 1988. Carbohydrate changes in tobacco systemically protected against blue mold by stem infection with *Peronospora tabacina.* Phytopathology 78:733-738.
Samish, Z., and Dimant, D. 1959. Bacterial population in fresh, healthy cucumbers. Food Manuf. 34:17-20.
Samish, Z., and Etinger-Tulczynska, R. 1963. Distribution of bacteria within the tissue of healthy tomatoes. Appl. Microbiol. 11:7-10.
Sasaki, T., Honda, Y., Umekawa, M., and Nemoto, M. 1985. Control of certain diseases of greenhouse vegetables with ultraviolet-absorbing vinyl film. Plant Dis. 69:530-533.
Sase, S., Takakura, T., and Nara, M. 1984. Wind tunnel testing on airflow and temperature distribution of a naturally ventilated greenhouse. Acta Hortic. 148:329-336.
Savage, S. D., and Sall, M. A. 1981. Radioimmunosorbent assay for *Botrytis cinerea.* Phytopathology 71:411-415.
Sayre, R. M., Patrick, Z. A., and Thorpe, H. J. 1965. Detection and identification of a nematicidal component in decomposing plant residues. Nematologica 11:263-268.
Schein, R. D. 1964. Comments on the moisture requirements of fungus germination. Phytopathology 54:1427.
Schenk, P. K. 1961. Biologie en bestrijding van *Urocystis gladiolicola* Ainsw. op gladiolen. Tijdschr. Plantenziekten 67:313-416.
Schepers, H. T. A. M. 1983. Decreased sensitivity of *Sphaerotheca fuliginea* to fungicides which inhibit ergosterol biosynthesis. Neth. J. Plant Pathol. 89:185-187.
_____. 1984. Resistance to inhibitors of sterol biosynthesis in cucumber powdery mildew. Pages 495-496 in: Proc. Br. Crop Prot. Conf. Pests Dis. 1984.
Scher, F. M., and Baker, R. 1980. Mechanism of biological control in a *Fusarium*-suppressive soil. Phytopathology 70:412-417.
_____. 1982. Effect of *Pseudomonas putida* and a synthetic iron chelator on induction of soil suppressiveness to Fusarium wilt pathogens. Phytopathology 72:1567-1573.
Scher, F. M., and Castagno, J. R. 1986. Biocontrol: A view from industry. Can. J. Plant Pathol. 8:222-224.
Schippers, B., and Bouman, A. 1973. Inhibition of germination and mycelial growth of *Fusarium solani* f. *cucurbitae* and *Aspergillus flavus* by volatiles from soil. Acta Bot. Neerl. 22:166.
Schippers, B. A., and de Weyer, W. M. M. M. 1972. Chlamydospore formation and lysis of macroconidia of *Fusarium solani* f. sp. *cucurbitae* in chitin-amended soil. Neth. J. Plant Pathol. 78:45-54.
Schippers, B., Schroth, M. N., and Hildebrand, D. C. 1967. Emanation of water from underground plant parts. Plant Soil 27:81-91.
Schippers, B., Lugtenberg, B., and Weisbeek, P. J. 1987. Plant growth control by fluorescent pseudomonads. Pages 19-39 in: Innovative Approaches to Plant Disease Control. I. Chet, ed. John Wiley & Sons, New York.
Schnathorst, W. C. 1959a. Spread and life cycle of the lettuce powdery mildew fungus. Phytopathology 49:464-468.
_____. 1959b. Resistance in lettuce to powdery mildew related to osmotic value. Phytopathology 49:562-571.
_____. 1962. Comparative ecology of downy and powdery mildews of lettuce. Phytopathology 52:41-46.
Schnathorst, W. C., and Weinhold, A. R. 1957. An osmotic mechanism for resistance to powdery mildew in lettuce and peach. (Abstr.) Phytopathology 47:533.
Schneider, R. W., ed. 1982. Suppressive Soils and Plant Disease. American Phytopathological Society, St Paul, MN.
_____. 1985. Suppression of Fusarium yellows of celery with potassium, chloride, and nitrate. Phytopathology 75:40-48.
Schneider, R. W., Grogan, R. G., and Kimble, K. A. 1978. Colletotrichum root rot of greenhouse tomatoes in California. Plant Dis. Rep. 62:969-971.
Schneiderhan, F. J. 1927. The black walnut (*Juglans nigra* L.) as a cause of the death of apple trees. Phytopathology 17:529-540.
Schoeneweiss, D. F. 1975. Predisposition, stress, and plant disease. Annu. Rev. Phytopathol. 13:193-211.

Schroeder, C. A. 1965. Temperature relationships of fruit tissues under extreme conditions. Proc. Am. Soc. Hortic. Sci. 87:199-203.

Schroeder, W. T., and Provvidenti, R. 1969. Resistance to benomyl in powdery mildew of cucurbits. Plant Dis. Rep. 53:271-275.

Schroth, M. N., and Hancock, J. G. 1982. Disease-suppressive soil and root-colonizing bacteria. Science (Washington, D.C.) 216:1376-1381.

Schuler, C., Biala, J., Bruns, C., Gottschall, R., Ahlers, S., and Vogtmann, H. 1989. Suppression of root rot on peas, beans, and beetroots caused by *Pythium ultimum* and *Rhizoctonia solani* through the amendment of growing media with composted organic household waste. J. Phytopathol. (Berlin) 127:227-238.

Schwarz, M. 1989. Oxygenating of nutrient solution in normal and stress conditions. Soilless Cult. 5:45-53.

Sciaroni, R. H., McCain, A. H., Lear, B., and Branson, R. L. 1972. Methyl bromide fumigation for carnations. Flower Nursery Rep. (Apr.):2-3.

Scott, J. W., and Jones, J. P. 1989. Monogenic resistance in tomato to *Fusarium oxysporum* f. sp. *lycopersici* race 3. Euphytica 40:49-53.

Seemann, J. 1962. Ein Betrag zur Problem der Kühlung von Gewächshäusern wahrend des Sommers mit einfachen Verfahrung. Gartenbauwissenschaft 27:33-44.

Seginer, I. 1980. Optimizing greenhouse operation for best aerial environment. Acta Hortic. 106:169-178.

Seginer, I., and Raviv, M. 1984. Selecting greenhouse night temperatures for tomato seedlings. Acta Hortic. 148:603-610.

Sharvelle, E. E. 1969. Chemical Control of Plant Diseases. Purdue University Press, Lafayette, IN.

Shawish, O., and Baker, R. 1982. Thigmomorphogenesis and predisposition of hosts to Fusarium wilt. Phytopathology 72:63-68.

Shiraishi, M., Fukutomi, M., and Akai, S. 1970. Effect of temperature on the conidium germination and appressorium formation by *Botrytis cinerea.* Ann. Phytopathol. Soc. Jpn. 36:297-303.

Short, T. H., Badger, P. C., and Roller, W. L. 1976. OARDC's solar-heated greenhouse. Agric. Eng. 57:30-32.

Shull, C. A. 1936. Rate of adjustment of leaf temperature to incident energy. Plant Physiol. 11:181-188.

Sidhu, G. S., and Webster, J. M. 1974. Genetics of resistance in the tomato to root-knot nematode: Wilt-fungus complex. J. Hered. 65:153-156.

______. 1979. Genetics of tomato resistance to the *Fusarium: Verticillium* complex. Physiol. Plant Pathol. 15:93-98.

Siebering, H., and Leistra, M. 1979. Computer simulation of fumigant behaviour in soil. Pages 135-161 in: Soil Disinfestation. D. Mulder, ed. Elsevier, Amsterdam.

Simay, E. I. 1988. In vitro occurrence of hyperparasitism of *Botrytis cinerea* Pers. by *Gliocladium catenulatum* Gilman & Abbott. Acta Phytopathol. Entomol. Hung. 23:133-135.

Simeoni, L. A., Lindsay, W. L., and Baker, R. 1987. Critical iron level associated with biological control of Fusarium wilt. Phytopathology 77:1057-1061.

Simons, J. N., and Cox, D. M. 1958. Transmission of pseudo-curly top virus in Florida by a treehopper. Virology 6:43-48.

Sinclair, W. A., Cowles, D. P., and Hee, S. P. 1975. Fusarium foot rot of Douglas-fir seedlings: Suppression by soil fumigation, fertility management and inoculation with spores of *Laccaria laccata.* For. Sci. 21:390-399.

Sironval, C. 1951. Une exemple de lutte physiologique contre l'infection. Lejeunia 15:51-54.

Sitterly, W. R. 1972. Breeding for disease resistance in cucurbits. Annu. Rev. Phytopathol. 10:471-490.

Sivan, A., and Chet, I. 1989a. Degradation of fungal cell walls by lytic enzymes of *Trichoderma harzianum.* J. Gen. Microbiol. 135:675-682.

______. 1989b. The possible role of competition between *Trichoderma harzianum* and *Fusarium oxysporum* on rhizosphere colonization. Phytopathology 79:198-203.

Skadow, K., Schaffrath, J., Gohler, F., Drews, M., and Lankow, H.-J. 1984. *Phytophthora nicotianae* und *Pythium aphanidermatum* als Schäderreger in NFT-Kultur. Nachrichtenbl. Dtsch. Pflanzenschutzdienstes (Braunschweig) 38:169-172.

Skylakakis, G. 1981. Effects of alternating and mixing pesticides on the buildup of fungal resistance. Phytopathology 71:1119-1121.

_____. 1982. The development and use of models describing outbreaks of resistance to fungicides. Crop Prot. 1:249-262.

_____. 1983. Theory and strategy of chemical control. Annu. Rev. Phytopathol. 21:117-135.

_____. 1984. Quantitative evaluation of strategies to delay fungicide resistance. Pages 565-572 in: Proc. Br. Crop Prot. Conf. Pests Dis. 1984.

Slusarski, C. 1983. The effects of methyl bromide fumigation and farm manure application of *Corynebacterium michiganense* on greenhouse tomatoes. Acta Hortic. 152:57-63.

Smart, R. E., and Sinclair, R. E. 1976. Solar heating of grape berries and other spherical fruits. Agric. Meteorol. 17:241-259.

Smiley, R. W. 1975. Forms of nitrogen and the pH in the root zone and their importance to root infections. Pages 55-62 in: Biology and Control of Soil-Borne Plant Pathogens. G. W. Bruehl, ed. American Phytopathological Society, St. Paul, MN.

Smith, A. M., and Cook, R. J. 1974. Implications of ethylene production by bacteria for biological balance of soil. Nature (London) 252:703-705.

Smith, I., ed. 1988. Growing greenhouse vegetables. Ont. Minist. Agric. Food Publ. 526.

Smith, I. M. 1988. *Puccinia pelargonii-zonalis* Doidge. Pages 491-492 in: European Handbook of Plant Diseases. I. M. Smith, J. Dunez, R. A. Lelliott, D. H. Phillips, and S. A. Archer, eds. Blackwell, Oxford.

Smith, J. H. 1923. The killing of *Botrytis cinerea* by heat, with a note on the determination of temperature coefficients. Ann. Appl. Biol. 10:335-347.

Smith, P. M. 1977. Control of *Phytophthora cinnamomi* in water. Page 121 in: Annu. Rep. Glasshouse Crops Res. Inst. 1977.

Smith, P. M., and Ousley, M. A. 1984. Epidemiology and control of Phytophthora root rot diseases of woody ornamentals. Pages 102-105 in: Annu. Rep. Glasshouse Crops Res. Inst. 1984.

Smith, W. H., Meigh, D. F., and Parker, J. C. 1964. Effect of damage and fungal infection on the production of ethylene by carnations. Nature (London) 204:92-93.

Smith, W. P. C., and Goss, O. M. 1946. Bacterial canker of tomatoes. J. Dep. Agric. West. Aust. 23:147-156.

Sneh, B. 1981. Use of chitinolytic bacteria for biological control of *Fusarium oxysporum* f. sp. *dianthi* in carnation. Phytopathol. Z. 100:251-256.

Snyder, R. S., and Shaw, R. H. 1984. Converting humidity expressions with computers and calculators. Univ. Calif. Coop. Ext. Leafl. 21372.

Soffer, H., and Berger, D. W. 1988. Studies on plant propagation using the aero-hydroponic method. Acta Hortic. 230:261-269.

Sommers, L. E., Harris, R. F., Dalton, F. N., and Gardner, W. R. 1970. Water potential relations of three root-infecting *Phytophthora* species. Phytopathology 60:932-934.

Sonneveld, C. 1969. De invloed van stomen op de stikstofhuishouding van de grond. Tuinbouw Meded. 32:197-203.

_____. 1979. Changes in chemical properties of soil caused by steam sterilization. Pages 39-50 in: Soil Disinfestation. D. Mulder, ed. Elsevier, Amsterdam.

Sonneveld, C., and Voogt, S. 1973. The effects of soil sterilization with air-steam mixtures on the development of some glasshouse crops. Plant Soil 38:415-423.

_____. 1975. Studies on the manganese uptake of lettuce on steam-sterilized glasshouse soils. Plant Soil 42:49-64.

Southey, J. F. 1978. Physical methods of control. Pages 302-312 in: Plant Nematology. 3rd ed. J. F. Southey, ed. Minist. Agric. Fish. Food (G.B.) Ref. Book 407. Her Majesty's Stationery Office, London.

Sparnaaij, L. D., Garretson, F., and Bekker, W. 1975. Additive inheritance of resistance to *Phytophthora cryptogea* Pethybridge & Lafferty in *Gerbera jamesonii* Bolus. Euphytica 24:551-556.

Speer, E. O. 1978. Beitrag zur Morphologie von *Ampelomyces quisqualis* Ces. Sydowia Ann. Mycol. Ser. II 31:242-246.

Spek, J. C. 1972. Het berekenen van kasenconstructies. Inst. Tuinbouwtech. Publ. 76.

Spencer, D. M. 1976. Pelargonium rust and its control by fungicides. Plant Pathol. 25:156-161.

_____, ed. 1978. The Powdery Mildews. Academic Press, London.

_____. 1980. Parasitism of carnation rust (*Uromyces dianthi*) by *Verticillium lecanii*. Trans. Br. Mycol. Soc. 74:191-194.

Spencer, D. M., and Atkey, P. T. 1981. Parasitic effects of *Verticillium lecanii* on two rust fungi. Trans. Br. Mycol. Soc. 77:535-542.

Spencer, D. M., and Ebben, M. H. 1981. Biological control of powdery mildew. Pages 128-129 in: Annu. Rep. Glasshouse Crops Res. Inst. 1981.

Stair, E. C., Brown, H. D., and Hienton, T. E. 1928. Forced ventilation as a means of controlling tomato *Cladosporium* and *Septoria* in hotbeds. Phytopathology 18:1027-1029.

Stall, R. E. 1963. Effects of lime on incidence of Botrytis gray mold of tomato. Phytopathology 53:149-151.

Stall, R. E., and Hall, C. B. 1969. Association of bacteria with graywall of tomato. Phytopathology 59:1650-1653.

Stall, R. E., Hortenstine, C. C., and Iley, J. R. 1965. Incidence of Botrytis gray mold of tomato in relation to a calcium-phosphorus balance. Phytopathology 55:447-449.

Stanghellini, M. E., and Burr, T. J. 1973. Effect of soil water potential on disease incidence and oospore germination of *Pythium aphanidermatum*. Phytopathology 63:1496-1498.

Stanghellini, M. E., and Kronland, W. C. 1986. Yield loss in hydroponically grown lettuce attributed to subclinical infection of feeder rootlets by *Pythium dissotocum*. Plant Dis. 70:1053-1056.

Stanghellini, M. E., Stowell, L. J., and Bates, M. L. 1984. Control of root rot of spinach caused by *Pythium aphanidermatum* in a recirculating hydroponic system by ultraviolet irradiation. Plant Dis. 68:1075-1076.

Stanghellini, M. E., White, J. G., Tomlinson, J. A., and Clay, C. 1988. Root rot of hydroponically grown cucumbers caused by zoospore-producing isolates of *Pythium intermedium*. Plant Dis. 72:358-359.

Stanghellini, M. E., Adaskaveg, J. E., and Rasmussen, S. L. 1990. Pathogenesis of *Plasmopara lactucae-radicis*, a systemic root pathogen of cultivated lettuce. Plant Dis. 74:173-178.

Staunton, W. P., and Cormican, T. P. 1980. The effects of pathogens and fungicides on tomatoes in a hydroponic system. Acta Hortic. 98:293-297.

Stepanov, K. M. 1935. Dissemination of infectious diseases of plants by air currents. (In Russian.) Bull. Plant Prot. (Leningrad) Ser. 2 8:1-68.

Stephens, C. T., and Stebbins, T. C. 1985. Control of damping-off pathogens in soilless container media. Plant Dis. 69:494-496.

Stern, V. M., Smith, R. F., van den Bosch, R., and Hagen, K. S. 1959. The integrated control concept. Hilgardia 29:81-101.

Stickler, M. P. 1975. The use of plastics for heat insulation in greenhouses. Plasticulture 25:41-53.

Still, S. M., Dirr, M. A., and Gartner, J. B. 1976. Phytotoxic effects of several bark extracts on mung bean and cucumber growth. J. Am. Soc. Hortic. Sci. 101:34-37.

Stoller, B. B. 1954. Principles and practices of mushroom culture. Econ. Bot. 8:48-95.

Stolzy, L. H., and Van Gundy, S. D. 1968. The soil as an environment for microflora and microfauna. Phytopathology 58:889-899.

Stolzy, L. H., Zentmyer, G. A., and Roulier, M. H. 1975. Dynamics and measurement of oxygen diffusion and concentration in the root zone and other microsites. Pages 50-54 in: Biology and Control of Soil-Borne Plant Pathogens. G. W. Bruehl, ed. American Phytopathological Society, St. Paul, MN.

Stout, G. L. 1962. Maintenance of "pathogen-free" planting stock. Phytopathology 52:1255-1258.

Strider, D. L. 1969a. Bacterial canker of tomato caused by *Corynebacterium michiganense*. N.C. Agric. Exp. Stn. Tech. Bull. 193.

_____. 1969b. Foliage blight phase of bacterial canker of tomato and survival of *Corynebacterium michiganense* in toxicants and in an air-dried condition. Plant Dis. Rep. 53:864-868.

_____. 1973. Damping-off of statice caused by *Botrytis cinerea* and its control. Plant Dis. Rep. 57:969-971.

_____. 1978. Alternaria blight of carnation in the greenhouse and its control. Plant Dis. Rep. 62:24-28.

Strong, M. C. 1946. The effects of soil moisture and temperature on Fusarium wilt of tomato.

Phytopathology 36:218-225.
Styer, D. J., and Chinn, C. K. 1983. Meristem and shoot-tip culture for propagation, pathogen elimination, and germplasm preservation. Hortic. Rev. 5:221-277.
Summers, R. W., Heaney, S. P., and Grindle, M. 1984. Studies of a dicarboximide resistant heterokaryon of *Botrytis cinerea.* Pages 453-458 in: Proc. Br. Crop Prot. Conf. Pests Dis. 1984.
Sun, S.-K., and Huang, J.-W. 1985. Formulated soil amendment for controlling Fusarium wilt and other soilborne diseases. Plant Dis. 69:917-920.
Sundheim, L. 1977. Attempts at biological control of *Phomopsis sclerotioides* in cucumber. Neth. J. Plant Pathol. 83(suppl. 1):439-442.
______. 1982. Control of cucumber powdery mildew by the hyperparasite *Ampelomyces quisqualis* and fungicides. Plant Pathol. 31:209-214.
Sundheim, L., and Amundsen, T. 1982. Fungicide tolerance in the hyperparasite *Ampelomyces quisqualis* and integrated control of cucumber powdery mildew. Acta Agric. Scand. 32:349-355.
Sussman, A. S. 1965. Dormancy of soil microorganisms in relation to survival. Pages 99-110 in: Ecology of Soil-Borne Plant Pathogens: Prelude to Biological Control. K. F. Baker and W. C. Snyder, eds. University of California Press, Berkeley.
Sutter, E. G. 1985. Morphological, physical, and chemical characteristics of epicuticular wax on ornamental plants regenerated in vitro. Ann. Bot. (London) 55:321-329.
Sutton, O. G. 1953. Micrometeorology. McGraw-Hill, New York.
Svedelius, G. 1989. Försök avseende *Trichoderma*-berikad kompost, Biobalans, mot gråmögel, *Botrytis cinerea,* på jordgubbar. Vaextskyddsnotiser 53:30-37.
Svedelius, G., and Unestam, T. 1978. Experimental factors favouring infection of attached cucumber leaves by *Didymella bryoniae.* Trans. Br. Mycol. Soc. 71:89-97.
Szaniszlo, P. J., Powell, P. E., Reid, C. P. P., and Cline, G. R. 1981. Production of hydroxamate siderophore iron chelators by ectomycorrhizal fungi. Mycologia 73:1158-1174.
Tachibana, K., and Minagawa, H. 1984. Thermal analysis of greenhouses with heat transfer matrix and energy efficiency. Acta Hortic. 148:369-373.
Tahvonen, R. 1982a. The suppressiveness of Finnish light coloured sphagnum peat. J. Sci. Agric. Soc. Finl. 54:345-356.
______. 1982b. Preliminary experiments into the use of *Streptomyces* spp. isolated from peat in the biological control of soil and seed-borne diseases in peat culture. J. Sci. Agric. Soc. Finl. 54:357-359.
Tahvonen, R., and Uoti, J. 1983. The use of *Streptomyces* sp. as a biological control agent. Page 795 in: Int. Congr. Plant Prot. Proc. Conf. 10th.
Takakura, T., Shono, H., and Hojo, T. 1984. Crop management by intelligent computer systems. Acta Hortic. 148:317-328.
Tamada, T. 1975. Beet necrotic yellow vein virus. Descriptions of Plant Viruses, No. 144. Commonwealth Mycological Institute and Association of Applied Biologists, Kew, England.
Tamietti, G., and Garibaldi, A. 1980. Il riscaldomento solare del terreno mediante pacciamatura con materiali nella lotta contro la radice suberosa del pomodoro in serra. Dif. Piante 3:143-150.
Tammen, J. F. 1973. Rose powdery mildew studied for epidemics. Sci. Agric. 20(2):10.
Tammen, J., Baker, R. R., and Holley, W. D. 1956. Control of carnation diseases through the cultured-cutting technique. Pages 72-76 in: Plant Dis. Rep. Suppl. 238.
Tantau, H. J. 1980. Climate control algorithms. Acta Hortic. 106:49-57.
______. 1984. Adaptive control of greenhouse climate. Acta Hortic. 148:251-257.
______. 1985. Greenhouse climate control using mathematical models. Acta Hortic. 174:449-459.
Taylor, C. E. 1980. Nematodes. Pages 375-416 in: Vectors of Plant Pathogens. K. F. Harris and K. Maramorosch, eds. Academic Press, New York.
Teakle, D. S. 1962. Transmission of tobacco necrosis by a fungus, *Olpidium brassicae.* Virology 18:224-231.
______. 1980. Fungi. Pages 417-438 in: Vectors of Plant Pathogens. K. F. Harris and K. Maramorosch, eds. Academic Press, New York.
Tezuka, N., and Kiso, A. 1970. Appearance of thiophanate-methyl resistant strains of *Botrytis* spp. in eggplant. (Abstr.) Ann. Phytopathol. Soc. Jpn. 43:303-304.

Thatcher, F. S. 1942. Further studies of osmotic and permeability relations in parasitism. Can. J. Res. Sect. C 20:283-311.

Thibodeau, P. O. 1983. Grafting on a resistant rootstock: An efficient control method against Fusarium crown and root rot of greenhouse tomato. (Abstr.) Can. J. Plant Pathol. 5:212.

Thinggaard, K., and Middelboe, A. L. 1989. *Phytophthora* and *Pythium* in pot plant cultures grown on ebb and flow bench with recirculating nutrient solution. J. Phytopathol. (Berlin) 125:343-352.

Thomas, C. S., Marois, J. J., and English, J. T. 1988. The effects of wind speed, temperature, and relative humidity on development of aerial mycelium and conidia of *Botrytis cinerea* on grape. Phytopathology 78:260-265.

Thompson, D. C., and Jenkins, S. F. 1985. Effects of temperature, moisture, and cucumber cultivar resistance on lesion size increase and conidial production by *Colletotrichum lagenarium.* Phytopathology 75:828-832.

Thornthwaite, C. W., and Holzman, B. 1939. The determination of evaporation from land and water surfaces. Mon. Weather Rev. 67:4-11.

Thorpe, H. J., and Jarvis, W. R. 1981. Grafted tomatoes escape Fusarium foot and root rot. Can. J. Plant Sci. 61:1027-1028.

Thresh, J. M. 1982. Cropping practices and virus spread. Annu. Rev. Phytopathol. 20:193-218.

Thut, H. F. 1939. The relative humidity gradient of stomatal transpiration. Am. J. Bot. 26:315-319.

Tichelaar, G. M. 1967. Studies on the biology of *Botrytis allii* on *Allium cepa.* Neth. J. Plant Pathol. 73:157-160.

Ting, K. C., Dijkstra, J., Fang, W., and Giniger, M. 1989. Engineering economy of controlled environment for greenhouse production. Trans. ASAE 32:1018-1022.

Tirilly, Y., Trique, B., and Maisonneuve, J. 1987. Perspectives d'utilisation de *Hansfordia pulvinata* contre la cladosporiose de la tomate. EPPO Bull. 17:639-643.

Tjamos, E. C. 1979. Induction of resistance to Verticillium wilt in cucumber (*Cucumis sativus*). Physiol. Plant Pathol. 15:223-227.

———. 1983. Control of *Pyrenochaeta lycopersici* by combined soil solarization and low dose of methyl bromide in Greece. Acta Hortic. 152:233-258.

Tjamos, E. C., and Faridis, A. 1980. Control of soilborne pathogens by solar heating in plastic houses. Pages 82-84 in: Proc. Congr. Mediterr. Phytopathol. Union 5th.

Tomlinson, J. A., and Faithfull, E. M. 1979. Effects of fungicides and surfactants on the zoospores of *Olpidium brassicae.* Ann. Appl. Biol. 93:13-19.

———. 1980. Studies on the control of lettuce big-vein disease in recirculated nutrient solutions. Acta Hortic. 98:325-331.

Tomlinson, J. A., and Garrett, R. G. 1964. Studies on the lettuce big vein virus and its vector *Olpidium brassicae* (Wor.) Dang. Ann. Appl. Biol. 54:45-61.

Tomlinson, J. A., and Thomas, B. J. 1986. Studies on melon necrotic spot virus disease of cucumber and on control of the fungus vector (*Olpidium radicale*). Ann. Appl. Biol. 108:71-80.

Tomlinson, J. A., Carter, A., Dale, W. T., and Simpson, C. J. 1970. Weed plants as sources of cucumber mosaic virus. Ann. Appl. Biol. 66:11-16.

Tomlinson, J. A., Faithfull, E. M., and Clay, C. M. 1980. Big vein disease of lettuce. Pages 82-83 in: Rep. Natl. Veg. Res. Stn. Wellesbourne 1980.

Tomlinson, J. A., Faithfull, E. M., Webb, M. J. W., Fraser, R. S. S., and Seeley, N. D. 1983. *Chenopodium* necrosis: A distinctive strain of tobacco necrosis virus isolated from river water. Ann. Appl. Biol. 102:135-147.

Tompkins, C. M., and Hansen, H. N. 1950. Flower blight of *Stephanotis floribunda,* caused by *Botrytis elliptica,* and its control. Phytopathology 40:780-781.

Towers, G. H. N., and Wat, C. K. 1978. Biological activity of polyacetylenes. Rev. Latinoam. Quim. 9:162-170.

Trappe, J. M. 1977. Selection of fungi for ectomycorrhizal inoculation in nurseries. Annu. Rev. Phytopathol. 15:203-222.

Traquair, J. A. 1984. Etiology and control of orchard replant problems: A review. Can. J. Plant Pathol. 6:54-62.

Traquair, J. A., Meloche, R. B., Jarvis, W. R., and Baker, K. W. 1983. The hyperparasitism

of *Cladosporium uredinicola* on *Puccinia violae*. Can. J. Bot. 62:181-184.
Traquair, J. A., Shaw, L. A., and Jarvis, W. R. 1988. New species of *Stephanoascus* with *Sporothrix* anamorphs. Can. J. Bot. 66:926-933.
Trenbath, B. R. 1984. Gene introduction strategies for the control of crop diseases. Pages 142-168 in: Pest and Pathogen Control: Strategic, Tactical, and Policy Models. G. R. Conway, ed. John Wiley & Sons, Chichester.
Trickett, E. S., and Goulden, J. D. S. 1958. The radiation transmission and heat conserving properties of glass and some plastic films. J. Agric. Eng. Res. 3:281-285.
Trigalet, A., and Trigalet-Demery, D. 1990. Use of avirulent mutants of *Pseudomonas solanacearum* for the biological control of bacterial wilt of tomato plants. Physiol. Mol. Plant Pathol. 36:27-38.
Trolinger, J. C., and Strider, D. L. 1984. Botrytis blight of *Exacum affine* and its control. Phytopathology 74:1181-1188.
Trutman, P., Keane, P. J., and Merriman, P. R. 1982. Biological control of *Sclerotinia sclerotiorum* on aerial parts of plants by the hyperparasite *Coniothyrium minitans*. Plant Pathol. 24:109-113.
Tu, C. M. 1972. Effect of 4 nematocides on activities of microorganisms in soil. Appl. Microbiol. 23:398-401.
Tu, J. C. 1980. *Gliocladium virens,* a destructive mycoparasite of *Sclerotinia sclerotiorum*. Phytopathology 70:670-674.
Tu, J. C., and Vaartaja, O. 1981. The effect of the hyperparasite *Gliocladium virens* on *Rhizoctonia solani* and on Rhizoctonia root rot of white beans. Can. J. Bot. 59:22-27.
Tukey, H. B. 1969. Implications of allelopathy in agricultural plant science Bot. Rev. 35:1-16.
Udink ten Cate, A. J. 1980. Remarks on greenhouse computer control models. Acta Hortic. 106:43-47.
Udink ten Cate, A. J., and Challa, H. 1984. On optimal control of the crop growth system. Acta Hortic. 148:267-276.
Udink ten Cate, A. J., and van de Vooren, J. 1984. New models for greenhouse climate control. Acta Hortic. 148:277-285.
Udink ten Cate, A. J., Bot, G. P. A., and van Dixhorn, J. J. 1978. Computer control of greenhouse climates. Acta Hortic. 87:263-272.
Uhlenbroek, J. H., and Bijloo, J. D. 1959. Isolation and structure of a nematicidal principle occurring in *Tagetes* roots. Pages 579-581 in: Proc. Int. Congr. Crop Prot. 4th.
Vaartaja, O. 1964. Survival of *Fusarium, Pythium* and *Rhizoctonia* in very dry soil. Bimon. Prog. Rep. Dep. For. Can. 20:3.
Vanachter, A. 1979. Fumigation against fungi. Pages 163-183 in: Soil Disinfestation. D. Mulder, ed. Elsevier, Amsterdam.
Vanachter, A., van Wambeke, E., and van Assche, C. 1983. Potential danger for infection and spread root diseases of tomatoes in hydroponics. Acta Hortic. 133:119-128.
Van Alfen, N. K. 1982. Biology and potential for disease control of hypovirulence of *Endothia parasitica*. Annu. Rev. Phytopathol. 20:349-362.
van Assche, C. 1979. Aims of soil disinfestation. Pages 9-15 in: Soil Disinfestation. D. Mulder, ed. Elsevier, Amsterdam.
van Asten, J., and Dorpema, J. W. 1982. A new approach to sterilization conditions: The IMO concept. Pharm. Weekbl. Sci. Ed. 4:49-56.
van Berkum, J. A., and Hoestra, H. 1979. Practical aspects of the chemical control of nematodes in soil. Pages 53-134 in: Soil Disinfestation. D. Mulder, ed. Elsevier, Amsterdam.
van Bruggen, A. H. C., and Bouchibi, N. 1989. Effect of sodium/calcium ratio and ionic concentration of a nutrient solution on Phytophthora root rot of tomato. (Abstr.) Phytopathology 79:912.
van den Ende, J., Koorneef, P., and Sonneveld, C. 1975. Osmotic pressure of the soil solution: Determination and effects in some glasshouse crops. Neth. J. Agric. Sci. 23:181-190.
van der Borg, H. H., ed. 1980. Symposium on computers in greenhouse climate control. Acta Hortic. 106:1-209.
Vanderplank, J. E. 1963. Plant Diseases: Epidemics and Control. Academic Press, London.
_____. 1984. Disease Resistance in Plants. 2nd ed. Academic Press, Orlando, FL.
_____. 1989. A paradox as an aid to understanding host-pathogen specificity. Plant Pathol. 38:144-145.

van der Post, C. J., Schie, J. J., and de Graf, R. 1974. Energy balance and water supply in glasshouses in the West-Netherlands. Acta Hortic. 35:13-22.

van der Vlugt, J. L. F. 1989. A literature review concerning root death in cucumber and other crops. Norw. J. Agric. Sci. 3:265-274.

van de Vooren, J. 1980. Data analyses. Acta Hortic. 106:39-41.

van de Vooren, J., and Strijbosch, T. 1980. Glasshouse ventilation control. Acta Hortic. 106:117-124.

van Dommelen, L., and Bollen, J. G. 1973. Antagonism between benomyl-resistant fungi on cyclamen sprayed with benomyl. Acta Bot. Neerl. 22:169-170.

van Dorst, H. J. M. 1975. Influence of temperature regime on virus diseases in cucumber. Acta Hortic. 51:329-332.

van Dorst, H. J. M., and Peters, D. 1974. Some biological observations on pale fruit, a viroid incited disease of cucumber. Neth. J. Plant Pathol. 80:85-96.

van Dorst, H. J. M., Huijberts, N., and Bos, L. 1983. Yellows of glasshouse vegetables, transmitted by *Trialeurodes vaporariorum*. Neth. J. Plant Pathol. 89:171-184.

van Eck, W. H. 1978. Autolysis of chlamydospores of *Fusarium solani* f. sp. *cucurbitae* in chitin and laminarin amended soils. Soil Biol. Biochem. 10:89-92.

van Leeuwen, F. X. R., and Sangster, B. 1987. The toxicology of the bromide ion. Crit. Rev. Toxicol. 18:189-213.

van Lenteren, J. C., and Woets, J. 1988. Biological and integrated pest control in greenhouses. Annu. Rev. Entomol. 33:239-269.

van Os, E. A., van de Braak, N. J., and Klomp, G. 1988. Heat treatment for disinfecting drainwater, technical and economic aspects. Pages 353-359 in: Proc. Int. Soc. Soilless Cult. 1988.

van Os, H. 1964. Production of virus-free carnations by means of meristem culture. Neth. J. Plant Pathol. 70:18-26.

van Peer, R., Xu, T., Rattink, H., and Schippers, B. 1988. Biological control of carnation wilt caused by *Fusarium oxysporum* in hydroponic systems. Pages 361-373 in: Proc. Int. Soc. Soilless Cult. 1988.

van Slogteren, D. H. M., Groen, N. P. A., and Muller, P. J. 1976. Mycoplasmaziekten van bol- en knolegewassen. Pages 50-51 in: Jaarversl. Lab. Bloembollenonderz. Lisse 1976.

van Steekelenburg, N. A. M. 1983. Epidemiological aspects of *Didymella bryoniae,* the cause of stem and fruit rot of cucumber. Neth. J. Plant Pathol. 89:75-86.

_____. 1984. Influence of ventilation temperature and low ventilation rates on incidence of *Didymella bryoniae* in glasshouse cucumbers. Acta Hortic. 156:187-197.

_____. 1985a. Influence of time of transition from night to day temperature on incidence of *Didymella bryoniae* and influence of the disease on growth and yield of glasshouse cucumbers. Neth. J. Plant Pathol. 91:225-233.

_____. 1985b. Influence of humidity on incidence of *Didymella bryoniae* on cucumber leaves and growing tips under controlled conditions. Neth. J. Plant Pathol. 91:277-283.

_____. 1986. Factors influencing internal fruit rot of cucumber caused by *Didymella bryoniae.* Neth. J. Plant Pathol. 92:81-91.

van Steekelenburg, N. A. M., and van der Sar, M. 1987. Epidemiology of pathogens in substrate crops. Page 82 in: Annu. Rep. Glasshouse Crops Res. Exp. Stn. Naaldwijk 1987.

van Steekelenburg, N. A. M., and van de Vooren, J. 1980. Influence of the glasshouse climate on development of disease in a cucumber crop with special reference to stem and fruit rot caused by *Didymella bryoniae.* Acta Hortic. 118:45-56.

van Wambeke, E. 1983. Efficiency increase of methyl bromide soil fumigation by admixture with methyl chloride or ameliorated tarps. Acta Hortic. 152:137-141.

van Wambeke, E., Wijsmans, J., and d'Hertefelt, P. 1983. Possibilities in microwave application for growing substrate disinfestation. Acta Hortic. 152:209-213.

van Weel, P. A. 1984. Bench heating for potplants or cutflower production. Acta Hortic. 148:57-61.

van Zinderen-Bakker, E. M., Molnar, J. M., MacKenzie, A., and Monk, G. 1984. Process control energy reduction and data acquisition of a fully programmable greenhouse computer. Acta Hortic. 148:309-316.

Vedie, R., and Le Normand, M. 1984. Modulation de l'expression du pouvoir pathogène de *Botrytis fabae* Sard. et de *Botrytis cinerea* Pers. par des bactéries du phylloplan de

Vicia faba L. Agronomie (Paris) 4:721-728.
Veenebos, J. A. J. 1984. The green corner: A pre-export inspection system for chrysanthemum cut flowers and pot plants in the Netherlands. EPPO Bull. 14:369-372.
Verhoeff, K. 1963. Voetrot en "kanker" bij tomaat, veroorzaakt door *Didymella lycopersici*. Neth. J. Plant Pathol. 69:298-313.
_____. 1965. Studies on *Botrytis cinerea* in tomatoes: Mycelial development in plants growing in soil with various nutrient levels, as well as in internodes of different age. Neth. J. Plant Pathol. 71:167-175.
_____. 1967. Studies on *Botrytis cinerea* in tomatoes: Influence of methods of deleafing on the occurrence of stem lesions. Neth. J. Plant Pathol. 73:117-120.
_____. 1968. Effect of soil nitrogen level and of methods of deleafing upon the occurrence of *Botrytis cinerea* under commercial conditions. Neth. J. Plant Pathol. 74:184-194.
_____. 1970. Spotting of tomato fruits caused by *Botrytis cinerea*. Neth. J. Plant Pathol. 76:219-226.
_____. 1974. Latent infections by fungi. Annu. Rev. Phytopathol. 12:99-110.
_____. 1980. The infection process and host-pathogen interactions. Pages 153-180 in: The Biology of *Botrytis*. J. R. Coley-Smith, K. Verhoeff, and W. R. Jarvis, eds. Academic Press, London.
Verhoeff, K., Bulit, J., and Dubos, B. 1988. *Botryotinia fuckeliana* (de Bary) Whetzel. Pages 432-435 in: European Handbook of Plant Diseases. I. M. Smith, J. Dunez, R. A. Lelliott, D. H. Phillips, and S. A. Archer, eds. Blackwell, Oxford.
von Zabeltitz, C. 1976a. Möglichkeiten zur Nutzung der Sonnenenergie für die Beheizung von Gewächshäusern. Gartenbauwissenschaft 41:213-220.
_____. 1976b. Temperaturverteilung bei einer Luft-vegetationsheizung aus PE-Släuchen. Dtsch. Gartenbau 30:7-8.
_____. 1976c. Gewächshausheizung mit Sonnenergie. Dtsch. Gartenbau 30:1221-1222.
Waaijenberg, D. 1985. Research on greenhouse cladding materials. Acta Hortic. 170:103-109.
Waggoner, P. E. 1984. Models for forecasting disease control. Pages 221-232 in: Pest and Pathogen Control: Strategic, Tactical, and Policy Models. G. R. Conway, ed. John Wiley & Sons, Chichester.
Waggoner, P. E., and Shaw, R. H. 1953. Stem and root temperatures. Phytopathology 43:317-318.
Waggoner, P. E., and Taylor, G. S. 1958. Dissemination by atmospheric turbulence: Spores of *Peronospora tabacina*. Phytopathology 48:46-51.
Wahl, V. 1989. Les systèmes-experts de diagnostic en protection des plantes de l'INRA: Un point de vue méthodologique. Phytoprotection 70:111-118.
Walker, J. C. 1950. Environment and host resistance in relation to cucumber scab. Phytopathology 40:1094-1102.
Walker, J. C., and Patel, P. N. 1964. Splash dispersal and wind as factors in epidemiology of halo blight of bean. Phytopathology 54:140-141.
Walker, J. N., and Duncan, G. A. 1973a. Estimating greenhouse ventilation requirements. Ky. Agric. Exp. Stn. AEN-9.
_____. 1973b. Air circulation in greenhouses. Ky. Agric. Exp. Stn. AEN-18.
_____. 1974a. Greenhouse ventilation systems. Ky. Agric. Exp. Stn. AEN-30.
_____. 1974b. Greenhouse location and orientation. Ky. Agric. Exp. Stn. AEN-32.
Walker, J. N., and Slack, D. C. 1970. Properties of greenhouse covering materials. Trans. ASAE 13:682-684.
Wallace, H. R. 1978. The diagnosis of plant diseases of complex etiology. Annu. Rev. Phytopathol. 16:379-402.
Walton, W. M., and Prewitt, W. C. 1949. The production of sprays and mists of uniform drop size. Proc. Phys. Soc. London Sect. B 62:341-350.
Ward, G. M. 1964. Observations on root growth in the greenhouse tomato. Can. J. Plant Sci. 44:492-494.
_____. 1967. Observations of root growth in the greenhouse cucumber. Can. J. Plant Sci. 47:215-217.
Water, J. K. 1981. Chrysanthemum white rust. EPPO Bull. 11:239-242.
Watkinson, A. 1975. Insulating fully creates other problems. Grower (London) 84:611-613.

Watson, A. G., and Koons, C. E. 1973. Increased tolerance to benomyl in greenhouse populations of *Botrytis cinerea.* (Abstr.) Phytopathology 63:1218-1219.
Weast, R. C., ed. 1968. Handbook of Chemistry and Physics. 62nd ed. Chemical Rubber, Cleveland, OH.
Weinhold, A. R. 1961. Temperature and moisture requirements for germination of conidia of *Sphaerotheca pannosa* from peach. Phytopathology 51:699-703.
Weinhold, A. R., and Bowman, T. 1968. Selective inhibition of the potato scab pathogen by antagonistic bacteria, and substrate influence on antibiotic production. Plant Soil 28:12-24.
Weiss, A., Hipps, L. E., Blad, B. L., and Steadman, J. R. 1980. Comparison of within-canopy microclimate and white mold disease (*Sclerotinia sclerotiorum*) development in dry edible beans as influenced by canopy structure and irrigation. Agric. Meteorol. 22:11-21.
Wellman, F. L. 1937. Control of southern celery mosaic in Florida by removing weeds that serve as a source of mosaic infection. U.S. Dep. Agric. Bull. 548.
Wenner, M. A. 1979. Advances in controlled droplet application. Agrichem. Age (Nov.-Dec.), pp. 18-25.
Wenzl, H. 1938a. *Botrytis cinerea* als Erreger einer Fleckenkrankheit der Cyclamen-Blüten. Phytopathol. Z. 11:107-118.
——. 1938b. *Botrytis cinerea* als Erreger einer Fleckenkrankheit der Knospen und Blüten der Rose ("Blütenfeuer"). Gartenbauwissenschaft 11:462-472.
Whalen, M. C. 1988. The effect of mechanical impedance on ethylene production by maize roots. Can. J. Bot. 66:2139-2142.
Wheeler, B. E. J. 1978. Rose powdery mildew: An analysis of its epidemiology. Pages 121-127 in: Plant Disease Epidemiology. E. Griffiths, ed. Blackwell, Oxford.
White, J. G., and Hague, N. G. M. 1971. Control of black root rot of cucumber (*Phomopsis sclerotioides*) with methyl bromide. Pages 237-242 in: Proc. Br. Insectic. Fungic. Conf. 6th.
Whitehead, A. G. 1978. Chemical control: (a) Soil treatment. Pages 283-296 in: Plant Nematology. 3rd ed. J. F. Southey, ed. Minist. Agric. Fish. Food (G.B.) Ref. Book 407. Her Majesty's Stationery Office, London.
Whittaker, R. H., and Feeny, P. P. 1971. Allelochemics: Chemical interactions between species. Science (Washington, D.C.) 171:757-770.
Whittle, R. M., and Lawrence, W. J. C. 1960. The climatology of glasshouses. II. Ventilation. J. Agric. Eng. Res. 5:36-41.
Wicks, T. J., Volle, D., and Baker, B. T. 1978. The effect of soil fumigation and fowl manure on populations of *Fusarium oxysporum* f. sp. *cucumerinum* in glasshouse soil and on the incidence of cucumber wilt. Agric. Rec. (S. Aust.) 5:4-8.
Wijetunga, C., and Baker, R. 1979. Modeling of phenomena associated with soil suppressive to *Rhizoctonia solani.* Phytopathology 69:1287-1293.
Wijnen, A. P., Volker, P., and Bollen, G. J. 1983. De lotgevallen van pathogene schimmels in een composthoop. Gewasbescherming 14:5.
Wilhelm, S. 1956. A sand-culture technique for the isolation of fungi associated with roots. Phytopathology 46:293-295.
——. 1965. *Pythium ultimum* and the soil fumigation growth response. Phytopathology 55:1016-1020.
Wilhelm, S., and Nelson, P. E. 1970. A concept of rootlet health of strawberries in pathogen-free field soil achieved by fumigation. Pages 208-215 in: Root Diseases and Soil-Borne Pathogens. T. A. Toussoun, R. V. Bega, and P. E. Nelson, eds. University of California Press, Berkeley.
Wilhelm, S., and Raabe, R. D. 1956. Culture-indexing of budwood to provide *Verticillium*-free greenhouse roses. Pages 85-87 in: Plant Dis. Rep. Suppl. 238.
Williams, P. F. 1978. Epidemiology and stratagems for the control of fungal diseases. Australas. Plant Pathol. 8:8-10.
Williams, P. H., Sheard, E., and Read, H. 1953. Didymella stem rot of the tomato. J. Hortic. Sci. 28:278-294.
Williamson, C. E. 1953. Methyl bromide injury to some ornamental plants. (Abstr.) Phytopathology 43:489.
Wilson, A. R. 1963. Some observations on the infection of tomato stems by *Botrytis cinerea.* (Abstr.) Ann. Appl. Biol. 51:171.

_____. 1964. Grey mould of tomato: Stem infection by *Botrytis cinerea*. Annu. Rep. Scott. Hortic. Res. Inst. 11:73-74.

Wilson, J. D., and Norris, M. G. 1966. Effect of bromine residues in muck soil on vegetable yields. Down to Earth 22:15-18.

Winoto Suatmadji, R. 1969. Studies on the Effect of *Tagetes* Species on Plant Parasitic Nematodes. H. Veenman & Zonen, Wageningen, Netherlands.

Winspear, K. W., Postlethwaite, J. D., and Cotton, R. F. 1970. The restriction of *Cladosporium fulvum* and *Botrytis cinerea*, attacking glasshouse tomatoes, by automatic humidity control. Ann. Appl. Biol. 65:75-83.

Wolcott, A. R., Maciak, F., Shepherd, L. N., and Lucas, R. E. 1960. Effects of Telone on nitrogen transformations and on growth of celery in organic soil. Down to Earth 16:10-14.

Wolfe, J. S. 1970. Feasibility and economics of conditioning recirculated greenhouse air by means of evaporative cooling. J. Agric. Eng. Res. 15:265-273.

Wolfe, J. S., and Cotten, R. F. 1975. Airflow and temperature distribution in greenhouses with fan ventilation. Acta Hortic. 46:34-56.

Wolfe, M. S. 1981. Integrated use of fungicides and host resistance for stable disease control. Philos. Trans. R. Soc. London B 295:175-184.

Wolffhechel, H. 1989. Fungal antagonists of *Pythium ultimum* isolated from a disease-suppressive *Sphagnum* peat. Vaextskyddsnotiser 53:1-2, 7-11.

Woltz, S. S., and Engelhard, A. W. 1973. Fusarium wilt of chrysanthemum: Effect of nitrogen source and lime on disease development. Phytopathology 63:155-157.

Woltz, S. S., and Waters, W. E. 1975. Chives production as affected by fertilizer practices, soil mixes and methyl bromide soil residues. Proc. Fla. State Hortic. Soc. 88:133-137.

Wood, R. K. S. 1951a. The control of diseases of lettuce by the use of antagonistic organisms. I. The control of *Botrytis cinerea* Pers. Ann. Appl. Biol. 38:203-216.

_____. 1951b. The control of diseases of lettuce by the use of antagonistic organisms. II. The control of *Rhizoctonia solani* Kühn. Ann. Appl. Biol. 38:217-230.

Woodruff, N. P. 1954. Shelterbelt and surface barrier effects on wind velocities, evaporation, house heating and snow drifting. Kans. Agric. Exp. Stn. Tech. Bull. 77.

Woolley, J. T. 1964. Water relations of soybean leaf hairs. Agron. J. 56:569-571.

Wymore, L. A., and Baker, R. 1982. Factors affecting cross-protection in control of Fusarium wilt of tomato. Plant Dis. 66:908-910.

Yamakawa, K., and Nagata, N. 1975. Three tomato lines obtained by the use of chronic gamma radiation with combined resistance to TMV and *Fusarium* race J-3. Tech. News Inst. Radiat. Breed. 16.

Yamamoto, I., Komada, H., Kuniyasu, K., Saito, M., and Ezuka, A. 1974. A new race of *Fusarium oxysporum* f. sp. *lycopersici* inducing root rot of tomato. Pages 17-29 in: Proc. Kansai Plant Prot. Soc. 16.

Yarwood, C. E. 1936. The tolerance of *Erysiphe polygoni* and certain other powdery mildews to low humidity. Phytopathology 26:845-859.

_____. 1939a. Control of powdery mildews with a water spray. Phytopathology 29:288-290.

_____. 1939b. Relation of moisture to infection with some downy mildews and rusts. Phytopathology 29:933-945.

_____. 1950. Water content of fungus spores. Am. J. Bot. 37:636-639.

_____. 1956. Humidity requirements of foliage pathogens. Plant Dis. Rep. 40:318-321.

_____. 1957. Powdery mildews. Bot. Rev. 23:235-301.

_____. 1959a. Microclimate and infection. Pages 548-556 in: Plant Pathology: Problems and Progress 1908-1958. C. S. Holton, G. W. Fischer, R. W. Fulton, H. Hart, and S. E. A. McCallan, eds. University of Wisconsin Press, Madison.

_____. 1959b. Predisposition. Pages 521-562 in: Plant Pathology: An Advanced Treatise. Vol. 4. J. G. Horsfall and E. B. Cowling, eds. Academic Press, New York.

_____. 1963. Heat adaptation in cucumber powdery mildew. Plant Dis. Rep. 47:824-825.

_____. 1978. Water stimulates *Sphaerotheca*. Mycologia 70:1035-1039.

Yarwood, C. E., and Gardner, M. W. 1964. Unreported powdery mildews. III. Plant Dis. Rep. 48:310.

Yarwood, C. E., and Hazen, W. E. 1944. The relative humidity at leaf surfaces. Am. J. Bot. 31:129-135.

Young, H. C., Jr., Prescott, J. M., and Saari, E. E. 1978. Role of disease monitoring in preventing epidemics. Annu. Rev. Phytopathol. 16:263-285.

Yuen, G. Y., and Raabe, R. D. 1984. Effects of small-scale aerobic composting on survival of some fungal plant pathogens. Plant Dis. 68:134-136.

Yunis, H., Elad, Y., and Mahrer, Y. 1990. The effect of air temperature, relative humidity and canopy wetness on gray mold of cucumbers in unheated greenhouses. Phytoparasitica 18:203-215.

Zadoks, J. C. 1985. On the conceptual basis of crop loss assessment: The threshold theory. Annu. Rev. Phytopathol. 23:455-473.

Zak, B. 1964. Role of mycorrhizae in root disease. Annu. Rev. Phytopathol. 2:377-392.

Zakaria, M. A., and Lockwood, J. L. 1980. Reduction in *Fusarium* populations in soil by oilseed meal amendments. Phytopathology 70:240-243.

Zentmyer, G. A. 1979. Effect of physical factors, host resistance and fungicides on root infection at the soil-root interface. Pages 315-328 in: The Soil-Root Interface. J. L. Harley and R. S. Russell, eds. Academic Press, London.

Zinnen, T. M. 1988. Assessment of plant diseases in hydroponic culture. Plant Dis. 72:96-99.

Index